MW01387687

Statistical and Managerial Techniques for Six Sigma Methodology

Statistical and Managerial Techniques for Six Sigma Methodology

Theory and Application

Stefano Barone

University of Palermo, Italy
and
Chalmers University of Technology, Sweden

Eva Lo Franco

University of Palermo, Italy

A John Wiley & Sons, Ltd., Publication

Library of Congress Cataloging-in-Publication Data

Barone, Stefano.
  Statistical and managerial techniques for six sigma methodology : theory and application / Stefano Barone and Eva Lo Franco.
      p. cm.
  Includes bibliographical references and index.
  ISBN 978-0-470-71183-5 (cloth)
 1. Total quality management. 2. Quality control – Management. 3. Six sigma (Quality control standard)
I. Franco, Eva Lo. II. Title.
  HD62.15.B367 2012
  658.4′013 – dc23

                                                                      2011039805

A catalogue record for this book is available from the British Library.

ISBN: 978-0-470-71183-5

Set in 10/12pt Times by Laserwords Private Limited, Chennai, India
Printed and bound in Malaysia by Vivar Printing Sdn Bhd

1   2012

To our families.

To Palermo, Napoli and Göteborg.

Contents

Preface   xi
About the Authors   xiii

1 Six Sigma methodology   1
  1.1 Management by process   1
    1.1.1 The concept of 'process'   1
    1.1.2 Managing by process   1
    1.1.3 The process performance triangle   2
    1.1.4 Customer satisfaction   3
    1.1.5 The success of enterprise   4
    1.1.6 Innovation and Six Sigma   5
  1.2 Meanings and origins of Six Sigma   5
    1.2.1 Variation in products and processes   5
    1.2.2 Meaning of 'Six Sigma'   6
    1.2.3 Six Sigma process   7
    1.2.4 Origins of Six Sigma   7
    1.2.5 Six Sigma: Some definitions   9
  1.3 Six Sigma projects   11
    1.3.1 Why implement Six Sigma projects?   11
    1.3.2 Six Sigma paths   12
  1.4 The DMARIC path   18
    1.4.1 Human resources and training   20
    References   21

2 Basic managerial techniques   23
  2.1 For brainstorming   23
    2.1.1 Cause–effect diagram   23
    2.1.2 Affinity diagram (KJ analysis)   26
  2.2 To manage the project   29
    2.2.1 Work breakdown structure   29
    2.2.2 Gantt chart   30
  2.3 To describe and understand the processes   30
    2.3.1 The SIPOC scheme   31
    2.3.2 The flow chart   32
    2.3.3 The ServQual model   33
  2.4 To direct the improvement   37

      2.4.1   The Kano model   37
      References   39

3 Basic statistical techniques   **41**
   3.1 To explore data   41
      3.1.1   Fundamental concepts and phases of the exploratory
              data analysis   41
      3.1.2   Empirical frequency distribution of a numerical variable   46
      3.1.3   Analysis by stratification   59
      3.1.4   Other graphical representations   60
   3.2 To define and calculate the uncertainty   62
      3.2.1   Definitions of probability   63
      3.2.2   Events and probabilities in the Venn diagram   64
      3.2.3   Probability calculation rules   66
      3.2.4   Dispositions, permutations and combinations   69
   3.3 To model the random variability   70
      3.3.1   Definition of random variable   70
      3.3.2   Probability distribution function   71
      3.3.3   Probability mass function for discrete random variables   71
      3.3.4   Probability density function for continuous variables   71
      3.3.5   Mean and variance of a random variable   72
      3.3.6   Principal models of random variables   74
   3.4 To draw conclusions from observed data   82
      3.4.1   The inferential process   82
      3.4.2   Sampling and samples   82
      3.4.3   Adopting a probability distribution model by graphical
              analysis of the sample (probability plot)   84
      3.4.4   Point estimation of the parameters of a Gaussian population   88
      3.4.5   Interval estimation   90
      3.4.6   Hypothesis testing   91
      References   93

4 Advanced managerial techniques   **95**
   4.1 To describe processes   95
      4.1.1   IDEF0   95
   4.2 To manage a project   98
      4.2.1   Project evaluation and review technique   98
      4.2.2   Critical path method   104
   4.3 To analyse faults   109
      4.3.1   Failure mode and effect analysis   110
      4.3.2   Fault tree analysis   114
   4.4 To make decisions   122
      4.4.1   Analytic hierarchy process   122
      4.4.2   Response latency model   129
      4.4.3   Quality function deployment   135
      References   143

5   Advanced statistical techniques   **145**
    5.1   To study the relationships between variables   145
            5.1.1   Linear regression analysis   145
            5.1.2   Logistic regression models   156
            5.1.3   Introduction to multivariate statistics   157
    5.2   To monitor and keep processes under control   171
            5.2.1   Process capability   172
            5.2.2   Online process control and main control charts   174
            5.2.3   Offline process control   183
    5.3   To improve products, services and production processes   189
            5.3.1   Robustness thinking   189
            5.3.2   Variation mode and effect analysis   200
            5.3.3   Systemic robust design   209
            5.3.4   Design of experiments   212
            5.3.5   Four case studies of robustness thinking   243
    5.4   To assess the measurement system   259
            5.4.1   Some definitions about measurement systems   259
            5.4.2   Measurement system analysis   260
            5.4.3   Lack of stability and drift of measurement system   262
            5.4.4   Preparation of a gauge R&R study   263
            5.4.5   Gauge R&R illustrative example   263
            References   265

**6   Six Sigma methodology in action: Selected Black Belt projects
    in Swedish organisations**   **267**
    6.1   Resource planning improvement at SAAB Microwave Systems   269
            6.1.1   Presentation of SAAB Microwave Systems   269
            6.1.2   Project background   269
            6.1.3   Define phase   270
            6.1.4   Measure phase   275
            6.1.5   Analyse phase   275
            6.1.6   Improve phase (ideas and intentions)   280
            6.1.7   Control phase (ideas and intentions)   282
    6.2   Improving capacity planning of available beds: A case study
            for the medical wards at Sahlgrenska and Östra Hospitals   283
            6.2.1   Presentation of Sahlgrenska and Östra Hospitals   284
            6.2.2   Project background   284
            6.2.3   Define phase   284
            6.2.4   Measure phase   286
            6.2.5   Analyse phase   288
            6.2.6   Improve phase (ideas and intentions)   293
            6.2.7   Control phase (ideas and intentions)   296
    6.3   Controlling variation in play in mast production process at ATLET   296
            6.3.1   Presentation of Atlet AB   297
            6.3.2   Project background   297
            6.3.3   Define phase   298

6.3.4	Measure phase	302
6.3.5	Analyse phase	307
6.3.6	Improve phase (ideas and intentions)	312
6.3.7	Control phase (ideas and intentions)	313
6.4	Optimising the recognition and treatment of unexpectedly worsening in-patients at Kärnsjiukhuset, Skaraborg Hospital	314
6.4.1	Presentation of Skaraborg Hospital	314
6.4.2	Project background	314
6.4.3	Define phase	315
6.4.4	Measure phase	321
6.4.5	Analyse phase (ideas and intentions)	328
6.4.6	Improve phase (ideas and intentions)	329
6.4.7	Control phase (ideas and intentions)	329
6.5	Optimal scheduling for higher efficiency and minimal losses in warehouse at Structo Hydraulics AB	330
6.5.1	Presentation of Structo Hydraulics AB	330
6.5.2	Project background	331
6.5.3	Define phase	332
6.5.4	Measure phase	335
6.5.5	Analyse phase	338
6.5.6	Improve phase (planning)	343
6.5.7	Control phase (planning)	348
6.6	Reducing welding defect rate for a critical component of an aircraft engine	350
6.6.1	Presentation of Volvo Aero Corporation	350
6.6.2	Project background	350
6.6.3	Define phase	351
6.6.4	Measure phase	354
6.6.5	Analyse phase	359
6.6.6	Improve phase (ideas and intentions)	364
6.6.7	Control phase (ideas and intentions)	365
6.7	Attacking a problem of low capability in final machining for an aircraft engine component at VAC – Volvo Aero Corporation	365
6.7.1	Presentation of Volvo Aero Corporation	365
6.7.2	Project background	366
6.7.3	Define phase	366
6.7.4	Measure phase	367
6.7.5	Analyse phase	371
6.7.6	Improve phase (ideas and intentions)	373
Index		**375**

Preface

The Six Sigma methodology is characterised by the concurrent and integrated use of managerial and statistical techniques. This special feature makes the methodology not always understandable or easily applicable. In this book we present the methodology through the illustration of the most widespread techniques and tools involved in its application.

When we decided to write the book, our first intention was to produce a reference textbook for a Six Sigma Black Belt university course, and in general a textbook that would be useful for university courses and specialist training on Six Sigma, Quality Management, Quality Control, Quality and Reliability Engineering, Risk Analysis and Business Management. Today, with the book finished, we can say that it is also addressed to business and engineering consultants, quality managers and managers of public and private organisations.

Our certainty that both managerial and statistical aspects of Six Sigma are fundamental for successful projects has permeated all parts of the book. Both aspects are equally considered, and enough room is given to preliminary notions to allow the reader acquiring the ability to apply the techniques and tools in real life.

The book aims to share the knowledge and the use of the Six Sigma methodology with everybody wishing to study its contents and having a basic knowledge of mathematics (at least secondary school level).

The structure of the book is essentially based on the main distinction between managerial and statistical techniques, and between basic and advanced ones. It gives special emphasis to variation and risk management, and to the structure and the organisational aspects of Six Sigma projects.

A clear and rigorous language characterises the book. The approach is basically pragmatic; in fact the topics are discussed with the support of many examples and case studies. To simplify the writing, we will use 'he' instead of 'he/she' when referring to a generic person. We note that we have met a lot of Six Sigma enthusiasts of both genders.

The book is based on a wide international scientific research and on the academic experience of the authors, both in research and in teaching. Furthermore, we relied on our current jobs as statisticians, and also on our backgrounds in engineering and economics, respectively.

The book is made up of six chapters.

Chapter 1 'Six Sigma Methodology' is divided into three sections: Management by process; Meanings and origins of Six Sigma; Six Sigma frameworks. This is essentially a conceptual chapter.

Chapter 2 to Chapter 5 illustrate the techniques. Their sections are titled according to the main scope of the techniques being presented.

Chapter 2 'Basic Managerial Techniques' has four sections: For brainstorming; To manage the project; To describe and understand the processes; To direct the improvement.

Chapter 3 'Basic Statistical Techniques' is divided into the following sections: To explore data; To define and calculate the uncertainty; To model the random variability; To draw conclusions from observed data.

Chapter 4 'Advanced Managerial Techniques' is divided into four sections: To describe processes; To manage a project; To analyse faults; To make decisions.

Chapter 5 'Advanced Statistical Techniques' is divided into four sections: To study the relationships between variables; To monitor and keep processes under control; To improve products, services and production processes; To assess the measurement system.

Essential bibliographic references conclude every chapter. We have intentionally referred only to the most authoritative books and articles relevant to the topics dealt with in each chapter.

Finally, Chapter 6 'Six Sigma methodology in action: Selected Black Belt projects in Swedish organisations' contains the reports of seven projects carried out within the last three editions of the Six Sigma Black Belt course held by one of the authors at the Chalmers University of Technology in Göteborg (Sweden) in the period 2009–2011.

This book, to which we have equally contributed, is the result of years of study and work, *but* also it is the result of precious collaborations with many key persons: our students, our colleagues, our teachers. We would like to thank all of them, but we prefer mentioning just one as a representative, Professor Bo Bergman, to whom we both feel profoundly linked by a sense of deep professional esteem and human friendship.

Stefano Barone and Eva Lo Franco

About the Authors

Stefano Barone received his PhD in Applied Statistics in 2000 from the University of Naples (Italy). After having worked for two years as a post doc researcher at ELASIS (FIAT Research Centre), at University of Naples and then at Chalmers University of Technology, he became Assistant Professor of Statistics at the University of Palermo, Faculty of Engineering. In 2005–2006 he served as council member and vice president of ENBIS, the European Network for Business and Industrial Statistics of which he was one of the founders. In 2008 he was awarded a Fulbright visiting research grant and worked at the Georgia Institute of Technology (Atlanta, USA). For the past three years he was Associate Professor (Docent) of Industrial Statistics at Chalmers University of Technology where he was responsible for the Six Sigma Black Belt education at Master level.

Eva Lo Franco actively collaborates to the chair of Statistical Quality Control at the University of Palermo (Italy) since 2003. She got her PhD in Economic Statistics in 2005. In 2006–2009 she was research fellow dealing with "education on quality and quality of education" especially at University. She had experience with the implementation of quality management systems and models both in public administration and SMEs. She is a senior assessor for the Italian Quality Award. She currently teaches Statistical Quality Control for the School of Economics and Basic Statistics for the bachelor program in Managerial Engineering.

1

Six Sigma methodology

1.1 Management by process

The application of Six Sigma always involves the implementation of a path: that is, the development of a coherent set of activities that together help to achieve one or more planned results (see Section 1.3.2). On the other hand, Six Sigma projects always involve one or more business processes and related performance. Therefore, an understanding of the methodology cannot ignore the study of the concept of 'process' and the deepening of methods and tools for its management.

1.1.1 The concept of 'process'

Any process in its simplest form can be illustrated as in Figure 1.1. A process is a logically consistent and repeatable sequence of activities that allow the transformation of specified inputs (or resources) into desired output (or results), and generating value. Some activities may also run in parallel.

Inputs, outputs and the value generated by the process must be measurable. Moreover, a process has a well-defined beginning and end, Finally, for each activity its manager is defined, that is the person who is responsible for its performance. Accordingly, defining a process is a way of answering the question: 'who does what?'

1.1.2 Managing by process

Managing by process is a principle of management of the organisation as a whole; it involves the design of the processes of the organisation, their realisation, their monitoring and evaluation, their improvement over time. It is an iterative method that can be synthesised by the four steps of the well-known Deming cycle (Figure 1.2): Plan, Do, Check, Act.

The first phase, 'Plan', provides for the establishment of the goals of the organisation and the planning of processes necessary to deliver results in line with the objectives. Phase

Statistical and Managerial Techniques for Six Sigma Methodology: Theory and Application, First Edition.
Stefano Barone and Eva Lo Franco.
© 2012 John Wiley & Sons, Ltd. Published 2012 by John Wiley & Sons, Ltd.

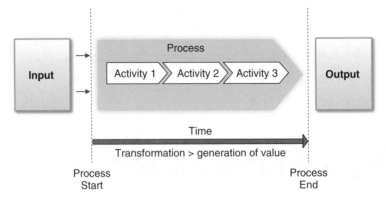

Figure 1.1   A generic process.

Figure 1.2   Deming cycle.

'Do' consists of implementing the processes. Next is the monitoring of the processes and the measurement of the results against the prefixed objectives: this is the 'Check' phase. The measurability of a process that is the object of a Six Sigma improvement project is a *conditio sine qua non* (i.e. a necessary condition) if you are looking to reduce the variation in the performance of the process. In the fourth phase, 'Act', it is time to take action to improve the process performance.

The Deming cycle is iterative and responds to the basic principle of continuous improvement.

1.1.3   The process performance triangle

In general, any process performance has some unique factors that need to be considered for its evaluation. However, we may imagine three basic dimensions that characterise each process performance:

(i) the variability of the performance with respect to a prefixed target (variation),

(ii) the mean time needed to obtain one unit of output from the allocated resources (cycle time), and

(iii) the return that the provision allows for in terms of the difference between costs and revenues, income and expenditure (yield).

In these basic dimensions we may trace a hierarchy of importance that arises from a cause–effect relationship: improving the cycle time rather than the yield may not affect the third dimension (variation). Conversely, an improvement in process variability will always reflect positively on both the cycle time (fewer minor alterations, less reworking, fewer controls, etc.) and the yield of the process (such as lower costs due to maintenance, reduced waiting times, etc. and/or increasing revenues through increased sales, etc.). The so-called process performance triangle (Figure 1.3) showing the three dimensions is a useful reference scheme for the analysis of a process along the temporal dimension and for the concomitant evaluation of different processes.

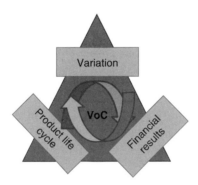

Figure 1.3   The triangle of process performance.

1.1.4 Customer satisfaction

Inside the process performance triangle, the role of the customer is of fundamental importance. First, the tolerable variation of product characteristics, both in terms of target value and limits, is or should be agreed with the client. The organisation's inability to adapt to the customer will always, sooner or later, result in incurring costs (e.g. returns, rework, penalties, etc.) and/or non-achievement of income (e.g. poor sales, rebates off the expected price, etc.). Therefore, it is necessary to know thoroughly the customer's expectations. These expectations can be explicit, since they are made public through, for example contract terms or expressed through legislation, and can be implicit, because they are normally taken for granted, or unexpressed.

A useful methodology to identify customer requirements and translate them into product/service characteristics is the quality function deployment (QFD, see Section 4.4.3). Once the characteristics critical for the customer (CTQ = critical to quality) are known, it will be possible to prioritise potential projects to improve the processes related to the achievement of those characteristics. In Six Sigma projects it is expected that, based on the collection of customer expectations (VOC = voice of customer), the expected value of the characteristic and its allowed range of variation (its tolerance) can be given.

The wishes of the customer, through the dimension of variability, are reflected both on the average length of the product life cycle and on the financial results of the process.

1.1.5   The success of enterprise

The evaluation of the performances of individual processes of an organisation is part of the more general and complex evaluation of its success – this latter evaluation is intended to be the full realisation of the raison d'être of the organisation. This means assessing the organisation's ability to meet the needs of customers in a competitive environment, economically, enhancing and developing its resources, first of all the human ones. Therefore, the success of an organisation is measured on three dimensions: cost-effectiveness, competitiveness and the satisfaction of the participants: first, workers and owners (Figure 1.4). This theoretical approach, typical of the business-institutionalist school of thought, leads us to ponder some basic management principles underlying the process of an organisation along these three dimensions of success: the economy (i.e. the ability of management to pay from its revenues all the costs of the inputs it needs), the logic of service to customer, and the promotion and development of resources. These three management principles must be present together and united by a logic of continuous improvement that qualifies the organisation, in one word, as 'innovative'. In summary, a successful organisation is always characterised by its innovativeness, that is the ability to continuously seek new opportunities for enhancement and development of resources to maintain a higher ability to economically serve the customer.

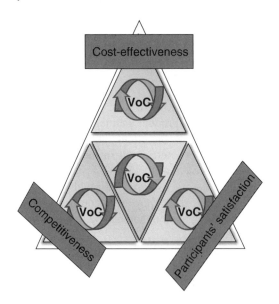

Figure 1.4   The size of the success of an organisation.

1.1.6   Innovation and Six Sigma

The descriptions given so far lead us to highlight a unique relationship between innovation (as understood in the previous section) and Six Sigma projects and, more generally, the philosophy of Six Sigma. The philosophy of Six Sigma is imbued with principles which underpin the innovative ability of an organisation, and encourages the organisation to innovate through the minimisation of the variability in processes and products. We can therefore say that the positive results achieved by a Six Sigma project (for example, the solution of a chronic problem of quality would permanently reduce the defect rate in a process) is always an innovation (Bisgaard, 2008). However, the opposite is not true, in the sense that innovation can be sought through routes that do not necessarily pass through the study of variation in product and processes.

1.2   Meanings and origins of Six Sigma

1.2.1   Variation in products and processes

No production process ever generates two products with exactly the same characteristics because a number of sources of variation always act on the process. The variation that is observed in the products can be traced back to two types of factors: the so-called **control factors**, that is inputs fed into the process on which a control activity can be made, and the so-called **noise factors**, that is inputs to the process that are not controllable.

The phenomenon of variation associated with a product characteristic can be represented graphically using the concept of a probability mass/density function (see Sections 3.3.3 and 3.3.4) that best fits the observed data. Such data may concern a variable, that is a quantitative characteristic (for example the temperature of a liquid, the tension of an electric component, etc.) or a qualitative characteristic (e.g. colour of a fabric, the type of manufactured wood, etc.).

If the observed characteristic can be modelled by a Gaussian random variable (Section 3.3.6.2), the values it can assume are symmetrically distributed around a central value (the mean) and the probability associated with them, somehow expressed by the probability density function (Figure 1.5), decreases as you move away from the central value. The distance measured on the horizontal axis between the mean ($\mu = 4$ in Figure 1.5) and the point of inflection of the curve is the standard deviation σ (in Figure 1.5, σ is equal to 1). The mean and standard deviation, respectively, give a location and a dispersion index for the random variable.

The probability density function represents the feasible production at time t, that is it is a kind of snapshot of the results obtained and obtainable by the process being analysed. By varying the instant of observation, the dispersion and shape of the distribution may vary (Figure 1.6). In such cases the phenomenon of variation in the product is also associated with a variation in the process.

The first variation (i.e. the variation in the product) is an inevitable phenomenon that we can try to limit by adjusting the control factors and the relations between them and the noise factors. The process variation, however, can be minimised by avoiding the so-called special causes of variation (see Section 5.2), but the process will always be subject to a

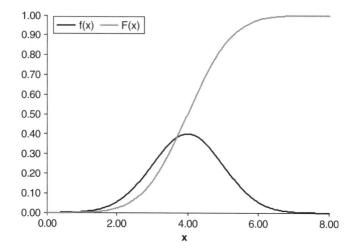

Figure 1.5   Gaussian distribution.

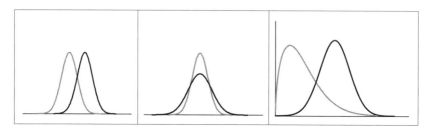

Figure 1.6   Possible process variations.

small number of causes – so-called common causes, in contrast to the special causes – that have a stable and repeatable distribution over time.

1.2.2   Meaning of 'Six Sigma'

'Six Sigma' literally means six times sigma. In statistical terminology the Greek letter σ usually indicates the standard deviation of a random variable (see Section 3.3.1). σ is an index of the dispersion of an empirical or theoretical distribution, and describes the variability of the characteristic under study compared with a central reference value. In the case of Gaussian distributed random variables (Section 3.3.6.2.1), if we consider all possible values in the range $(\mu - 6\sigma, \mu + 6\sigma)$ they represent almost the entire population. In fact, for a Gaussian random variable with mean μ and standard deviation σ, the probability of observing values outside the range $(\mu - 6\sigma, \mu + 6\sigma)$ is equal to 1.98×10^{-9}.

Therefore, supposing that the random variable in question represents a process in control for which a lower tolerance limit and a higher tolerance limit are set, if the mean of the process coincides with the centre of the tolerance interval and if the standard deviation of the process is such that the tolerance range $\Delta = 12\sigma$, then only two times in a billion can we expect the variable to have a value outside the tolerance limits. That is to say that the faulty units per million we can expect (so-called DPMO, defects per million opportunities) will amount to no more than 0.002. This situation is illustrated in Figure 1.7.

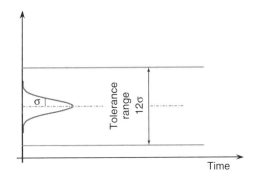

Figure 1.7   Process in statistical control within a tolerance bandwidth equal to 12σ.

However, the standard deviation of the process is rarely such that the tolerance range is about 12 times σ. On the contrary, in Six Sigma programmes this situation represents a goal to be achieved. In fact, the variation of the process is measured against the specification limits required by the customer and the organisation must work on the process to reduce it.

1.2.3   Six Sigma process

When a process is such that its mean coincides with the centre of the tolerance interval and the tolerance range is equal to 12σ we talk about a Six Sigma process. However, we must consider that while the process in control, at time t, provides a value for the DPMO equal to 0.002, systematic causes of action may lead to a shift of the central value of the distribution and/or a change in the dispersion of values around it. For this reason, at the instant $t + \Delta t$, the number of sigmas between the central value of the distribution and the nearest specification limit will be lower than that at time t.

To account for such phenomena in Six Sigma processes observed over time, a measure of the shift equal to 1.5σ has been empirically established. By performing again the calculation of probability, we can expect values outside the tolerance limits only 3.4 times in a million (Figure 1.8). Table 1.1 shows the value of DPMO for different levels of σ.

We will see later that a Six Sigma process can also be expressed in terms of process capability (Section 5.2.3).

1.2.4   Origins of Six Sigma

The term Six Sigma rose from being a mere statistical concept to become a problem-solving methodology in the mid 1980s at Motorola, under the direction of Robert W. Galvin. According to many people it is due to Bill Smith, chief engineer of the

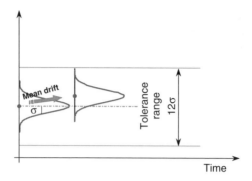

Figure 1.8   A drift of the mean equal to 1.5σ determines a defect rate equal to 3.4 DPMO.

Table 1.1   DPMO versus values of σ.

Level of quality σ^a	DPMO	
	with μ_0	with $\mu_1 = \mu_0 + 1.5\sigma$
σ	317 311	697 672
2σ	45 500	308 770
3σ	2 700	66 811
4σ	63.40	6210
5σ	0.57	233
6σ	0.002	3.4

[a]Distance between the central value of the tolerance ($= \mu_0$) and the nearest tolerance limit.

communications division. He had the idea of reducing the failures of the product during its manufacturing process in order to improve its performance after its delivery to the customer. (The evidence that products repaired along the production line had a higher probability of failure on the field, in contrast to non-repaired products, which had higher expectations of life and better performances, were shown by Smith in an internal report in 1985) However, it must be to a young engineer from Motorola's Governmental Electronics Group, Michael Harry, in the early years of his career at Motorola alongside Smith, that the consolidation of Six Sigma as a problem-solving methodology is really due. Harry developed the so-called MAIC approach (measure, analyse, improve and control) in the report entitled 'The Strategic Vision for Accelerating Six Sigma Within Motorola' and its development strategy for quality. In 1987, Galvin, as part of an ambitious long-term programme called 'The Six Sigma Quality Program' established for 1992 the goal of 3.4 DPMO (Bhote, 1989). Only two years later, Galvin asked Harry to lead Motorola's Six Sigma Research Institute. Harry had soon defined the hierarchy of experts in Six Sigma from the Champion to the Green Belt.

The improvement programme adopted by Motorola during the years 1987–1997 achieved cost savings amounting to 13 billion dollars and increased the productivity of employees by 204 % (Park, 2003). In 1988 Motorola's participation in the Malcolm Baldrige American Quality Award and its victory greatly contributed to spreading knowledge of Six Sigma programmes and successes achieved thanks to them. Already in the early 1990s other big companies like IBM, Texas Instruments and DEC launched Six Sigma programmes.

In 1993 Harry moved to AlliedSignal. In the same year its CEO, Bossidy, started adopting Six Sigma. The following year, Michael Harry and Richard Schroeder, a former Motorola manager, founded the Six Sigma Academy. Among the first customers were AlliedSignal and General Electric. It was in 1996 that Jack Welch, CEO of GE, espoused the Six Sigma strategy. He gave an important contribution to the development of Six Sigma and its dissemination in the world. He believed so much in this strategy that he tied the incentive system to the objectives of Six Sigma and declared that to be a manager one should be at least a Green Belt. GE obtained a cost reduction of 900 million dollars in two years. Also at GE there was the evolution of the standard approach to Six Sigma projects from MAIC to DMAIC, by including the 'Define' phase as the first step.

It was after 1995 that Six Sigma began to spread to industries other than electronics: Toyota, Siemens, Honeywell, Microsoft, Whirlpool, ABB, Polaroid, Sony and Nokia are just some examples. The twenty-first century is seeing the spread of Six Sigma in the service sector, particularly healthcare and financial services (Hoerl et al., 2004).

Along with the spread between organisations, including sectors other than manufacturing, Six Sigma has spread within organisations, involving processes other than production (e.g. design for Six Sigma, DFSS).

Smith (2001) developed a very interesting insight of the evolution of Six Sigma in product development. In organisations where Six Sigma has become established as an approach to manage processes, people tend to identify it with a management philosophy that permeates all activities. In this sense it is considered the paradigm for innovation in the strategic management of organisations of the new millennium (Park, 2003), the successor to the Total Quality Management (TQM), which in turn had replaced the Total Quality Control (TQC), In fact, Six Sigma can certainly be considered as a fruit of earlier decades of philosophical and methodological evolution of the concept of quality. The evolution of Six Sigma is continuing, and one of the most recent developments is the Lean Six Sigma methodology combining the two approaches of Lean Production and Six Sigma.

The lean approach aims to 'rapidly respond to changing customer demands and to create more value at a lower cost' (Womack and Jones, 1996). It was established in manufacturing (lean production) and has gradually spread to other areas of management. The lean methodology is characterised by five steps: (i) to identify what the customer really perceives as value; (ii) to line up value-creating activities for a specific product/service along a value stream; (iii) to eliminate activities that do not add value; (iv) to create a flow condition in which the product/service advances smoothly and rapidly at the pull of the customer and (v) to speed up the cycle of improvement in pursuit of perfection (Su, Chiang and Chang, 2006).

1.2.5   Six Sigma: Some definitions

This section gives a definition of Six Sigma developed by the authors, and then provides definitions that appear significant for the attention given to certain aspects of a concept, which is undoubtedly complex to summarise.

Today the term Six Sigma, in addition to clearly evoking a measure of aspired defect/fault reduction, identifies a strategic programme of continuous improvement characterised by defined stages and the use of statistical and managerial tools. Its aim is to reduce the undesired variability of process and product performances, and the costs associated with it, in order to increase customer satisfaction and increase market share. This definition emphasises that different aspects are equally relevant.

1. The term programme shows that the design of Six Sigma has to follow the process logic through the clear identification of all the elements that characterise it as such (input, output, customers, etc.), and also follow the execution of a clear set of phases (e.g. DMAIC).

2. The strategic nature of the Six Sigma programmes. In fact, the decision on their adoption is to be taken by the top management and it has long-term effects on the entire organisation.

3. The tendency of the programmes towards the 'continuous' improvement of business performance, implemented by minimising non-value added activities (the so-called MUDA, 'muda' in Japanese means waste), is inherent in each of the possible paths suggested by Six Sigma (Section 1.3.2) and pushes towards increasingly ambitious goals.

4. The use of statistical techniques is fundamental. It is not accidental that the term itself, identifying the programmes in question, expresses a statistical concept. This results from the use of the so-called principle of management by facts.

5. Increased customer satisfaction and, consequently, greater market share, is achieved through the improvement of the products and/or services offered and therefore the optimisation of the processes.

The following definitions are, in some cases, compiled by distinguished representatives of big companies, and in other cases by internationally well-known academics.

> ... is a new strategic paradigm of management innovation for company survival in this 21st century, which implies three things: statistical measurement, management strategy and quality culture. (Park, Lee and Chung, 1999)

> ... is a company-wide strategic initiative for the improvement of process performance with the core objectives to reduce costs and increase revenue – suitable in both manufacturing and service organizations. At the core of Six Sigma is a formalized, systematic, heavily result oriented, project-by project improvement methodology tailor-made to achieve improvements on variation first of all, but also in cycle time and yield. (Magnusson, Kroslid and Bergman, 2003)

> Motorola definition: ... a disciplined method of using extremely rigorous data gathering and statistical analysis to pinpoint sources of errors and ways of eliminating them. (Harry and Schroeder, 2000)

General electric definition: Six Sigma is a highly disciplined process that helps us focus on developing and delivering near-perfect products and services' (www.ge.com).

It is a business strategy based on objective decision making and problem solving, relying on meaningful and real data to create actionable goals, analysing root cause(s) of defects, and thus suggesting the ways to eliminate the gap between existing performance and the desired level of performance (Kumar *et al.*, 2008).

By now Six Sigma is a mature framework for quality improvement; an overall approach consisting of a systematic alignment and application of statistical tools for customer satisfaction and business competitiveness (Goh, 2010).

What appears clear is that both Six Sigma programmes and its philosophy, while responding to a series of principles of TQM, including continuous improvement (achieved by continuing efforts to reduce the variability of processes), management by processes (Section 1.1.2), leadership and involvement of the entire organisation, management by fact, focus on customer (Section 1.1.4), is thus representing a continuity with the past; while, on the other side, Six Sigma programmes and its philosophy are characterised by: the indispensable use of statistical methods and concepts, the focus on economic and financial returns (cost reductions and profit increases), and the need for ad hoc trained human resources (Section 1.3.3). What it does mean is that, although according to some scholars (e.g. the well-known expert on quality Joseph M. Juran), Six Sigma does not represent a novel idea (Paton, 2002), according to others (Walters, 2005), Six Sigma is a new recipe made of already known ingredients, a recipe that until today has proved absolutely winning.

1.3   Six Sigma projects

1.3.1   Why implement Six Sigma projects?

The company's productivity is the ratio between revenues and incurred costs. In general, the costs can be grouped into two categories: in 'productive costs' (i.e. those able to generate revenues) and 'unproductive costs': scrap, waste, not required performance (e.g. 'over quality'–'over production'), inspections, maintenance, stock, waiting time, rework, transfers/transport, and so on. The elimination of unproductive costs (the so-called 'Muda') translates into more profit. A winning company will be the one that can provide the market with the products/services with the minimum level of quality that the market is willing to pay for, through processes that naturally produce products/services at Six Sigma.

In particular, in accordance with Goh (2010), Six Sigma when effectively applied:

- helps achieve customer satisfaction with continuous process improvement;

- assesses and compares performances via simple metrics;

- sets concrete targets for impactful improvement projects;

- leads to selection of appropriate problem-solving tools;

- marks progress via verifiable project achievements;

- sustains improved performances throughout the organisation;

- enhances the overall business bottom line and return to stakeholders.

In addition, Six Sigma tends to generate a long-term success because it involves a cultural change and when this happens the results of the change remain for a long time. John Chambers, CEO of Cisco Systems, one of the companies that have experienced tremendous growth in the 1990s, said that there are products that can end up being out of the market in just three years. The only way to continue to grow rapidly and firmly hold market share is to continue to innovate and rethink the entire organisation. Six Sigma develops the necessary culture and skills to support this approach (Pande, Neuman and Cavanagh, 2000).

However, we must be aware of the fact that Six Sigma is not applicable at all times – for a thorough reflection on the possible limits of its application see Goh (2002).

1.3.2   Six Sigma paths

1.3.2.1   DMAIC

As mentioned earlier (Section 1.1) a Six Sigma project and, therefore, a Six Sigma strategy, is characterised by the succession of predefined phases. Originally, the Six Sigma methodology, as it was developed within Motorola, included four steps: measure, analyse, improve and control. The acronym used to describe such a path was MAIC.

Later, at General Electric, a fifth phase was put before the previous four: define, from which was derived DMAIC.

The DMAIC path was born to be applied in the production area of an organisation; however, *ad hoc* paths were developed for the improvement of activities belonging to design, as well as other services. These will be discussed later (Sections 1.3.2.2 and 1.4).

'DMAIC is a closed loop that eliminates unproductive steps, often focuses on measurements and applies technology for continuous improvement' (Kwak and Anbari, 2006). The flow chart in Figure 1.9 shows the DMAIC process and highlights the decision nodes that characterise this path. Here are the steps of the DMAIC process, invoking the most appropriate managerial and statistical techniques in support of each phase.

- *Phase D – define*. The first phase of the DMAIC process aims to identify the products or processes that need a priority intervention through Six Sigma projects. This is a very important phase because all subsequent stages depend on it. Measures of DPMO (Section 1.2.2), customer complaints, suggestions from company people, reports on non-compliance, and so on are useful data for this purpose. Once several products/processes on which it is evident some improvement potential exists are identified, the prioritisation is done through the use of tools such as the Pareto diagram (Section 3.1.4.4), the cause and effect diagram (Section 2.1.1), and so on. Variables to which to refer in order to prioritise possible projects are, for example, the benefits that would accrue to the customers, the benefits to the organisation, the complexity of the concerned process/es, the potential in terms of cost savings, and so on.

The identification of the products/processes on which to act primarily includes the following activities:

— define the requirements and expectations of the customers;

— define the project boundaries;

— define the processes by mapping the business flow.

The execution of this phase should include benchmarking with other international competitors in order to highlight the key product/process features.

- *Phase M – measure*. The first activity included in the measure phase is represented by the selection of one or more characteristics of the chosen product/process ('response' variables), as well as the corresponding resources that have influence on these characteristics ('input' variables). The response variables are usually represented by those features most important to the customers (the so-called 'Critical To Quality', CTQ, characteristics). In this sense, a widely used technique to select the CTQ characteristics is the 'Quality Function Deployment' (QFD, Section 4.4.3).

  After that, all preparatory activities for the collection of data on selected input variables and response variables should follow, such as sampling techniques, measurement intervals, operational modalities for data recording, roles and responsibilities, and so on. Finally in this second phase, all data collection is performed. It should be emphasised that the measurements made at this stage have a definite purpose, namely providing the necessary information to make decisions in terms of changes to the product/process. Therefore, with respect to situations where the measuring activity is aimed at the evaluation of the performance of products and processes, the data collection is much more detailed and is related to the improvement project. Among the measurements to be carried out should be an estimation of the 'capability' of each involved process (Section 5.2.3).

  The main activities consist of:

— measure the process to satisfy customer needs;

— develop a data collection plan;

— collect and compare data to determine issues and shortfalls.

- *Phase A – analyse*. The analyse phase consists essentially in the evaluation of data collected in the previous phase, assessing this through statistical methods, evaluating the process centring and variation, the process stability, the trend of process/product performance, and also making evaluations in terms of DPMO (defects per million opportunities). Various basic managerial and statistical tools can be used to support this phase. Often at this stage a gap analysis is also carried out to identify those common factors that determine best performance. In some cases it might be appropriate to modify the process/product target, in which case the next step would again be represented by 'measurement'.

The execution of this phase, as with the define phase, should include benchmark activities on the performance of response variables for similar products/processes of competing and/or best-in-class organisations. This activity could provide an important basis of information for setting improvement targets.

The main activities consist of:

— analyse the causes of defects and sources of variation;

— prioritise the opportunities for future improvement.

- *Phase I – improve.* The improve phase starts with the decision about the product characteristics on which to intervene, if it is considered that, through their improvement, we can achieve the target set. To establish improvement measures, a set of simple tools is represented by the so-called '7QC tools' (see Section 2.1.1). These tools should allow one to verify the presence of special causes of variation, from which an activity aimed at their elimination and reduction should follow. However, if by the empirical evidence special causes of variation are not found, the design of experiments (see Section 5.4) will be adopted, voluntarily introducing variations in the process and analysing their effect. In addition, we appeal to the robust design method, discussed in Section 5.3.1.

  The main activities consist of:

— improve the process to eliminates variation;

— develop creative alternatives and implement enhanced solutions.

- *Phase C – control.* Initially, the control phase was thought to ascertain the actual improvement in terms of values of the response variables, following the measures adopted in the previous phase. For this it uses statistical process control tools, in particular the control charts (see Section 5.2.2), to monitor and provide evidence of the results arising from the new process conditions. After a period of adjustment, process capability is evaluated again, and, depending on the evidence arising from this analysis, it may be necessary to review all or part of the DMAIC path. Another common activity at this stage is to 'institutionalise' the achieved results. It could be, for example a need to update the flow-chart, the procedure or the process design/product that has been affected by the Six Sigma project, or produce such documents from scratch. Institutionalisation can also be made through an estimate of annual savings, for example in terms of costs due to the implemented improvement. Finally, it is important to disseminate the results of Six Sigma within the organisation, through the production of short reports that describe the case study.

  The main activities consist of:

— control process variations to meet requirements;

— develop a strategy to monitor and control the improved processes;

— implement the improvements of systems and structures.

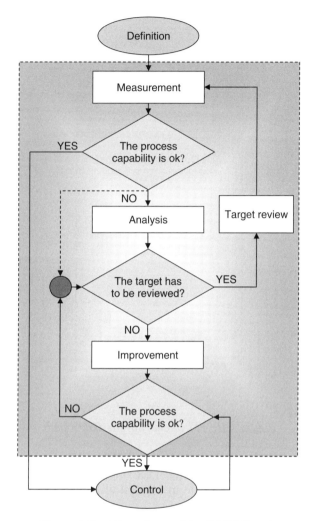

Figure 1.9   Flow chart of the DMAIC path.

1.3.2.2   DFSS – Design for Six Sigma

If the Six Sigma logic is applied to research and development (R&D) we use to talk about Design for Six Sigma (DFSS). The DFSS approach is characterised by the integration of statistical and managerial techniques, and enables organisations to more effectively manage their development process of new products through the optimisation of several key factors such as costs, time to market, and so on (Mader, 2002). To carry out a Six Sigma project in this area some paths have been developed, such as DMADV (define, measure, analyse, design and verify) and IDOV/ICOV (identify, design, optimise and validate or identify, characterise, optimise and validate).

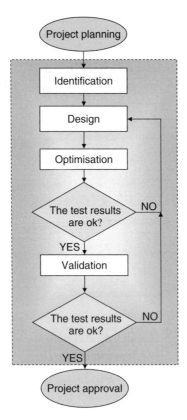

Figure 1.10   Flow chart of the IDOV path.

The flow chart in Figure 1.10 represents the IDOV path, highlighting the decisional nodes that characterise it. Below are the main activities included in each step of the IDOV path, highlighting the most appropriate managerial and statistical techniques to support each phase.

- *Phase I – identify*. Main activities are:

  — identify customer requirements (CTQ variables);

  — consider financial goals;

  — translate the identified customer requirements into quantitative performance measures (and specification limits);

  — consider all needed inputs and establish a business plan;

  — identify roles, responsibilities and milestones.

Useful techniques for the development of this phase are: QFD (Section 4.4.3), Kano analysis (Section 2.4.1), suppliers, inputs, process, outputs and customers (SIPOC) (Section 2.3.2), gauge repeatability & reproducibility (GR&R) (see Section 5.4.4).

- *Phase D – design*. The Design phase emphasises CTQs and consists of developing alternative concepts or solutions, evaluating alternatives and selecting a best-fit concept, deploying CTQs and predicting process capability. Main activities are:

  — formulate several competing design concepts;

  — determine the measurements systems that will allow one to test the performance of the system for important design attributes;

  — implement the concept testing process;

  — evaluate the risk with regard to the CTQs and decide for proceeding or revisiting the design concepts.

Useful techniques for the development of this phase are: flow-charts, cause–effect diagrams (see Section 2.1.1), failure mode and effect analysis (FMEA) (see Section 4.3.1), design of experiment (DOE) (see Section 5.4), correlation and regression analysis (see Section 5.1).

- *Phase O – optimise*. The Optimise phase requires use of process capability information and a statistical approach to tolerancing. Developing detailed design elements, predicting performance and optimising design take place within this phase. Main activities are:

  — identify the key process/product output variables (KPOVs) for each CTQ variable, that is those variables the desired performance depends on;

  — identify the key process/product input variables (KPIVs) that can be controlled in order to optimise the KPOVs;

  — establish criteria for assessing if the process/product fulfils the customer's requirements;

  — estimate the risks related to the CTQs desired performance on the basis of the predicted performance of the KPOVs.

Useful techniques for the development of this phase are: robust design (Section 5.4.3), parameter estimation (Section 3.4) and process capability models (Section 5.2).

- *Phase V – validate*. The Validate phase consists of testing and validating the optimised design. As increased testing using formal tools occurs, feedback of requirements should be shared with manufacturing and sourcing and future manufacturing and design improvements should be noted. Main activities are:

  — prototype test and validation;

  — assess performance, failure modes, reliability and risks;

  — design iteration;

  — final phase review.

Useful techniques for the development of this phase are FMEA (Section 4.3.1), statistical process control (Section 5.2).

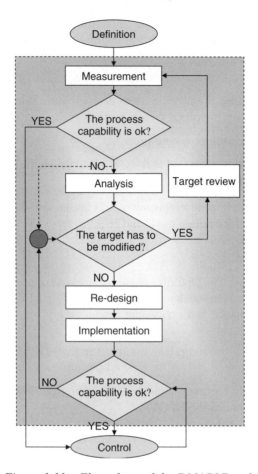

Figure 1.11   Flow chart of the DMARIC path.

1.4   The DMARIC path

If the Six Sigma logic is applied to services then we talk about TSS: transactional Six Sigma. The path that is typically followed to develop a Six Sigma project in this area is the DMARIC: define, measure, analyse, redesign, implement and control.

The flow chart in Figure 1.11 represents a DMARIC process highlighting its characteristic decision nodes. Below are listed the main activities included in each step of the DMARIC process, recalling the most appropriate managerial and statistical techniques in support of each phase.

- *Phase D – definition*
  - define the purpose and context conditions for conducting the project;
  - identify the CTQ characteristics;
  - check your competitiveness with reference to the CTQ's through benchmarking;

— determine the impact of the project in terms of business.

Tools such as the discrete event simulation to evaluate allow one to identify, even at modest cost, which systems do not yet exist and which would be prohibitively expensive to build.

- *Phase M – measurement*
  - identify metrics for the project CTQ's;
  - measure the performance of the CTQ's over time to draw conclusions on the variability in the long term;
  - look after the aspects of the financial arrangements relating to the project.

Tools such as Conjoint Analysis (Section 5.3.5.2) can be used for assessing customer preferences.

- *Phase A – analysis*
  - draw a flow-chart of the process at a level of detail that gives the possibility to see improvement opportunities;
  - draw a cause and effect diagram or a matrix that clarifies the relationships between input variables and outcome variables (CTQ's);
  - use the Pareto chart to sort out the input variables;
  - perform regression, correlation and variance analysis to investigate the cause and effect relationships between input and outcome variables.

- *Phase R – redesign*
  - use DOE to assess the impact of possible changes to the process;
  - evaluate eventual changes to the work standard or process flow in terms of improvement of the quality and/or productivity of the process;
  - use DOE and other tools to determine the optimal operating scenarios of input variables (operating windows).

- *Phase I – implement*
  - set optimal work standards and/or the process flow;
  - check that the optimal operating scenarios of input variables are suitable and then implement them;
  - verify process improvements, process stability and its performance.

- *Phase C – control*
  - update the monitoring plan;
  - monitor the relevant input and output variables through the control charts;
  - prepare a final project report highlighting the benefits deriving from it;
  - disseminate the report among staff within the organisation;

— monitor results 3 and 6 months after the end of the project to ensure that process improvements are kept.

1.4.1 Human resources and training

The implementation of Six Sigma projects is possible only if the organisation has adequate human resources. In general we can say that the main required skills relate to the Total Quality Management (TQM) and Statistics. In large companies, where the Six Sigma methodology is adopted as a management philosophy, several figures are identified within the organisation that constitute a kind of human 'infrastructure' to guarantee the implementation of Six Sigma over time through the succession of Six Sigma projects.

Such an infrastructure is characterised by five hierarchical levels:

- *The Executive Leadership* – it is usually granted to the CEO and other key members of senior management to have the role of creating a corporate vision of Six Sigma and to sustain it over time. They also have to ensure that those in other roles have the freedom of action and resources needed to be able to explore new areas for improvement.

- *The Champions* – they are usually selected by the executive leadership from among the top-level management positions. The main task of a champion is to develop a deployment strategy of Six Sigma in an integrated manner among the various business functions, and monitor the implementation over time.

- *The Master Black Belts (MBBs)* – they are usually internal resources for full-time leadership for the organisation of Six Sigma, in principle, limited to one or more areas or business functions; they are selected from the Champions to accompany the Champions themselves and to carry on training and guidance for the Black Belts and the Green Belts. The MBB are experts in Six Sigma, able to identify, select and approve projects to oversee their development and completion, ensuring the correct and rigorous application of statistical and managerial techniques.

- *The Black Belts* – are employees of the organisation, full time in this role. However, this role can be temporary (e.g. two years). Hoerl (2001) defines Black Belts in this way: 'I view them to be the technical backbone of successful Six Sigma initiatives – the folks who actually generate the savings.' They work under the guidance of Master Black Belts to apply Six Sigma on specific projects; they are therefore the material executors of Six Sigma projects and are often in charge of a team of Green Belts to work on solving a problem.

- *The Green Belts* – are also employees of the organisation but they work for the implementation of Six Sigma projects only partially, in addition to their specific roles and responsibilities. They are running projects under the guidance of Black Belts.

In some companies, such as Honeywell, GE and 3M, GB (Green Belt) certification, and sometimes the BB certificate, is necessary in order to obtain promotions to management roles. In these cases, Six Sigma is also considered a tool for leadership development, a key leadership skill necessary to lead change in organisations (Snee in Hoerl, 2001).

References

Bhote, K.R. (1989) Motorola's long march to the Malcolm Baldrige national quality award. *National Productivity Review*, **8** (4), 365–376.

Bisgaard, S. (2008) Innovation and Six Sigma, *ASQ Six Sigma Forum Magazine*, **3**, 33–35.

Goh, T.N. (2002) A strategic assessment of Six Sigma. *Quality and Reliability Engineering International*, **18**, 403–410.

Goh, T.N. (2010) Six Sigma in industry: some observations after twenty-five years. *Quality and Reliability Engineering International*, doi: 10.1002/qre.1093.

Harry, M. and Schroeder, R. (2000) *Six Sigma: The Breakthrough Management Strategy Revolutionizing the World's Top Corporations*, Doubleday, New York.

Hoerl, R.W. (2001) Six Sigma black belts: what do they need to know? *Journal of Quality Technology*, **33** (4), 391–406.

Hoerl, R.W., Snee, R.D., Czarniak, S. and Parr, W.C. (2004) The future of Six Sigma. *Six Sigma Forum Magazine*, August, 38–43.

Kumar, M., Antony, J., Madu, C.N., Montgomery, D.C. and Park, S.H. (2008) Common myths of Six Sigma demystified. *International Journal of Quality and Reliability Management*, **25** (8), 878–895.

Kwak, Y.H. and Anbari, F.T. (2006) Benefits, obstacles, and future of six sigma approach. *Technovation*, **26** (5–6), 708–715.

Mader, D.P. (2002) Design for Six Sigma. *Quality Progress*, July, 82–84.

Magnusson, K., Kroslid D. and Bergman, B. (2003) *Six Sigma: The Pragmatic Approach*, Studentlitteratur, Sweden.

Pande, P.S., Neuman, R.P. and Cavanagh R.R. (2000) *The Six Sigma Way: How GE, Motorola, and Other Top Companies are Honing their Performance*, McGraw-Hill.

Park, S.H. (2003) Six Sigma for Quality and Productivity Promotion, Productivity Series 32, Asian Productivity Organization, Tokio.

Park, S.H., Lee, M.J. and Chung, M. (1999) *Theory and Practice of Six Sigma*, Publishing Division of Korean Standards Association, Seoul.

Paton, S.M. (2002) Juran: a lifetime of quality. *Quality Digest*, **22** (8), 19–23.

Smith, L.R. (2001) Six Sigma and the evolution of quality in product development'. *Six Sigma Forum Magazine*, November, 82–84.

Su, C., Chiang, T. and Chang, C. (2006) Improving service quality by capitalising on an integrated Lean Six Sigma methodology. *International Journal of Six Sigma and Competitive Advantage*, **2** (1), 1–22.

Walters, L. (2005) Six Sigma: is it really different? *Quality and Reliability Engineering International*, **21** (6), 221–224.

Womack, J. and Jones, D. (1996) *Lean Thinking: Banish Waste and Create Wealth in your Corporation*, Free Press, New York.

2

Basic managerial techniques

2.1 For brainstorming

'Brainstorming' is a technique developed around the 1940s by F. A. Osborn to facilitate the creation of innovative advertising. It consists in generating the largest number of possible ideas regarding a certain object of interest by a specially constituted team, under the guidance of a so-called 'facilitator'. The facilitator is responsible for ensuring that each group member can state his opinion freely, succinctly and without being subject to criticism.

First of all, the object of interest must have been well defined to discriminate against any irrelevant ideas. In each round everyone has the right to express one single idea, and in turn and in a fixed time, and the ideas generated should be recorded. The brainstorming session then ends. Normally, at the end of it, when all group members feel they have no more ideas to be presented, the ideas are screened individually for the necessary clarifications and possible redefinition (amalgamations of similar ideas, etc.). The final objective is to find the most shared ideas. To this end, typically some voting system is called for.

Both the cause–effect diagram (Section 2.1.1) and the affinity diagram (Section 2.1.2) use brainstorming.

2.1.1 Cause–effect diagram

The cause–effect diagram is a graphical tool that shows all possible causes that could lead to a certain effect, the relationships between those causes and among them and the effect itself. The effect is most frequently a problem. However the cause–effect diagram can also be used to study the causes of a positive effect. The cause–effect diagram is also described as a 'fishbone' diagram, because of its shape resembling a fishbone, or an Ishikawa diagram after its inventor Kauru Ishikawa, who developed it in the 1960s. It is considered to be one of the seven basic tools of quality management (7 QC tools),

Statistical and Managerial Techniques for Six Sigma Methodology: Theory and Application, First Edition.
Stefano Barone and Eva Lo Franco.
© 2012 John Wiley & Sons, Ltd. Published 2012 by John Wiley & Sons, Ltd.

together with the histogram (Section 3.1.2.2), Pareto chart (Section 3.1.4.4), check sheet (Section 3.1.1.2.1), control chart (Section 5.2), analysis by stratification (Section 3.1.3), and scatter plot (Section 3.1.4.3).

The identification of the effect is the basis of a successful analysis, as the better defined the effect, the more focused and effective the analysis will be. Next, you need to list the potential causes of the defined effect. To carry out this second phase three methods can be followed:

1. causes classification;

2. process steps;

3. cause listing.

Whichever method is chosen, the search for all possible causes is an activity to be conducted by a team of people who have a strong interest in the effect being studied and normally a brainstorming is adopted.

The classification of the causes involves, first, the identification of general categories for an orderly analysis of detail. One criterion may be the so called '5 M': Machines, Men, Methods, Materials and Measurements. The chosen categories will form the 'bone' of the diagram. This classification serves to organise ideas and stimulate creativity. The causes identified through an initial round of brainstorming should be placed on the diagram in the relevant category (e.g. Men). When you think you have exhausted the enunciation of the direct causes of the effect, you may continue brainstorming to identify sub-causes or indirect causes, that is causes of the already stated causes, and so on (see Figure 2.1). In general, this process goes on until the level of detail is considered sufficient for the analysis.

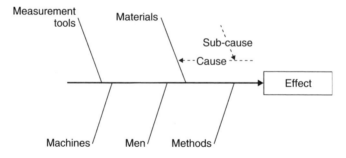

Figure 2.1   Cause–effect diagram made according to classification of causes.

For example, in Figure 2.2 the effect of 'variation in play in mast section' has been traced back to a possible cause relating to 'method', and the problems associated with it can be traced back to the 'assembly process' that in turn depends on 'estimation of play' (see the 'Atlet' case study in Chapter 6).

Once the diagram has been constructed, the team will discuss the effect with more knowledge of the facts (3). The group will examine in an orderly manner the identified root causes by analysing the consequences of a possible intervention on each of them.

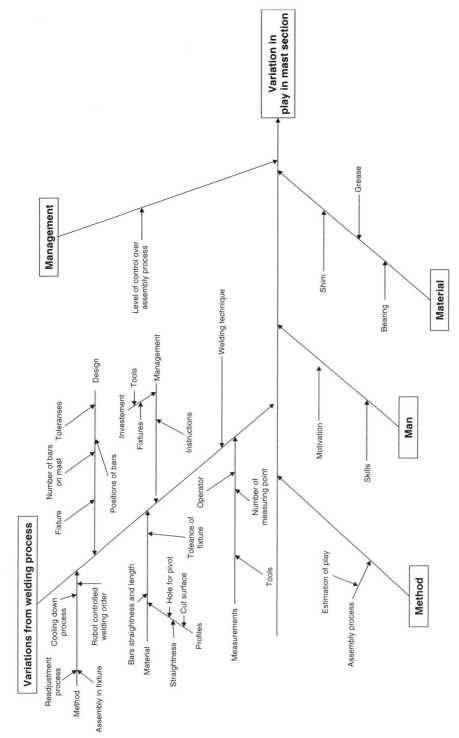

Figure 2.2   Example of cause–effect diagram with classifcation of causes.

We need to come to a ranking of importance of them, also through a voting system. An activity leading to the verification of the most likely causes may follow.

We use the 'process steps' method when the effect to be examined can be thought of as the result of a sequence of stages. This method requires a clear and precise definition of the generating process of the effect being studied.

With the 'cause listing' method, we start from a simple listing of causes. The list should be as broad and comprehensive as possible. Only in a second phase must the causes be ordered and hierarchised, highlighting their reciprocal relationships. This third method has the advantage of not forcing the operators into preconceived schemes, referring the drawing of the scheme to a later date.

2.1.2   Affinity diagram (KJ analysis)

The affinity diagram or KJ analysis, was proposed by Jiro Kawakita, a Japanese anthropologist, who developed it around the 1960s as a 'means of decision-making that can be utilised in all societies throughout the world, to implement social and economic devc.-opment'. It is a structured method to discuss, analyse and have a coherent vision of a complex and hard to define problem.

Also in this case, as for the cause–effect diagram, a team of people is involved who represent a strong interest in the problem being analysed through a procedure akin to brainstorming (Section 2.1). The team nominates a leader who will lead the analysis and manage the time. Also needed is a large white board (or otherwise a large paper sheet) on which to post notes (sticky notes are good for the task) and on which to write normally using felt-tip pens.

The problem under study must be well defined and reported, and this is written on the top left of the board. The main steps of the analysis are as follows.

1. After a preliminary discussion, each participant provides his opinions about the problem.

2. In turn, each team member writes his thoughts by succinct explanatory sentences, one concept/opinion on each note. Duplications should be avoided. One by one, the notes are neatly posted on the left-hand side of the board.

3. All sentences written on the notes are reviewed so that they are well understood by everybody: the leader focuses on a note at a time; the person who wrote it will illustrate its meaning. If the leader deems it necessary, during the debate on its meaning, he can correct and/or add terms. When he considers the final sentence clear and well expressed, he moves the note to the right-hand side of the board, and so on.

4. The notes should be organised into logical first-level groups: the leader does this by grouping them and continues until all team members feel satisfied. At that point, circles are drawn (usually in black colour) including the groups of notes.

5. A title is given to each first level group, that is a summary sentence clarifying the relationship between the notes in that group (usually a red pen is used). For isolated notes no title is given (see Figure 2.3).

6. First-level groups are in turn organised into second-level groups and each of them, marked with a circle (usually green) is assigned an explanatory sentence (or title, this time using a blue pen). At this level of aggregation any isolated note should be reconsidered and may be included within a second level group; otherwise they remain isolated.

7. Cause–effect relationships among second level groups are set out, by distinguishing relations of dependence (using the arrow symbol), contrast relations (using a segment). Relationships of mutual influence should not be considered. The clusters should be created so that they leave free the right top and bottom corners of the whiteboard.

8. Now each team member can vote on the first-level groups considered most important. Everyone has three votes, of different weights: a red vote is worth three points, a blue vote is worth two points and a green vote is worth one point.

9. Based on the results of the vote, the group/s of notes which is recognised to be of highest importance will be found and therefore the conclusions drawn should be placed on the top right-hand corner of the board.

10. Finally, in the lower right-hand corner of the board, the place, date and names of the participants are reported (see Figure 2.4).

In the example shown in Figure 2.5, the issue analysed through the KJ Analysis was poor behaviour of some hospital departments when looking at problems encountered in their daily work (see the 'Sahlgrenska and Östra Hospitals' case study in Chapter 6).

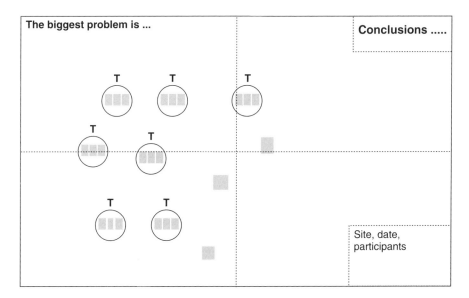

Figure 2.3   Affinity diagram – KJ analysis, grouping at first level.

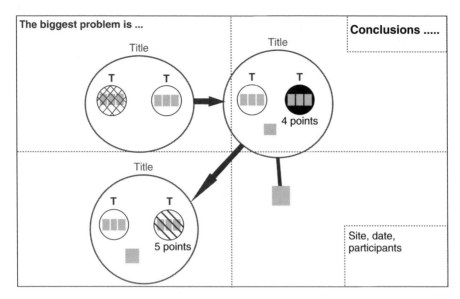

Figure 2.4   Affinity diagram – KJ analysis.

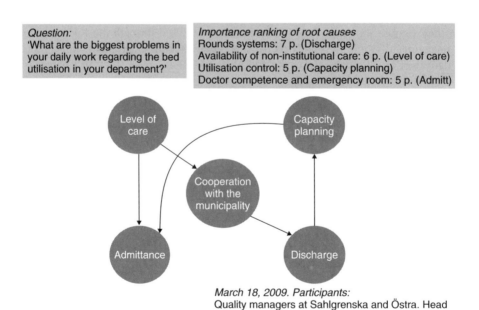

Figure 2.5   Example of affinity diagram ('Sahalgrenska-Östra Hospitals' project).

2.2 To manage the project

2.2.1 Work breakdown structure

A complex project can be more easily managed if it is broken down into individual components according to independent targets and following a hierarchal policy. The resulting structure, known as 'work breakdown structure' (WBS), favours the allocation of resources, accountability, measurement and control of the project. The project WBS is usually illustrated by a block diagram, in which are highlighted: the tasks, sub-tasks and work packages (see Figure 2.6). Since the WBS is a hierarchical structure it can also be represented through a table (see Table 2.1).

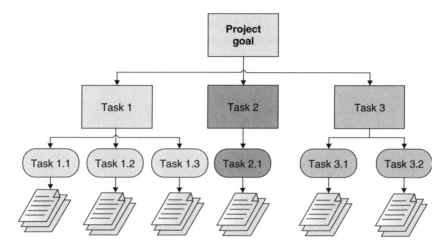

Figure 2.6   Work breakdown structure diagram. Pictorial format.

Table 2.1   Work breakdown structure: tabular format.

Level 1	Level 2	Level 3
Task 1		
	Subtask 1.1	
		Work Package 1.1.1
		Work Package 1.1.2
		Work Package 1.1.3
	Subtask 1.2	
		Work Package 1.2.1
		Work Package 1.2.2
		Work Package 1.2.3
Task 2		
	Subtask 2.1	
		Work Package 2.1.1
		Work Package 2.1.2
		Work Package 2.1.3

Each organisation may use specific terminology to classify the components of the WBS based on their level in the hierarchy. For example, some organisations use the terms task, sub-task and work-packages; others use phases, entries and activities.

The WBS can be organised on the basis of 'deliverables' or lifecycle stages of the project. The highest levels in the general structure are usually carried out by teams. Conversely, the lowest level in the hierarchy often includes activities of individuals. A WBS based on deliverables does not necessarily specify the activities to be carried out.

In creating the WBS it is important to choose an appropriate level of detail. We must avoid the extremes: that is a very high level of detail that may cause so-called micro-management, or a level of detail which is not sufficient to ensure that the relevant tasks are effectively handled. Usually the tasks should be defined so that their lifetime spans between several days and a few months. The WBS is the foundation of project management. Accordingly, it should be used as a preliminary tool to identify (e.g. along with the critical path method, CPM, Section 4.2.2) and to estimate the duration of the project activities (e.g. with the programme evaluation and review technique, PERT, Section 4.2.1).

2.2.2   Gantt chart

Every project requires a lead time, because the start of some activities of a project can be independent or dependent on the end of the previous activities. Making the schedule for a project means determining the exact time line for the implementation of the whole project.

As part of project management this activity can be supported by tools such as the Gantt chart and PERT (see Chapter 4), which allow one to better visualise the evolution of an ongoing project.

Henry Laurence Gantt, an American mechanical engineer, is credited with the invention of the Gantt chart in the early 1900s. The Gantt chart is a graphical representation of the duration of work compared to the progression of time. It is a useful tool for project planning and for monitoring the progress of a project.

Project planning and scheduling: the Gantt chart is used to plan how long a project should last. It explains the order in which tasks must be carried out. Older versions of Gantt charts did not show the interdependence between tasks; however, modern Gantt charts made with the use of computer software include this feature.

Project monitoring: the Gantt chart allows one to immediately see what needs to be executed at a certain time. It allows one to understand what remedial actions may bring the project within the set time line. Most of the current Gantt charts include 'milestones', which technically should not appear in a Gantt chart. However, to represent deadlines and other important events, it is useful to include this feature in the chart.

The example in Figure 2.7 is taken from the 'Sahlgrenska-Östra Hospitals' case study (see Chapter 6).

2.3   To describe and understand the processes

The study of a process – useful for its monitoring and improvement – requires a thorough knowledge of it, of the elements that constitute it and the relationships characterising it, within it and with other activities external to it. Accordingly, the mapping of a process,

ID	Task name	Start	Finish	Duration	Feb 2009				Mar 2009				Apr 2009				May 2009		
					2/1	2/8	2/15	2/22	3/1	3/6	3/15	3/22	3/29	4/5	4/12	4/19	4/26	5/3	5/10	5/17	5/24
1	Project definition	2/3/2009	2/12/2009	1.6w																
2	Cost of poor quality	2/3/2009	2/12/2009	1.6w																
3	Critical to quality	2/3/2009	2/12/2009	1.6w																
4	Process maps	2/3/2009	2/25/2009	3.4w																
5	SIPOC	2/3/2009	2/25/2009	3.4w																
6	Ishikawa	2/25/2009	3/4/2009	1.2w																
7	P-diagram	2/25/2009	3/4/2009	1.2w																
8	KJ shiba analysis	3/4/2009	3/18/2009	2.2w																
9	Data collection	2/25/2009	3/25/2009	4.2w																
10	Pareto	3/25/2009	3/30/2009	.8w																
11	Graph data	3/25/2009	4/24/2009	4.6w																
12	Analyse data	3/25/2009	4/24/2009	4.6w																
13	Improvement suggestions	4/20/2009	5/6/2009	2.6w																
14	Control procedure	5/6/2009	5/14/2009	1.4w																
15	1st Draft	4/24/2009	5/15/2009	3.2w																
16	Final Draft	5/15/2009	5/20/2009	.8w																
17	PPT presentation	5/15/2009	5/20/2009	.8w																

Figure 2.7   Example of Gantt chart ('Sahalgrenska-Östra Hospitals' project).

namely the description of activities and/or representation thereof, is a fundamental step. Sometimes, when no preliminary description/representation of the process under study is available, the P diagram (process diagram) can be used. It is a basic scheme, mostly used in the robust design methodology (see Section 5.3.1).

2.3.1   The SIPOC scheme

SIPOC stands for: Suppliers–Inputs–Process–Outputs–Customers (Figure 2.8). This is a process mapping tool. Through the conceptual framework suggested by the acronym, the fundamental elements of a process are defined, being described in a clear and essential way. This method of process analysis can assist in the identification stage of the processes of an organisation and in the first phases of their design.

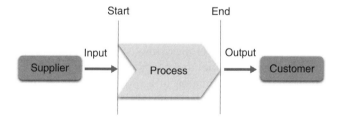

Figure 2.8   SIPOC scheme.

The activities suggested by the SIPOC scheme (see Figure 2.8) are:

• identify the process and its purpose;

• define the beginning and the end of the process;

- make a list of the main outputs and their customers;

- make a list of main inputs and their suppliers.

The example in Figure 2.9 is taken from the 'Saab' case study (see Chapter 6). The process in it has been expressed through a flow chart (Section 2.3.2).

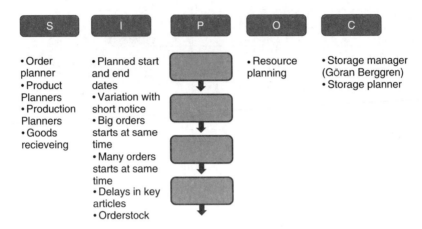

Figure 2.9   Example of application of the SIPOC scheme ('Saab' project).

2.3.2   The flow chart

The flow chart is a graphical representation of a process. The use of the flow chart was developed in the software industry around the 1950s. Over time it spread to all sectors where principles of process management are applied.

The flow chart is a graphical tool requiring the identification of the activities that make the process which is being represented. It uses some symbols to graphically represent all key elements of the process, but also shows their relations, which is why the term flow chart is used.

The symbols used are, among others, rectangle, rhombus and arrow. The rectangle represents an activity/action to be performed, the rhombus is a decision or condition, which may result in alternative actions, while the arrow is a line characterised by a direction. An arrow shows a link and at the same time a temporal/logical order between two actions (Figure 2.10). The graphical representation of a process simplifies its understanding by employees, customers, and so on, and facilitates the monitoring and the identification of possible critical points.

The graphical description of a process can be more or less detailed; the level of detail must be functional to the purpose for which the flow chart is used. This graphical tool can be used after or in conjunction with the SIPOC scheme (Section 2.3.1) to analyse the activities that constitute the process and/or with the IDEF method (Section 4.1.1) to represent timing and roles of the identified activities.

The example in Figure 2.11 is taken from the 'Sahlgrenska-Östra Hospitals' case study (Chapter 6).

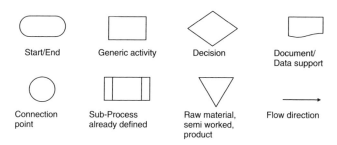

Figure 2.10   Basic symbols of a flow chart.

2.3.3   The ServQual model

A service is the result of an activity or a process, and it implies an interaction between a supplier and a customer (e.g. a medical examination, room cleaning, etc.). The main characteristics of a service are:

- intangibility;
- heterogeneity;
- inseparability of production and consumption.

A service is 'intangible'. This feature makes it difficult to define and understand production specifications, experimentation of prototypes before supply, the awareness of customer wants and needs and the measurement of customer perceptions.

Service 'heterogeneity' is a direct consequence of its unrepeatability; heterogeneity is mostly due to the prevalence of the human factor (suppliers and receptors of services) and to the difficulty of performance standardisation.

Finally, the 'inseparability' is a reference to the phases of production and consumption, which are mostly linked and reciprocally influencing each other. It is these three characteristics that show the big difference between a service and a product.

The ServQual model was developed by A. Parasuraman, L.L. Berry and V.A. Zeithaml in the 1980s. The scope of their research was to develop a conceptual model which identifies factors influencing the service quality and the sources of non-quality. To pursue the model, they carried out an explorative survey on managers and some focus groups of customers in four American banking services companies.

From the focus groups, they discovered that the perceived quality of a service is the result of a comparison between expectations and perceptions. This comparison is made by every customer in a more or less explicit manner. They called the possible difference between perceptions and expectations the **main gap** (or gap 5, see Figure 2.12). They also highlighted that expectations for services are primarily formed through three channels:

1. communication through 'word of mouth';
2. personal needs;
3. past experience.

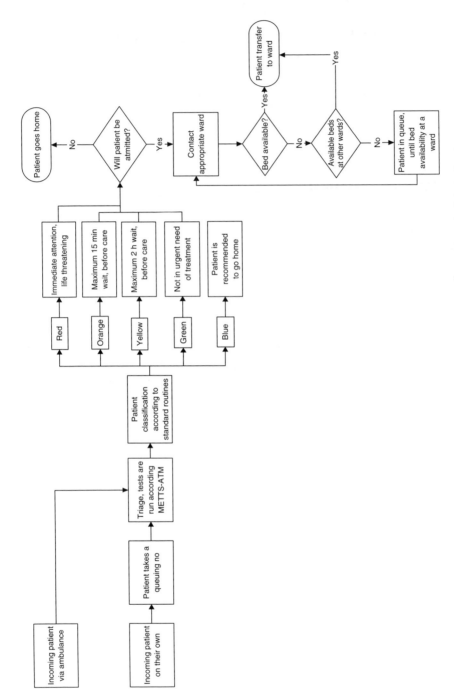

Figure 2.11   Example of flow chart ('Sahalgrenska-Östra Hospitals' project).

Finally, from the focus groups they classified the main aspects of the quality of a generic service into five so-called 'determinants':

- assurance (knowledge and courtesy of employees, their ability to inspire trust and confidence);

- empathy (care, individualising attention the firm provides its customers);

- reliability (ability to perform the promised service dependably and accurately);

- tangibles (physical facilities, equipment, appearance of personnel);

- responsiveness (knowledge, skills, flexibility, ...).

For example, in the case of the 'University teaching' service, we can define the five determinants as:

- *Responsiveness*. This refers to the willingness and readiness of the teacher and his staff to respond to student expectations during the course. It also refers to their capacity to face unforeseen events of an organisational nature without disappointing students, their willingness to support students in their learning process during the course and their willingness to transfer all necessary information about the course.

- *Tangibles*. This includes the physical aspects of the course, such as the textbooks suggested by the teacher for studying the subject, the material tools that help the teacher explain the lesson (traditional blackboard, computer, slides, etc.) and the room conditions in which lessons take place.

- *Assurance*. Assurance covers the teacher's competence (i.e. the abilities and knowledge necessary to teach the course contents) and credibility (i.e. the ability to inspire loyalty and honesty). It also means courtesy (kindness and respect towards students) and safety (freedom from doubt and uncertainty).

- *Empathy*. Empathy refers to a teacher's ability to transfer clearly his own knowledge to students, and the ability to retain interest during a lesson. The teacher is committed to understanding student needs and pays attention to them individually. It also means access to lessons and contact with the teacher (timetable and place where lessons and student reception are delivered).

- *Reliability*. The reliability implies an agreement between the perceived service and the trust that the student places on the course as a result of the information found (e.g. at the department, on websites, etc.) and received (e.g. by the teacher, by other students who attended the same course, etc.). It means the ability of the course to keep its promises, in the sense that everything that was initially announced regarding the course (its contents, class schedule, examination modalities, recommended study materials, etc.) occurred without any unwanted change.

As a result of the interviews with the managers, Parasuraman *et al.* understood that when the main gap (gap 5) occurs, it is as the result of one or more of the following other gaps:

- *Gap 1*: difference between customer expectations and management perceptions.
  Main factors:

  — inappropriate market research, insufficient market research, misuse of the survey results, lack of interaction between managers and customers;

  — inadequate upward communication, presence of too many hierarchical levels.

- *Gap 2*: difference between management perceptions and service quality specifications.
  Main factors:

  — inadequate commitment of managers to the quality of service;

  — poor confidence on objectives that need to be achieved;

  — inadequate standardisation of tasks;

  — lack of definition of objectives.

- *Gap 3*: deviation of actual performances from service quality specifications.
  Main factors:

  — ambiguity of roles, conflict of roles, poor suitability to the employee role;

  — poor suitability of the technology, inadequate supervisory systems and controls;

  — feeling of lack of control, lack of teamwork.

- *Gap 4*: no-correspondence between the supplier's promises about a service characteristic and the actual performance.
  Main factors:

  — inadequate horizontal communication (e.g. between the advertising department and operational functions, between vendors and operational functions, between human resources, marketing and operating department);

  — differences of policy and procedures between different branches or departments;

  — tendency to make exaggerated promises.

  Possible applications within Six Sigma projects are:

  — measuring customer's satisfaction (five quality determinants, service quality definition);

  — facilitating the mapping of a new process (i.e. design) or existing process (i.e. improvement) (five gaps, expectation sources).

Figure 2.12 is an illustration of the ServQual model.

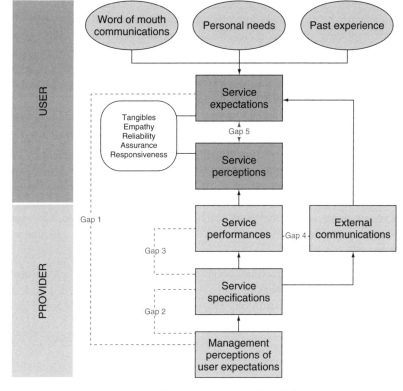

Figure 2.12   The ServQual model.

2.4   To direct the improvement

2.4.1   The Kano model

Developed in the 1980s by Prof. Noriaki Kano, the Kano model helps one to understand the relationship between the degree of achievement of objective quality and the satisfaction level experienced by a customer. While working on social science theories of satisfaction, Kano concluded that the relationship between the fulfilment of a need and the experienced satisfaction is not necessarily linear. He found that a product or a service can be thought of as being made of elements or attributes having different relations with satisfaction.

The Kano diagram is a Cartesian graph showing the degree of achievement on the horizontal axis and the satisfaction level on the vertical axis (Figure 2.13). In this model, Kano distinguishes five categories of attributes:

- *Must-be elements*. The 'must-be' elements are the basic elements of a product. If these elements are not fulfiled, the customer will be extremely dissatisfied. On the other hand, their fulfilment does not increase customer satisfaction. Customers consider must-be elements as prerequisites; they take them for granted and therefore do not explicitly demand them. Must-be requirements are in any case decisive competitive factors; if they are not fulfiled, customers will not be interested in the product/service at all.

For example, when going to a restaurant, the customer expects that there will be a place to sit. If there is no seat, the customer will be dissatisfied. If there is a seat, no credit will be given because it is supposed to be there. Alongside this, the availability of many seats does not give additional satisfaction.

- *One-dimensional elements*. One-dimensional quality elements result in satisfaction when fulfiled and dissatisfaction when not fulfiled. These elements are usually explicitly demanded by the customer. A customer in a restaurant expects his order to be taken promptly and the food delivered in a reasonable time. The better the restaurant meets these needs, the more satisfied the customer will be. One-dimensional attributes are product elements that have great influence on how satisfied a customer will be with the product.

- *Attractive elements*. Attractive quality attributes can be described as 'surprise and delight' attributes; they provide satisfaction when fully achieved, but do not cause dissatisfaction when not fulfiled. These are attributes normally not expected and, therefore, they are often unspoken.

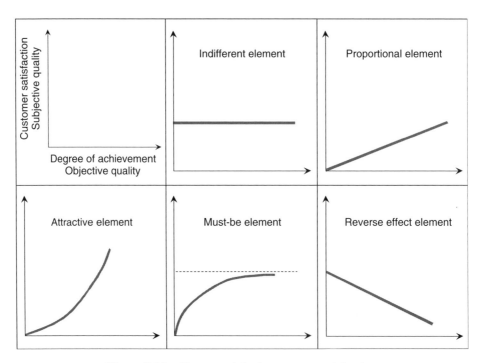

Figure 2.13   Kano model of customer satisfaction.

For example, if the restaurant gives a glass of wine for free the customer will be pleasantly surprised. Yet the absence of this additional service will not result in customer dissatisfaction or loss of clients.

- *Indifferent elements*. Indifferent quality elements refer to aspects that are neither good nor bad. Consequently, their achievement does not result in either customer satisfaction or dissatisfaction.

- *Reverse requirements*. Reverse quality requirements refer to a high degree of achievement resulting in dissatisfaction (and vice versa, a low degree of achievement resulting in satisfaction). For example, too many 'extra features' in a technological device may dissatisfy some customers.

Meeting customers' basic quality needs provides the foundation for the elimination of dissatisfaction and complaints. Exceeding customers' performance expectations creates a competitive advantage, and innovations differentiating the product and the organisation creates an excited customer.

It has also been observed that customer requirements change over time. Over time, an 'attractive' can become 'one-dimensional', and, with the passing of time, become a 'must-be' requirement.

References

Barone, S. and Lo Franco, E. (2009) Design of a university course quality by teaching experiments and student feedback (TESF). *Total Quality Management*, **20** (7), 687–703.

Barone S. and Lo Franco E. (2010) TESF methodology for statistics education improvement. *Journal of Statistics Education*, **18** (3).

Griffin, A. and Hauser, J.R. (1993) The voice of the customer. *Marketing Science*, **12** (1), 1–27.

Hauser, J.R. and Clausing, D. (1988) The house of quality. *Harvard Business Review*, **66**, 63–73.

Ishikawa, K. (1991) *Introduction to Quality Control*, 3A Corporation, Tokyo, ISBN: 4-906224-61-X.

Kano, N., Seraku, N., Takahashi, F. and Tsuji, S. (1984) Attractive quality and must-be quality. *Hinshitsu (Quality, The Journal of the Japanese Society for Quality Control)*, **14** (2), 39–48.

Parasuraman, A., Zeithaml, V.A. and Berry, L.L. (1985) A conceptual model of service quality and its implication for future research. *Journal of Marketing*, **49** (4), 41–50.

Parasuraman, A., Zeithaml, V.A. and Berry, L.L. (1988) SERVQUAL: a multiple-item scale for measuring consumer perceptions of service quality. *Journal of Retailing*, **64** (1), 12–40.

Scupin, R. (1997) The KJ method: a technique for analyzing data derived from Japanese ethnology. *Human Organization*, **56** (2), 233–237.

3

Basic statistical techniques

We will use the following notation to improve the readability of this chapter:

- an uppercase italic letter will either denote a random event (in this case we will use letters from early in the alphabet, e.g. A) or a random variable (in this case we will use one of the later letters of the alphabet, e.g. X);

- a lowercase italic letter will denote either a deterministic variable or a constant;

- a lowercase Greek letter will denote a parameter of a random variable model or an analytic function;

- r.v. stands for random variable;

- s – stands for 'stochastically' (i.e. from a probabilistic viewpoint).

3.1 To explore data

3.1.1 Fundamental concepts and phases of the exploratory data analysis

Exploratory data analysis (EDA) aims to rapidly acquire a summary of the salient characteristics of the observed phenomenon, mainly using graphical tools. The EDA is usually the start-up phase of an investigation and is generally characterised by four steps:

- data collection and arrangement;

- data analysis (or processing);

- presentation of the results of the analysis;

- interpretation and discussion of the results.

Statistical and Managerial Techniques for Six Sigma Methodology: Theory and Application, First Edition.
Stefano Barone and Eva Lo Franco.
© 2012 John Wiley & Sons, Ltd. Published 2012 by John Wiley & Sons, Ltd.

Data collection requires one to define the objectives of the analysis itself. The objectives should, if possible, be clear, precise and unambiguous. They should be explicit in the temporal and spatial dimensions. Of course, the degree of complexity of the phenomenon to be analysed, its nature and the objectives of the analysis affect the collection modalities, which must be clearly specified.

The data collection arrangement is followed by a phase of methodological development. This phase involves the choice of statistical methods to be applied. We can use existing methods or develop new ones. The choice of the method does not always take place immediately after the collection of information. Indeed in some cases it may precede and be held together with the definition of research objectives. Again, the nature of the collected data and the intent of the analysis (simply descriptive and/or inferential) influence the choice.

The presentation of results of an exploratory analysis represents a very important activity because often the people who use the results of the exploratory analysis are not the same as those who carried out the data collection. For this reason the presentation of results must be effective, that is it must provide evidence of more substantial results in relation to the objectives of the analysis and allow for easy interpretation. Finally, the use of the results should be as balanced and scientific as possible.

We may unconsciously use EDA in dealing with real problems every day; here are some examples:

- to know the orientation of public opinion, in particular when elections are approaching;

- to understand a failure mechanism in a production process that determines certain defects;

- to evaluate the status and the progress of our company in the market, for example concerning sales results;

- to assess whether the number of accidents on the highways of a certain region should be considered normal or should require special attention.

The EDA, as a branch of statistics, contains all its typical aspects, that is:

- the discovery of 'new' information, currently unknown, about the phenomenon under study;

- the synthesis of information through analytical and graphical tools and graphs that express the 'relevant information' contained in the data;

- the dialectic – in fact, the acceptance of possible conclusions, proposals for new analysis and new hypotheses and decisions all may stem from the results of EDA;

- the support to science in the sense that EDA produces results in terms of new knowledge to support any other science or technology;

- the iterative aspect, since each result is often a prerequisite for further questions;

- the interactive and often multidisciplinary research;

- the two co-existing interpretations of methodological science and scientific method – it is often the EDA itself that gives inputs for the revision of existing methods or the development of new methods.

3.1.1.1 Fundamental definitions

- Population is a set (limited or unlimited, existing or hypothetical) of elements under study. Any subset of the population is a sample.

- Statistical survey is a set of operations to gain information about a population. A survey can provide responses (e.g. preferences, opinions, attitudes) or measures (e.g. length in metres, weight in grams, etc.). A statistical survey may be conducted on the entire population or on a sample.

- Statistical unit (or subject) is an element of the population or sample from which to gather one or more aspects under investigation.

- Character is any aspect gathered on the statistical units.

- Modality is the expression of the character, that is the number (in the case of a quantitative character) or attribute (in the case of a qualitative character) with which the character appears in the statistical unit.

- Frequency is the number of times a particular modality occurs in the examined sample or population.

- Categorical variable is a qualitative character that takes non-numerical modalities (often called attributes) of the most varied nature. For example, the geographic region of residence, the occupation, the business role, eye colour, and so on.

- Numerical variable is a quantitative character that takes numbers as modalities. A numerical variable can be discrete or continuous.

  — *Discrete variable*: a variable that can take a discrete (conceptually finite or infinite) number of modalities. The number of defects found on an artefact and an exam score are examples of discrete variables.

  — *Continuous variable*: a continuous variable can take any value in an interval (limited or unlimited) of real numbers. The weight of a valve engine, the diameter of a drive shaft section, the body temperature of a patient are examples of continuous variables.

  The distinction between discrete and continuous variables should be made at a conceptual level. For example, the length of an object is conceptually a continuous variable, although we must often rely on measuring instruments (e.g. a metre rule) with a limited resolution. So in reality it can happen that observed data from continuous variables are apparently treated as discrete variables. On the other hand, it may happen that a conceptually discrete variable (e.g. the number of bacteria present in a volume of sea water) will be treated as a continuous variable.

Characters can be distinguished based on a **measurement scale**.

In the case of categorical variables, the measurement scale can be nominal or ordinal.

In a **nominal scale**, modalities do not have a predetermined order (for example in a study on the ergonomics of a motorcycle seat, the gender of the potential user can be considered as a categorical variable with 'female' and 'male' modalities but with no predefined order). Conversely, in an **ordinal scale**, modalities have a predetermined order (examples of categorical variables with an ordinal scale are military rank, the severity level in a risk analysis, the customer satisfaction level, etc.).

For numerical variables, the scale of measurement can be an interval or a ratio scale.

An interval scale allows comparisons only by differences between modalities. It has a conventional zero. Examples of numerical variables measured on interval scales are the temperature in Celsius degrees, the geographic latitude and longitude.

A ratio scale allows comparisons both by differences and by ratios. It has an absolute zero, which means absence of the measured character. Examples of numerical variables measured on ratio scales are the weight of an object, the driving time on a highway, the length of a table, a patient's age, and so on.

3.1.1.2   Data collection

Data collection can be done with traditional methods, such as a datasheet, or with the aid of a computer, so widespread today. In the case of data collection with the aid of a computer, the data will be collected in digital format and stored in so-called databases from which it will be possible to extract and arrange the data of interest for the exploratory analysis. Below the traditional way is illustrated, as it is still widely used, especially when data are collected for a specific purpose within a Six Sigma project.

3.1.1.2.1   Data collection sheet   The data collection sheet or check sheet is a form for recording data about a certain phenomenon through the use of symbols or signs. Table 3.1 shows an example of check sheet about defects found on 2000 trays of ceramics produced by a factory in Sicily over six months.

The check sheet allows for data aggregation in order to make them ready for further processing. This facilitates the collection of information, finalises the gathered information and helps to minimise costs. There are different types of data collection sheets depending on the type of data to be collected and the main objectives of the analysis. Then there may be summary sheets that include information on data already collected and checklists that facilitate the checking tasks.

3.1.1.2.2   The data array for analysis with statistical software   The data array or worksheet is the easiest way to record the collected data. For analysis with statistical software (e.g. MINITAB, which has been mostly used during the preparation of this book), reporting the data in an array is the first and fundamental step.

Before putting data into the array we must follow a coding procedure to determine the location of each variable and its type in the data array.

The example in Table 3.2 shows the sex, height, eye colour and hair colour of 20 navy cadets. By the coding procedure we assign to each modality of each variable a number or a symbol. For example, for the categorical variable 'gender' the initials M (= male) and

Table 3.1   Example of data collection sheet for defects on ceramic trays.

Type of defect	January	February	March	April	May	June	Total
Damaged bottom	/	/	/	//			5
Non-flat surface	////	/////	///	/////	///	//	22
Non-uniform colour	////////////////////	///////////////////////	///////////////////////	///////////////////////	///////////////////////	///////////////////////	147
Damaged handles	//////	////	//////	//////	//////	//////	34
Defective decoration	/////	//		//////		///	16
Rough surface	/////////////	///////////////	/////////////	//////////////	///////////////	////////////////////	112
Missing colour	//	/					3
Broken handles	//////						6
Colour spots	////						4
Broken tray	/						1
Total defects	71	63	52	62	50	51	350

Table 3.2   Example of data array.

No.	Gender	Height (cm)	Eye colour	Hair colour
1	F	158		
2	F	156		
3	F	166		
4	F	168		
5	F	166		
6	F	164		
7	F	151		
8	F	160		
9	M	176		
10	M	187		
11	M	181		
12	M	176		
13	M	173		
14	M	174		
15	M	178		
16	M	177		
17	M	175		
18	M	179		
19	M	175		
20	M	174		

F (= female) are used; for the numeric variable 'height' the data recorded are expressed in centimetres.

3.1.2   Empirical frequency distribution of a numerical variable

3.1.2.1   Case of discrete variable

Let $x_1, x_2, \ldots, x_n$ be a set of observations of a discrete variable X. Suppose that the observations assume k modalities ($k \leq n$). The observed modalities are ordered from the smallest to the largest: $x_1^* \leq x_2^* \leq \ldots \leq x_k^*$ and the number of observations for each modality are counted: say that the modality x_1^* appears n_1 times, the modality x_2^* appears n_2 times and so on till the modality x_k^* which appears n_k times.

The result is a set of integers $n_1, n_2, \ldots, n_k$ which are the absolute frequencies for each modality. The ratios of the absolute frequencies to the total number of observations n are called relative frequencies: $f_i = n_i/n$, being $n = \sum_{i=1}^{k} n_i$.

An empirical frequency distribution can be represented by a table or a graph. In the case of discrete variable it is possible to use a dot plot. In it, for each of the ordered modalitities shown on the horizontal axis, the graph shows a number of points corresponding to their absolute frequencies.

EXAMPLE

For the purpose of basin sizing, the number of rainfall days (days when there was at least one episode of precipitation) were recorded in 30 Sicilian locations in the month of December (statistically the most 'wet'). The data collected are presented in Table 3.3.

Table 3.3   Rainfall days in December in 30 Sicilian locations.

10	11	10	14	13	15	18	16	8	15
9	15	12	13	11	15	11	14	12	9
13	12	11	14	15	9	10	17	12	14

The distribution of frequencies (absolute, relative and cumulated) is presented in Table 3.4. In Figure 3.1 the graphical representation (dot plot) is given.

Table 3.4   Empirical frequency (absolute, relative and cumulative) distribution of the rainfall days.

x_i^*	n_i	f_i	F_i
8	1	0.03	0.03
9	3	0.10	0.13
10	3	0.10	0.23
11	4	0.13	0.37
12	4	0.13	0.50
13	3	0.10	0.60
14	4	0.13	0.73
15	5	0.17	0.90
16	1	0.03	0.93
17	1	0.03	0.97
18	1	0.03	1.00
	30	1	

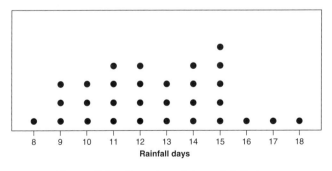

Figure 3.1   Dot plot of the rainfall days.

3.1.2.2 Case of continuous variable

For a continuous variable, having to deal with a hypothetically infinite number of modalities, the observed data will be conveniently grouped into classes of values. Each class is represented by an interval $(x_i, x_i + \Delta)$ with $i = 1, \ldots, k$, with k being the number of adopted classes. The class width Δ is often the same for all classes. The number of classes can be set in relation to the characteristics of the phenomenon or by reference to conventional criteria. Below we provide two conventional criteria.

- *First criterion*. The number of classes is chosen equal to the integer part of the square root of the number of observations, $k = \text{int} (\sqrt{n})$.

- *Second criterion*. The number of classes is chosen in accordance with the suggestions provided in Table 3.5.

Table 3.5   A criterion for the choice of the number of classes.

n	k
<50	5–7
50–100	6–10
100–250	7–12
>250	10–20

Once one has chosen the number of classes, the interval $(x_{\min}, x_{\max})$, is split into k intervals of equal width $\Delta = (x_{\max} - x_{\min})/k$. Let us say n_i is the number of observations falling in the i-th class and call it the absolute frequency. It should be noted that a given x is included in the i-th class if $x_i < x \leq x_{i+1}$. The only exception to this rule is $x_{\min}$ which is included in the first class.

The empirical distribution function corresponding to the i-th class is the fraction of modalities less than or equal to $x_i + \Delta$, upper limit of the class.

To graph the empirical distribution of a continuous variable we usually use a histogram. A histogram is a graph made on a Cartesian plane (see Figure 3.2): the horizontal axis is used for the observed variable, while the vertical axis is used for the frequency (absolute or relative). A histogram is a bar chart, each bar representing one class: the base is equal to the width and the height of the bar is equal to the frequency. If we want to make sure that the total area of the histogram is equal to 1 (a normalised histogram) the height of each bar is equal to the relative frequency divided by the class width.

EXAMPLE

In 50 stations located in various areas of the town, the average concentrations (over 8 hours of detection) of carbon monoxide in the air (legal limit 10 mg/m^3) have been detected on the same day. The data are presented in Table 3.6.

Table 3.6 Concentrations of carbon monoxide in 50 locations of the town.

6.2	6.6	2.6	3.1	9.5	4.8	7.8	6.4	8.0	4.0
6.8	2.8	8.3	5.0	5.6	1.1	4.2	5.0	5.0	6.3
6.8	3.1	6.8	3.2	9.1	6.3	6.2	5.9	5.5	6.2
3.2	6.9	6.6	6.8	8.0	4.8	3.1	2.7	1.4	5.7
3.7	7.4	6.7	2.8	7.5	3.7	6.2	5.8	6.1	7.9

To represent the frequency distribution, we calculate first: $x_{min} = 1.1$, $x_{max} = 9.5$, $k = \text{int}(\sqrt{50}) = 7$, $\Delta = 1.2$. Table 3.7 shows the tabular representation of the empirical distribution. Figure 3.2 shows the frequency histogram.

Table 3.7 Empirical distribution of carbon monoxide data.

x_i	$x_i + \Delta$	n_i	f_i	F_i
1.1	2.3	2	0.04	0.04
2.3	3.5	9	0.18	0.22
3.5	4.7	4	0.08	0.30
4.7	5.9	10	0.20	0.50
5.9	7.1	16	0.32	0.82
7.1	8.3	7	0.14	0.96
8.3	9.5	2	0.04	1.00
Total		**50**	**1**	

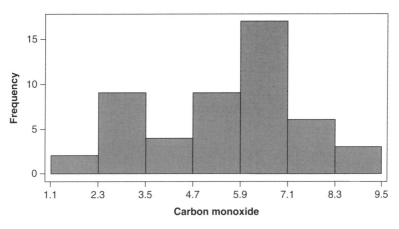

Figure 3.2   Histogram of frequencies of carbon monoxide data.

In the construction of histograms we should be aware that:

• the presence of anomalous values (so called *outliers*) can lead to an incorrect splitting into classes;

- the choice of a too large or too small number of classes can result in poor picture of the distribution of the variable.

Based on the two previous observations we shall realise that the choice of classes is a sensitive issue. It is based on established rules, but also on the care and experience of the person making the analysis.

Each empirical distribution has three main features:

- its location in reference to the assumed modalities and their respective frequencies;
- its dispersion, that is the expression of the attitude of the variable to assume different modalities;
- its shape, that is the overall appearance compared to reference configurations.

These characteristics of the distribution can be summarised by indices.

- As location indices we will define the average, the weighted average, the median, the mode and the quantiles.
- As indices of dispersion we will define the range, the interquartile range, the variance and the standard deviation.
- As shape indices we will define some skewness and kurtosis indices.

We must always remember that when we provide a summary of the distribution by using these indices, it also leads to a loss of information. Therefore characterising a distribution only through an average is less informative than characterising it by both average and variance, which in turn is less informative than characterising it by average, variance and skewness.

3.1.2.3   Location of the empirical distribution

3.1.2.3.1   The average   The average is the ratio between the sum of the values assumed by the variable and the number of observed data:

$$\bar{x} = \frac{1}{n} \sum_{i=1}^{n} x_i$$

If the data are grouped according to their modality (in the case of a discrete variable) the average is equal to the sum of the products of the modalities times their relative frequencies:

$$\bar{x} = \sum_{i=1}^{k} x_i^* f_i$$

For example: consider the data of Table 3.3.

First of all, we sort the data from the smallest to the largest: 8, 9, 9, 9, 10, 10, 10, 11, 11, 11, 11, 12, 12, 12, 12, 13, 13, 13, 14, 14, 14, 14, 15, 15, 15, 15, 15, 16, 17, 18. Then the average is calculated:

$$\bar{x} = \frac{1}{30}(8 + 9 + \cdots + 18) = \frac{1}{30}(8 \times 1 + 9 \times 3 + \cdots + 18 \times 1)$$

$$= 8 \times \frac{1}{30} + 9 \times \frac{3}{30} + \cdots + 18 \times \frac{1}{30} = 12.6$$

- *Properties of the average*

  — The average lies between the minimum and maximum observed modalities.

  — The sum of the differences from the average is zero. The average is the 'barycentre' or centre of mass of the observations, provided that all observations have the same 'weight'.

  — Linearity: if a new variable is so defined: $y = ax + b$, then $\bar{y} = a\bar{x} + b$.

  — Associative property: if we create groups of observations, the average of all observations can be calculated as the average of the group averages.

  — The average is the value that minimises the sum of the squared differences. In fact, considering a generic value h

$$\sum_{i=1}^{n}(x_i - h)^2 = \sum_{i=1}^{n}(x_i - \bar{x} + \bar{x} - h)^2 = \sum_{i=1}^{n}(x_i - \bar{x})^2 + n(\bar{x} - h)^2$$

  which is a minimum for $h = \bar{x}$.

3.1.2.3.2 The weighted average The weighted average is calculated whenever each observation is given a weight. It is calculated as the ratio between the sum of products of the values of the variable and their respective weights, and the sum of the weights.

$$\bar{x}_w = \frac{\sum_{i=1}^{n} x_i w_i}{\sum_{i=1}^{n} w_i}$$

This index is used if we want to weigh the data differently based on exogenous considerations.

3.1.2.3.3 The median The median is the value that divides the ordered observations $x_{(1)} \leq x_{(2)} \leq \cdots \leq x_{(n)}$ (note the use of brackets at the subscript indicating that the observations were ordered from lowest to highest) into two groups of equal size. By definition, at the median, the cumulated frequency is equal to 0.5.

For discrete numerical variables, the calculation of the median is quite simple:

$$
Me = \begin{cases} x_{\left(\frac{n+1}{2}\right)} & \text{if } n \text{ is odd} \\ \dfrac{x_{(n/2)} + x_{(n/2+1)}}{2} & \text{if } n \text{ is even} \end{cases}
$$

For example, for the rainfall data (Table 3.3): $n = 30$, $x_{(15)} = 12$, $x_{(16)} = 13$ and $Me = 12.5$ days.

For continuous variables, if the data were previously grouped into classes, one can determine the median class as the class at which the cumulated frequency reaches the value 0.5. Moreover, in the absence of details of individual values, it is also possible to calculate the approximate median assuming that the distribution function grows linearly in each class, and making a linear interpolation:

$$
Me = LL_{Me} + (UL_{Me} - LL_{Me}) \frac{0.5 - F(LL_{Me})}{F(UL_{Me}) - F(LL_{Me})} \tag{3.1}
$$

where LL_{Me}, UL_{Me} are respectively the lower and upper limits of the median class. For example, for the data in Table 3.6, using the division in classes shown in Table 3.7, there is no doubt that the median class is the fourth ($4.7 \div 5.9$) and furthermore, without having to resort to Equation 3.1, a point value of the median is $Me = 5.9$.

- *Properties of the median*

  — The number of positive differences from the median is equal to the number of negative differences (by definition).

  — The median is representative of the centre of the distribution also in the presence of anomalous values (outliers).

  — It can be demonstrated that the median is the value that minimises the sum of the absolute values of the differences.

3.1.2.3.4 The mode The mode of a frequency distribution is the most frequent modality, that is the modality with the largest frequency (both absolute and relative).

For example for the rainfall data in Table 3.3, the mode is the modality '15', corresponding to the largest observed frequency equal to 5 (see Table 3.4).

- *Properties of the mode*

  — The mode, unlike the average, is certainly an observed value (by definition).

  — The maximum number of zero differences is achieved at the mode (by definition).

— The mode is not necessarily unique: an empirical distribution can have more than one mode (unimodal, bimodal, trimodal, . . . , multimodal distribution).

— For continuous numerical variables we may define a modal class, that is the class which corresponds to the maximum frequency. See, for example the histogram in Figure 3.2, where the modal class is the fifth one.

The choice of the location indices mentioned above is mainly guided by the aims of the research. In summary we can say that:

1. taking the mode as the location index is somehow rewarding the most frequent;

2. the median, by minimising the sum of absolute values of differences, minimises the total cost when costs are related to either positive or negative gaps;

3. finally, the average minimises the overall risks because it gives weight to the extreme values more than the median; in this sense it is an index of general equilibrium.

3.1.2.3.5 Quantiles, percentiles and quartiles The *quantile of order q* is the value x_q for which the cumulative frequency is equal to q: $F(x_q) = q$.

If the quantile is to be expressed in percentage terms, it is called percentile. The *quartiles* are special cases of percentiles.

The first quartile Q_1 is the quantile of order 0.25, that is the value for which cumulative frequency is 0.25. The second quartile Q_2 is the quantile of order 0.5, which is the median. The third quartile Q_3 is the quantile of order 0.75.

Similarly to the calculation of the median, for the quartiles Q_1 and Q_3, in case of discrete variable, it is possible to use the formulas in Table 3.8.

Table 3.8   Formulas for the calculation of quartiles in case of discrete variable.

n odd	$\frac{n-1}{2}$ odd	$Q_1 = x_{\left(\frac{n+1}{4}\right)}$	$Q_3 = x_{\left(\frac{3}{4}(n+1)\right)}$
	$\frac{n-1}{2}$ even	$Q_1 = \dfrac{x_{((n-1)/4)} + x_{(((n-1)/4)+1)}}{2}$	$Q_3 = \dfrac{x_{((3(n-1)/4)+1)} + x_{((3(n-1)/4)+2)}}{2}$
n even	$\frac{n}{2}$ odd	$Q_1 = x_{\left(\frac{n+2}{4}\right)}$	$Q_3 = x_{\left(\frac{3n+2}{4}\right)}$
	$\frac{n}{2}$ even	$Q_1 = \dfrac{3 \times x_{(n/4)} + x_{((n/4)+1)}}{4}$	$Q_3 = \dfrac{x_{(3n/4)} + 3 \times x_{((3n/4)+1)}}{4}$

In the case of a continuous variable (with data arranged in classes), if we want to determine point values of Q_1 and Q_3, we need to firstly determine the class containing the quartile of interest, then apply the following formulas, which are similar to

Equation 3.1:

$$Q_1 = LL_{Q_1} + \left(UL_{Q_1} - LL_{Q_1}\right) \frac{0.25 - F\left(LL_{Q1}\right)}{F\left(UL_{Q_1}\right) - F\left(LL_{Q_1}\right)}$$

$$Q_3 = LL_{Q_3} + \left(UL_{Q_3} - LL_{Q_3}\right) \frac{0.75 - F\left(LL_{Q_3}\right)}{F\left(UL_{Q_3}\right) - F\left(LL_{Q_3}\right)}$$

3.1.2.3.6 Box-Whiskers plot and outliers detection   The box-whiskers plot is a graphical representation with the shape of a box and segments at the ends called whiskers (see Figure 3.3). The left and right sides of the box are placed at Q_1 and Q_3 respectively. We may see a line inside the box placed in correspondence of the median Q_2 and sometimes a mark in correspondence of the average. The whiskers are usually extended to the minimum and maximum values of the data set. However, when we want to be alerted of the presence of outliers, the lengths of the two whiskers are sized based on the size of the box.

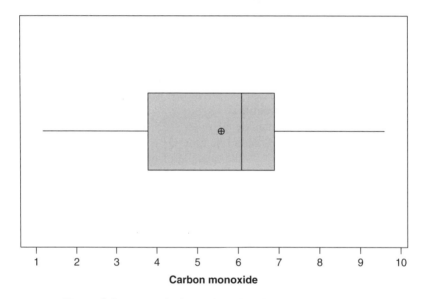

Figure 3.3   Box-whiskers plot of carbon monoxide data.

This graphical representation is more convenient than the histogram when we want to compare two or more distributions of data. In fact it gives the possibility to make comparisons with respect to location, dispersion, skewness and presence of outliers. Figure 3.4 depicts the double box-whiskers plot, separately by gender, of the height data of Table 3.2. The graph allows us to see clearly that the distributions of the heights of males and females are quite different. In addition, the graph highlights an outlier corresponding to a male cadet of height 187 cm. This simple example makes us understand that an outlier is not a datum to be deleted, because its presence can be considered natural in the distribution under study. On the other hand it could happen that the outlier, highlighted by the

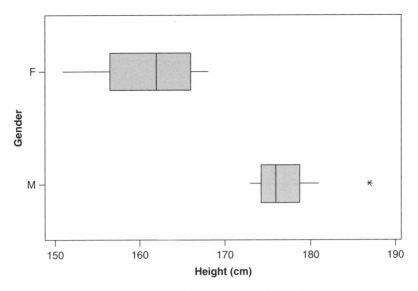

Figure 3.4   An example of double box-whisker plot and outlier.

box-whiskers plot, is the result of an error (e.g. an error of the measurement system or a transcription error); then since it should not belong to the distribution, we may decide to remove it from the dataset.

3.1.2.4   Dispersion of the empirical distribution

3.1.2.4.1   Range   The *range of variation* (more simply *range*) is the difference between the maximum and the minimum value of the observed modalities of the numerical variable.

$$\text{Range}\,(X) = \max(X) - \min(X) = x_k^* - x_1^*$$

The range is the simplest dispersion index, but it is strongly influenced by the possible presence of outliers.

3.1.2.4.2   The interquartile range (IQR)

$$\text{IQR} = Q_3 - Q_1$$

Unlike the range, the interquartile range of variation is not influenced by the presence of outliers. Note that the IQR is equal to the length of the box in the box-whiskers plot.

3.1.2.4.3   The variance   The variance (s^2) is the average of the squared differences between the individual observations of the numerical variable and the average $\bar{x}$, chosen as the reference value:

$$s^2 = \frac{1}{n} \sum_{i=1}^{n} (x_i - \bar{x})^2 \tag{3.2}$$

For ease of calculation, as an alternative to Equation 3.2, the following equivalent formula can also be adopted:

$$s^2 = \frac{1}{n} \sum_{i=1}^{n} x_i^2 - \bar{x}^2$$

If the data are grouped according to modality (discrete variable), we have:

$$s^2 = \frac{\sum_{i=1}^{k} n_i \left(x_i^* - \bar{x} \right)^2}{\sum_{i=1}^{k} n_i}$$

- *Properties of the Variance*
  - The variance is not affected by 'shifts'. If c is a constant and X is the observed variable, the variance of $X + c$ is equal to the variance of X.
  - The variance of cX is equal to c^2 times the variance of X.
  - The variance is a non-negative number (it is 0 only if all modalities are identical).
  - The variance is expressed in the squared measurement units of the variable; for example if the variable is expressed in metres, the variance will be expressed in square metres.

3.1.2.4.4   The standard deviation   The obvious inconvenience arising from the final property of the variance listed above, gave rise to the formulation of an index of variability that seems more natural:

$$s = \sqrt{s^2} = \sqrt{\frac{1}{n} \sum_{i=1}^{n} (x_i - \bar{x})^2} \tag{3.3}$$

This index is called the standard deviation and it is expressed in the same units as the variable under study.

EXAMPLE

For the rainfall data of Table 3.3 we have:

$$s^2 = \frac{1}{30} \sum_{i=1}^{30} (x_i - \bar{x})^2 = \frac{(8 - 12.6)^2 + \cdots + (18 - 12.6)^2}{30} = 6.31 \text{ days}^2$$

$$s = 2.51 \text{ days}$$

In some cases it is preferred to use $n - 1$ in place of n as the denominator of Equation 3.2. This aspect will be explained below (Section 3.4.4.2).

3.1.2.4.5   The coefficient of variation   The coefficient of variation is the ratio between
the standard deviation and the average:

$$C_V = \frac{s}{\bar{x}}$$

The coefficient of variation is a dimensionless index, that is it is independent from
the measurement unit of the variable. It is a good index if we want to make comparisons
between datasets with different measurement units or different orders of magnitude.

3.1.2.5   Skewness of the empirical distribution

The skewness can be defined as a lack of specularity of the empirical distribution with
respect to a hypothetical axis of symmetry. In fact, if the frequency distribution of the
variable is unimodal and symmetrical, then the average and the median of this distribution
tend to coincide. When the average exceeds the median, which in turn exceeds the mode,
we talk about positive skewness. Graphically the distribution appears with a tail to the
right (Figure 3.5a). Conversely, if the mode exceeds the median, which in turn exceeds
the average we talk about negative skewness (Figure 3.5b).

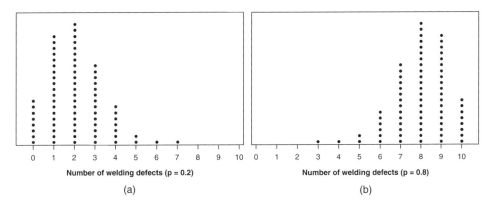

Figure 3.5   Example of (a) positive and (b) negative skewness.

Figure 3.5 shows dot plots of welding defects on produced units: the number of
weldings per unit is equal to 10, the number of welding defects appears to follow a
binomial distribution (see Section 3.3.6.1). In the left pane, the defect rate is low and the
empirical distribution is positively skewed. In the right pane, the defect rate is high and
the empirical distribution is negatively skewed.

Figure 3.6 shows the histogram of measurements of shaft diameters: the empirical
distribution appears to have no skewness and to follow a Gaussian distribution (see Section
3.3.6.2) with a mean of about 100 mm and standard deviation about 3 mm.

A skewness index normalised in the interval $[-1, +1]$ is the index of Hotelling-
Solomon:

$$Sk_1 = \frac{\bar{x} - Me}{s}$$

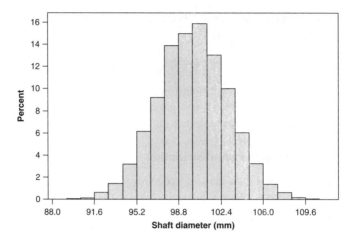

Figure 3.6   Example of symmetry of the empirical distribution.

If Sk_1 is higher than zero we say that there is positive skewness, if it is lower than zero there is negative skewness. If $Sk_1 = 0$ the empirical distribution is said to be symmetrical.
Another skewness index is the Fisher index:

$$Sk_2 = \frac{1}{n} \sum_{i=1}^{n} \left(\frac{x_i - \bar{x}}{s} \right)^3$$

It is positive, negative or null for a positively skewed, negatively skewed or symmetric distribution, respectively.

3.1.2.6   Kurtosis of the empirical distribution

The *kurtosis* of the empirical distribution measures the degree of 'sharpness' of the distribution, therefore the weight of the tails with respect to the central part of the distribution (Figure 3.7).

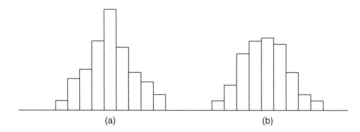

(a)   (b)

Figure 3.7   Examples of leptokurtic (a) and platykurtic (b) distributions.

The Pearson kurtosis index is defined as:

$$Ku = \frac{1}{n} \sum_{i=1}^{n} \left(\frac{x_i - \bar{x}}{s} \right)^4$$

It is equal to 3 if the distribution is bell-shaped and symmetrical respect to the centre, with the centre being coincident with the average, median and mode. In such case the distribution is said mesokurtic. For sharp distributions Ku is greater than 3 (leptokurtic distribution), while for flat distributions Ku is lower than 3 (platykurtic distribution).

3.1.3 Analysis by stratification

We use the analysis by stratification when we feel that we can have a better understanding of the phenomenon under study by splitting the available data into homogeneous groups.

Suppose we analyse the distribution of the cadet heights reported in Table 3.2. Suppose that the ultimate goal is to establish the sizes of a batch of uniforms to be ordered. Figure 3.8 shows the histogram of data for all cadets. The summary made by the histogram is not helpful for the intended goal, as it would be much more useful to conduct the analysis by distinguishing male from female data. So, it is convenient to stratify by gender, as depicted in Figure 3.9.

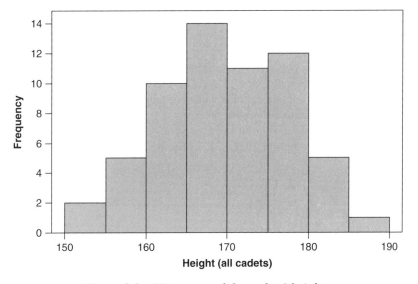

Figure 3.8   Histogram of the cadets' heights.

From the separate reading of the two graphs, we see more clearly what are the characteristics of the uniforms to be ordered.

The logical partitions according to which the data are grouped are said to be stratification factors. Examples of stratification factors in industry could be:

- time (morning/night shift, season of the year);
- operators (age, experience);
- machines or tools (model, type, age, technology);
- material (supplier, composition);
- measurement method (type of instrument, operator).

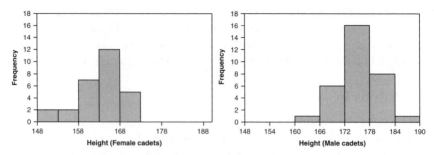

Figure 3.9   Histogram of the cadets' heights distinguished by gender.

3.1.4   Other graphical representations

3.1.4.1   Time series plots

A series is a dataset organised according to a criterion, for example chronological, in the case of time series, geographical, in the case of spatial series. These series can be graphically represented in a Cartesian plane placing the sorting criterion on the horizontal axis and the values taken by the variable on the vertical axis. Often, for better readability of the plot, the points are connected by line segments to form a continuous line. Examples of time series are: the MIB index (the volume of shares traded) calculated at the end of each day at the Milan Stock Exchange, the values of maximum daily temperature recorded in Palermo in one year, the exam grades taken by a student in his academic career, and so on. For example, Figure 3.10 shows the time series of the number of road accidents in Sicily for every year in the period 1978–2007. The trend did not look promising at least until year 2000.

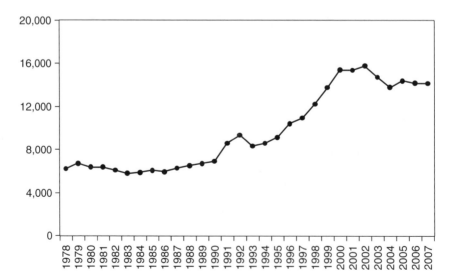

Figure 3.10   Number of road accidents in Sicily in 1978–2007.

3.1.4.2 Spatial series plots

The spatial series can also be graphically represented through a Cartesian plane, placing the spatial unit chosen on the horizontal axis and the values of the variable of interest on the vertical axis. The example of Figure 3.11 shows the values of the index accidents/kilometre (total number of accidents divided by the total length of the roads) for the 20 Italian regions in the year 2007.

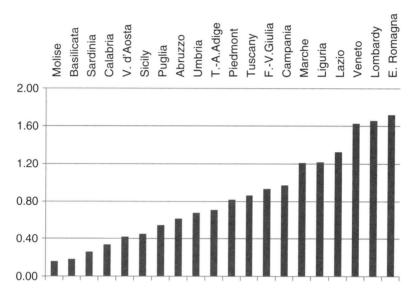

Figure 3.11   Index accidents per kilometre in the Italian regions in 2007.

3.1.4.3 Scatter plot

The scatter plot is a two-dimensional graph used to study the relationship between two quantitative variables. Two variables X and Y are said to be positively correlated if high values of X correspond to high values of Y and low values of X correspond to low values of Y (a direct relationship). Conversely, the two variables X and Y are said negatively correlated when high values of X correspond to low values of Y and low values of X correspond to high values of Y (an inverse relationship). If none of the two trends can be seen, we say that there is no correlation. Figure 3.12 shows examples of (a) a direct relationship, (b) an inverse relationship and (c) and no correlation. This topic will be further discussed in Chapter 5.

3.1.4.4 Pareto chart

The Pareto chart is a graphical technique aimed at identifying and prioritising problems. Consider the data of Table 3.1, related to defects on 2000 ceramic trays produced over a six-month period in a factory. Table 3.9 gives a tabular representation of this data. The frequencies are sorted from the highest to the lowest.

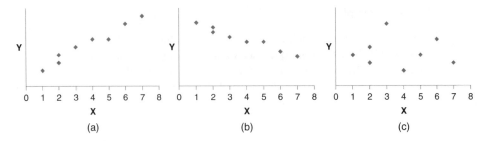

Figure 3.12   Examples of scatter plot with positive, negative and no correlation.

Table 3.9   Tabular representation of defects in ceramic trays.

Type of defect	n	$f\,(\%)$	$F\,(\%)$
Non-uniform colour	147	42	42
Rough surface	112	32	74
Damaged handles	34	9.7	83.7
Non-flat surface	22	6.3	90
Defective decoration	16	4.6	94.6
Broken handles	6	1.7	96
Damaged bottom	5	1.4	98
Colour spots	4	1.1	99
Missing colour	3	0.9	100
Broken tray	1	0.3	–
Total	350	100	

Figure 3.13 shows the histogram of the absolute frequencies for each category of defect. The red line in the graph is the cumulative frequency in percent.

The first type of defect accounts for 42 % of the total number of defects; the first two types account for the 74 % of the total. This analysis tells us how many and what types of defects we should give priority to and which defects we could initially neglect.

3.2   To define and calculate the uncertainty

The phase of probability calculation that we will discuss in this section can be used together with the EDA for statistical inference purposes, as we'll see in the last section of this chapter. In fact, inference is the branch of Statistics in which, starting from the observation of a part of reality, and using the laws of probability, we aim to 'infer' from the particular to the general with a calculated and controlled level of uncertainty.

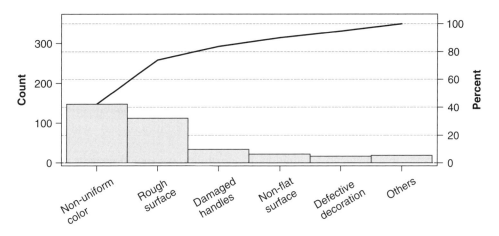

Figure 3.13   Example of a Pareto chart.

The term probability usually identifies two areas, which will be briefly examined: the calculus of probability and the random variables with their models.

3.2.1   Definitions of probability

An *event* is an assertion that we can neither say is true nor false, but we can say that it is possible. This assertion relates to something *random*, that is it is not known but it is well-specified. Unlike the event, the *fact* is something true or false, often has already occurred and therefore is not random.

Probability refers to events. We may say that the probability is a primitive concept that evolves into a human being from the moment of its birth.

There are three definitions given historically to the concept of probability: classical, frequentist and subjectivist. Each definition is tautological as we shall see, that is repetitive and/or redundant, and can be contradicted.

- *Classical definition*: the probability of occurrence of an event is the ratio of cases favourable to the occurrence of the event and all cases equally possible.

- *Frequentist definition*: the probability of occurrence of an event is given by the ratio of cases in which the event historically occurred and the total number of cases in which the event could occur.

- *Subjective definition*: the probability of occurrence of an event is the degree of confidence that a consistent subject places in the occurrence of the event. 'Consistent' means that the subject is properly using the information in its possession and makes evaluations which do not contradict each other.

Independently from the assumed definition, the probability will be expressed as a real number between 0 and 1.

EXAMPLE

When referring to an experiment whose outcome is uncertain (a test) the event may be a possible outcome of the experiment. For example, the toss of a coin is a test. For this test two events may be associated, namely two equally possible outcomes: the event 'head' and the event 'tail'.

According to the classical definition of probability, being two equally possible outcomes, the probability of 'head' as well as the probability of 'tail' will be equal to 1/2.

According to a frequentist definition of probability, we should imagine repeating the test of the coin toss an indefinite number of times, say n, and calculate the probability of 'heads' as the ratio between the number of times that there was 'head' and n. It is not said that the probability calculated in this way coincides with the probability defined by the classical approach. In order to give a confident judgement, we should consider a very large value of n, tending to infinity.

Finally, according to the subjective definition, the probability that the event 'head' occurs is the degree of confidence that an individual has in the occurrence of it. Two consistent individuals with a different state of knowledge may evaluate the desired probability differently.

The three different definitions of probability have been conceived because there are situations where one or two of them can not be used, either because we do not know the number of possible cases, or because the event has never occurred historically, or for other reasons. Also note that each of the definitions contains and uses more or less directly the concept of probability itself, from which we see the unresolved tautological problem.

With reference to a well-defined test and to a definite state of knowledge we denote as $\bar{A}$ (read 'no-A') the complementary or the negation of the event A.

The events A and $\bar{A}$ are *incompatible* because the occurrence of one excludes the occurrence of the other. We can say: $A \cap \bar{A} = \emptyset$ (read 'A intersection no-A is the impossible event').

On the other hand, if we join the event A and its negation, we obtain the totality of possible events, that is the so-called event space, denoted by Ω. We therefore say that: $A \cup \bar{A} = \Omega$ (read 'A union no-A is equal to Ω').

The impossible event has zero probability to occur: $\Pr\{\emptyset\} = 0$. The event space is instead a certain event: $\Pr\{\Omega\} = 1$. To any event A is assigned a probability $0 \leq \Pr\{A\} \leq 1$.

3.2.2 Events and probabilities in the Venn diagram

The Venn diagram is used to visualise events and probability.

In Figure 3.14 the rectangle Ω denotes the event space and the set A a generic event. In the Venn diagram it is assumed that Ω has unit area and the probability of an event contained in Ω is equal to its area. Figure 3.15 shows the event A and its negation. It is easy to see that $\Pr\{\bar{A}\} = 1 - \Pr\{A\}$. The event A and its negation constitute a *partition of the event space* since their union gives rise to the event space.

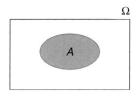

Figure 3.14   Event space and generic event in a Venn diagram.

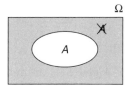

Figure 3.15   Complementary events.

There are situations in which if an event A occurs, it necessarily implies that another event B occurs (not necessarily the opposite). In this case we say that A is contained in B: $A \subset B$ (Figure 3.16).

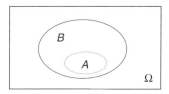

Figure 3.16   The event A *is contained in the event* B *(A ⊂ B).*

if A is contained in B and B is contained in A, then A and B are the same event.

The union of two events, $A \cup B$, is the event which means 'either only A, or only B, or both A and B occur' (Figure 3.17).

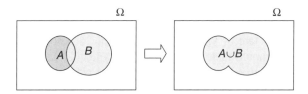

Figure 3.17   Union of two events.

The *concomitance* of two events A and B, $A \cap B$, is the event that occurs when both A and B occur (Figure 3.18).

Ω

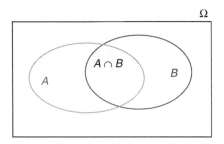

Figure 3.18   Concomitance of two events.

3.2.3   Probability calculation rules

If the concomitance of two events A and B is an impossible event then $A \cap B = \emptyset \Leftrightarrow$ $\Pr\{A \cap B\} = 0$. The events A and B are *incompatible* (see Figure 3.19).

Ω

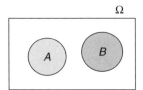

Figure 3.19   Incompatible events.

In the case of two incompatible events we have that the probability of their union is equal to the sum of their probabilities:

$$A \cap B = \emptyset \Leftrightarrow \Pr\{A \cup B\} = \Pr\{A\} + \Pr\{B\}$$

This is a special case because in general the probability of the union of two events is the sum of their probabilities minus the probability of their concomitance:

$$\Pr\{A \cup B\} = \Pr\{A\} + \Pr\{B\} - \Pr\{A \cap B\}$$

To understand this, just look carefully at Figure 3.17.

3.2.3.1.1   Conditional probability   As mentioned, the assessment of a probability is conditional to a precise state of knowledge. If the state of knowledge changes, the calculation of probability should take account of it. A typical situation might be like 'what would happen if the event A occurs'. In this case the calculation of the probability of another event B shall consider this additional information. This probability, so-called conditional, is indicated as $\Pr\{B|A\}$, and expresses the probability of the occurrence of B 'conditionally upon the occurrence of A' (read: 'given A').

The occurrence of B conditionally to A is equivalent to the occurrence of both, but in the event space consisting of the event A. Therefore the probability of $B|A$ is related to that of the concomitance $B \cap A$, see Figure 3.20.

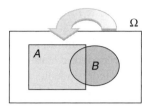

Figure 3.20   Conditional probability: A *becomes the new event space.*

The conditional probability is therefore given by the formula:

$$\Pr\{B|A\} = \frac{\Pr\{B \cap A\}}{\Pr\{A\}}$$

For example, the probability of getting a 'three' (event B) in the roll of a die is definitely 1/6, but the probability of getting a three conditionally to 'odd number' (event A) is:

$$\Pr\{B|A\} = \frac{\Pr\{B \cap A\}}{\Pr\{A\}} = \frac{\Pr\{'3' \cap' \text{odd number}'\}}{\Pr\{'\text{odd number}'\}} = \frac{1/6}{3/6} = \frac{1}{3}$$

From the conditional probability rule immediately follows the rule of composite probability, stating that:

$$\Pr\{A \cap B\} = \Pr\{A|B\}\Pr\{B\} = \Pr\{B|A\}\Pr\{A\} \tag{3.4}$$

3.2.3.1.2   Stochastic independence   Two events are said to be stochastically independent (hereinafter we will say s-independent) if the occurrence of one does not alter the probability of the occurrence of the other:

$$\Pr\{B|A\} = \Pr\{B\} \text{ and } \Pr\{A|B\} = \Pr\{A\}$$

If two events A and B are s-independent, the rule of composite probability Equation 3.4 becomes:

$$\Pr\{A \cap B\} = \Pr\{A\} \cdot \Pr\{B\}$$

Figure 3.21 shows how to represent two s-independent events in the Venn diagram.

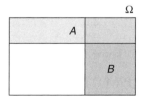

Figure 3.21   Stochastically independent events in the Venn diagram.

It must be stressed that if two events are incompatible, they are necessarily s-dependent (Figure 3.19). In fact, the occurrence of one affects the probability of the occurrence of the other because if one occurs then, by definition, it excludes the occurrence of the other and vice versa. Therefore, two s-independent events have to be compatible.

If the events A and B are s-independent, also $(\bar{A}, B)$, $(A, \bar{B})$, $(\bar{A}, \bar{B})$ are s-independent. Only the last case is shown in full here:

$$\Pr\{\bar{A} \mid \bar{B}\} = \frac{\Pr\{\bar{A} \cap \bar{B}\}}{\Pr\{\bar{B}\}} = \frac{1 - \Pr\{A \cup B\}}{1 - \Pr\{B\}} = \frac{1 - [\Pr\{A\} + \Pr\{B\} - \Pr\{A\}\Pr\{B\}]}{1 - \Pr\{B\}}$$

$$= \frac{1 - \Pr\{B\} - \Pr\{A\}[1 - \Pr\{B\}]}{1 - \Pr\{B\}} = 1 - \Pr\{A\} = \Pr\{\bar{A}\}$$

3.2.3.1.3   Rule of total probability   Consider an event space Ω and a partition $\{B, \bar{B}\}$. Then consider an event A that we assume to be compatible with B. Necessarily $(A \cap B)$ and $(A \cap \bar{B})$ will be incompatible events (see Figure 3.22).

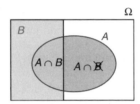

Figure 3.22   Rule of total probability.

It follows that:

$$\Pr\{A\} = \Pr\{A \cap B\} + \Pr\{A \cap \bar{B}\} = \Pr\{A|B\}\Pr\{B\} + \Pr\{A|\bar{B}\}\Pr\{\bar{B}\} \qquad (3.5)$$

The rule shown in Equation 3.5 is also called the theorem of total probability, and it allows one to calculate the probability of an event A by knowing the probability of another event B and the conditional probabilities of A given B and given $\bar{B}$.

3.2.3.1.4   Bayes' theorem   From the rule of the total probability follows directly the so-called *Bayes' theorem*. If we consider a partition of the event space Ω consisting of n events B_i, and consider a generic event A in the same space of events, by extending the previous rule Equation 3.5, we have:

$$\Pr\{A\} = \sum_{i=1}^{n} \Pr\{A|B_i\}\Pr\{B_i\}$$

However, if it is of interest to know the conditional probability $\Pr\{B_k|A\}$, it can be expressed as:

$$\Pr\{B_k|A\} = \frac{\Pr\{A \cap B_k\}}{\Pr\{A\}} = \frac{\Pr\{A|B_k\}\Pr\{B_k\}}{\sum_{i=1}^{n} \Pr\{A|B_i\}\Pr\{B_i\}}$$

If the events B_i interpret the possible 'causes' of an effect A, knowing the prior probabilities of the causes and the conditional probabilities of the effect given the individual causes (likelihoods), then Bayes' theorem allows an 'inversion' by which we can calculate the probability of a particular cause conditional upon the occurrence of the effect.

3.2.4   Dispositions, permutations and combinations

Sometimes we wonder about how many ways we have to dispose n objects in k places. See Figure 3.23 and imagine that the box contains n objects and the cabinet contains k places. To fill the first place we have n possibilities, for the second place we have $n - 1$ possibilities, and so on up to $n - (k - 1)$ possibilities for the last place.

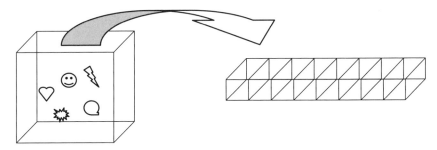

Figure 3.23   Scheme for the calculation of the dispositions of n objects in k places.

So, the number of possible dispositions of n objects in k places is:

$$n \times (n - 1) \times \cdots \times (n - k + 1) = \frac{n!}{(n - k)!}$$

Now consider a single disposition of k objects in the k places (Figure 3.24).

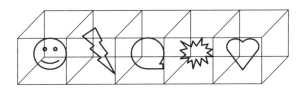

Figure 3.24   Permutations of k objects in k places.

If we are wanting to change the ordering of the k objects, how many possibilities do we have? Imagine removing the items and then puting them back one by one. For the first place we have k possibilities, $k - 1$ for the second place, ..., and finally only one for the last place. So the number of possible *permutations* of k objects in k places is:

$$k \times (k - 1) \times \cdots \times 1 = k!$$

The number of dispositions of n objects in k places 'contains' all permutations of k objects in k places. If we are only interested in the objects allocated to the places, but we are not interested in their specific order and therefore we want to exclude from counting all permutations, then we divide the number of dispositions by the number of permutations, so obtaining the so-called *combinations* of n objects in k places:

$$\frac{n!}{k!(n-k)!} = \binom{n}{k}$$

The right-hand side of the formula above is denoted by binomial coefficient and we will find it in the binomial model of a discrete random variable (Section 3.3.6.1).

3.3   To model the random variability

3.3.1   Definition of random variable

Defining a random variable (hereinafter we will use the acronym r.v.) means formulating a rule that establishes a correspondence between a real number and each event of an event space partition where the probabilities of all the events are known (Figure 3.25).

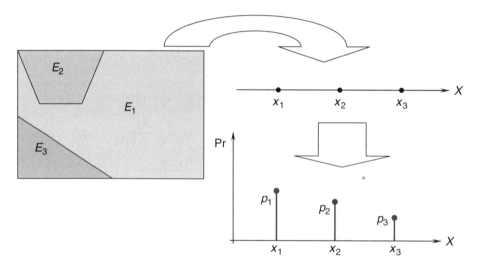

Figure 3.25   Definition of a random variable.

A r.v. is usually indicated by italic capital letters such as X. The set of values of the r.v. is said to be the *support* and is here indicated by Ξ. We call x_{min} and x_{max} the lower and upper bounds of the support.

The r.v. are classified into discrete and continuous. A discrete random variable can assume a finite or infinite, but countable, number of values. A continuous random variable can assume any value in a limited or unlimited interval of real numbers.

3.3.2   Probability distribution function

The probability distribution function $F_X(x)$ of a r.v. X is a mathematical function that assigns to each value $x \in \Xi$ the probability that the r.v. X gets values less than or equal to x:

$$F_X(x) = \Pr\{X \le x\} \text{ for } x \in \Xi$$

3.3.2.1   Properties of the probability distribution function

For any couple of values $x_1 \le x_2$, it is:

$$F_X(x_2) - F_X(x_1) = \Pr\{x_1 < X \le x_2\} \ge 0 \tag{3.6}$$

This means that the distribution function is non-decreasing. Demonstration hint:

$$\Pr\{X \le x_1\} + \Pr\{x_1 < X \le x_2\} + \Pr\{X > x_2\} = 1 \iff$$

$$F_X(x_1) + \Pr\{x_1 < X \le x_2\} + 1 - F_X(x_2) = 1$$

hence the assertion (Equation 3.6).

Furthermore it must be: $\lim_{x \to x_{\min}} F_X(x) = 0$ and $\lim_{x \to x_{\max}} F_X(x) = 1$

3.3.3   Probability mass function for discrete random variables

For a discrete r.v. X we define a *probability mass function*:

$$P_X(x) = \Pr\{X = x\}$$

For a probability mass function, it must be:

$$P_X(x_i) \ge 0 \qquad \forall x_i \in \Xi \qquad \sum_{x_i \in \Xi} P_X(x_i) = 1$$

The relation between the probability mass and the distribution functions is:

$$F_X(x_i) = \sum_{x \le x_i} P_X(x)$$

The probability distribution function of a discrete random variable is therefore a step function, with the steps raising from 0 to 1. Figure 3.26 shows an example of probability mass function and its corresponding distribution function.

3.3.4   Probability density function for continuous variables

For a continuous random variable we define a probability density function $f_X(x)$ so that the probability that the r.v. X gets values into the interval (a, b) is equal to:

$$\Pr\{a \le X \le b\} = \int_a^b f_X(x)\, dx$$

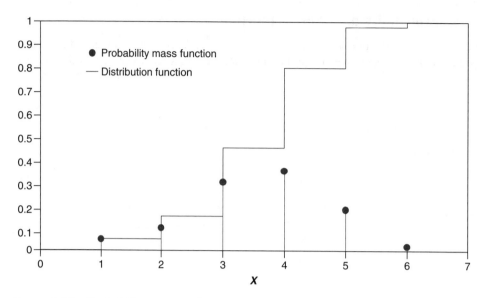

Figure 3.26   Probability mass and probability distribution functions for a discrete r.v.

For a probability density function it must be:

$$f_X(x) \geq 0 \quad \int_\Xi f_X(x)\,dx = 1$$

For a continuous r.v. X, the relationship between the probability density function and the distribution function is:

$$F_X(x) = \Pr\{X \leq x\} = \int_{x_{min}}^{x} f_X(\xi)\,d\xi$$

For a continuous r.v. the $F_X(x)$ is a continuous non-decreasing curve rising from 0 to 1 (Figure 3.27).

3.3.5   Mean and variance of a random variable

The mean or expectation of a r.v. is an index of the centre of the probability distribution. For a discrete r.v. it is given by:

$$E\{X\} = \sum_{x_i \in \Xi} x_i P_X(x_i) \tag{3.7}$$

As shown in Equation 3.7 it is an average of the values of the r.v. weighted by their corresponding probabilities. However, the expectation can be calculated for any function

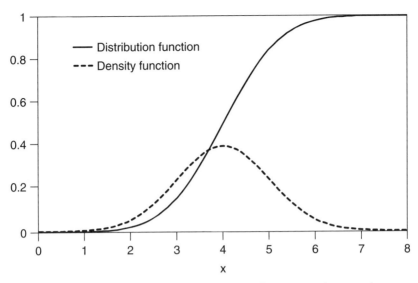

Figure 3.27   Probability density and distribution functions of a continuous random variable.

of the r.v.:

$$E\{\varphi(X)\} = \sum_{x_i \in \Xi} \varphi(x_i)\, P_X(x_i)$$

The expectation possesses the linearity property: $E\{aX + b\} = aE\{X\} + b$ where a and b are any two constants.

Another index that provides summary information on the probability law is the variance, which for a discrete r.v. is defined as:

$$V\{X\} = \sum_{x_i \in \Xi} (x_i - E\{X\})^2\, P_X(x_i)$$

The variance provides information on the dispersion of values around the mean. The variance, unlike the mean, does not possess the linearity property: $V\{aX + b\} = a^2 V\{X\}$. Furthermore, it is possible to demonstrate that: $V\{X\} = E\{X^2\} - (E\{X\})^2$.

The square root of the variance is called the standard deviation.

For a continuous r.v. it is:

$$E\{X\} = \int_{x_{min}}^{x_{max}} x f_X(x)\, dx \quad V\{X\} = \int_{x_{min}}^{x_{max}} [x - E\{X\}]^2\, f_X(x)\, dx$$

Moments of a r.v. and moment generating function

In general, all the characteristics of a probability distribution are defined by a set of constants called moments.

Let X be a r.v. and $\varphi(X) = (X - c)^r$.

We define the *moment of order r from c* using the expression:

$$E[\varphi(X)] = E[(X - c)^r] = \sum_i (x_i - c)^r P_X(x_i) \quad \text{if } X \text{ is a discrete r.v.}$$

$$E[\varphi(X)] = E[(X - c)^r] = \int_{x_{min}}^{x_{max}} (x - c)^r f_X(x)\, dx \quad \text{if } X \text{ is a continuous r.v.}$$

where r can get the values 0, 1, 2, ... and c is a constant. For $c = 0$ we have the so-called *absolute moments*. The *absolute moment of order* 1 is the mean $E\{X\}$.

If we pose $c = E\{X\}$ we have the *central moments*. It is easy to demonstrate that the *central moment of order* 1 is equal to zero for the linearity property of the expectation.

The *central moment of order* 2 is the variance.

Given a r.v. with its probability law, we can define a function through which we can obtain the absolute moments of the r.v. using simple mathematical operations. This function is called the *moment generating function*:

$$\Phi_X(t) = E\left\{e^{tX}\right\} = \sum_{x_i \in \Xi} e^{tx_i} P_X(x_i) \quad \text{if } X \text{ is a discrete r.v.}$$

$$\Phi_X(t) = E\left\{e^{tX}\right\} = \int_{x_{min}}^{x_{max}} e^{tx} f_X(x)\, dx \quad \text{if } X \text{ is a continuous r.v.}$$

In the two formulas above t is a deterministic variable. It can be shown that, for example if we want to calculate the first two absolute moments of the r.v. X, they can be obtained by derivative:

$$\frac{d}{dt}\Phi_X(t)\,|_{t=0} = E\{X\}$$

$$\frac{d^2}{dt^2}\Phi_X(t)\,|_{t=0} = E\{X^2\}$$

3.3.6   Principal models of random variables

The models of random variables are characterised by a specific form of the probability mass/density function and certain parameters.

3.3.6.1   Principal models of discrete random variables

3.3.6.1.1   The Bernoulli model   A 'Bernoulli test' has two possible outcomes that we can call *success*, with probability π and *failure*, with probability $1 - \pi$. The r.v. associated with this test is usually defined as follows:

$$X = \begin{cases} 0 & \text{in case of failure} \\ 1 & \text{in case of success} \end{cases}$$

The Bernoulli model is completely defined by the probability mass function:

$$P_X(x) = \begin{cases} 1 - \pi \text{ for } x = 0 \\ \pi \text{ for } x = 1 \end{cases}$$

A graphical representation of the probability mass function is given in Figure 3.28. The mean and variance of a Bernoulli r.v. can be simply calculated from their respective definitions:

$$E\{X\} = \sum_{x_i \in \Xi} x_i \, P_X(x_i) = 1 \times \pi + 0 \times (1 - \pi) = \pi$$

$$V\{X\} = \sum_{x_i \in \Xi} (x_i - E\{X\})^2 \, P_X(x_i) = (1 - \pi)^2 \pi + (0 - \pi)^2(1 - \pi) = \pi(1 - \pi)$$

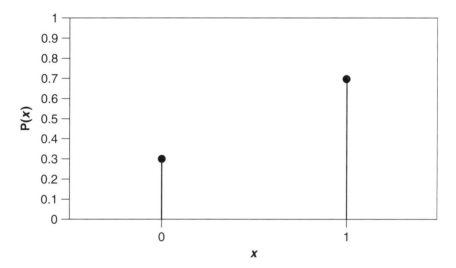

Figure 3.28   *Probability mass function for the Bernoulli model ($\pi = 0.3$).*

3.3.6.1.2   The binomial model   A sequence of v Bernoulli tests with the same success probability π and each other s-independent, leads to another model, the binomial model. The r.v. X that counts the number of successes in such a sequence is defined as a binomial r.v. with parameters π and v. The probability mass function is:

$$P_X(x) = \binom{v}{x} \pi^x (1 - \pi)^{v-x} \qquad (x = 0, 1, \ldots, v)$$

It is a probability mass function since:

$$\sum_{x=0}^{v} \binom{v}{x} \pi^x (1 - \pi)^{v-x} = [\pi + 1 - \pi]^v = 1$$

To calculate the mean and variance of a binomial r.v. we can profitably use the moment generating function:

$$\Phi_X(t) = \mathrm{E}\left\{e^{tX}\right\} = \sum_{x=0}^{\nu} e^{tx} \binom{\nu}{x} \pi^x (1-\pi)^{\nu-x} = \sum_{x=0}^{\nu} \binom{\nu}{x} (e^t \pi)^x (1-\pi)^{\nu-x}$$

$$= [\pi(e^t - 1) + 1]^\nu$$

So we get:

$$\mathrm{E}\{X\} = \left.\frac{d}{dt}\Phi_X(t)\right|_{t=0} = \nu\pi$$

$$\mathrm{E}\{X^2\} = \left.\frac{d^2}{dt^2}\Phi_X(t)\right|_{t=0} = \nu\pi + \nu^2\pi^2 - \nu\pi^2$$

$$\mathrm{V}\{X\} = \mathrm{E}\{X^2\} - (\mathrm{E}\{X\})^2 = \nu\pi(1-\pi)$$

Figure 3.29 shows some examples of binomial r.v. with $\nu = 10$ and three different values of the parameter π.

3.3.6.1.3   The Poisson model   Suppose we set a reference time interval and we count how many events of a certain type (say *arrivals*) occur in this time interval. In general, we may also refer to a spatial interval.

A Poisson r.v. X counts the number of arrivals that occur in the time interval, so it is a discrete r.v. which can take an infinite, but countable, number of values 0, 1, 2, ... ∞.

The probability mass function of a Poisson r.v. (see Figure 3.30) is defined by the following expression:

$$P_X(x) = \frac{\theta^x}{x!} e^{-\theta} \qquad (x = 0, 1, 2 \ldots, \infty)$$

Only one parameter θ defines the distribution model. It is easily shown that:

$$\sum_{x=0}^{\infty} \frac{\theta^x}{x!} e^{-\theta} = e^{-\theta} \sum_{x=0}^{\infty} \frac{\theta^x}{x!} = e^{-\theta} e^{\theta} = 1$$

The moment generating function of a Poisson r.v. is:

$$\Phi_X(t) = \mathrm{E}\left\{e^{tX}\right\} = \sum_{x=0}^{\infty} e^{tx} \frac{\theta^x}{x!} e^{-\theta} = e^{-\theta} \sum_{x=0}^{\infty} \frac{(e^t\theta)^x}{x!} = e^{-\theta} e^{e^t\theta} = e^{\theta(e^t - 1)}$$

Using the moment generating function we obtain the mean and variance:

$$\mathrm{E}\{X\} = \theta$$

$$\mathrm{V}\{X\} = \theta$$

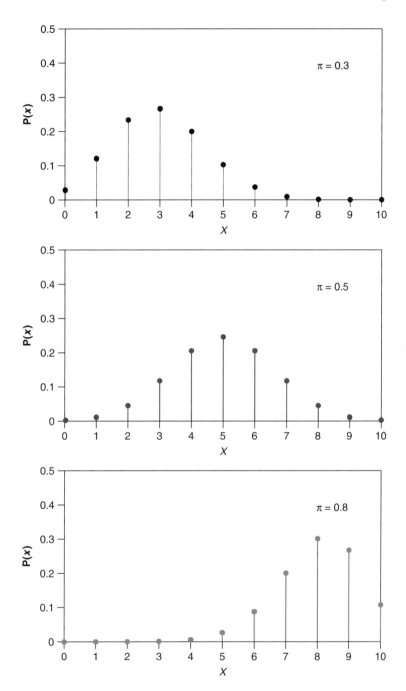

Figure 3.29   Examples of Binomial r.v. with $v = 10$.

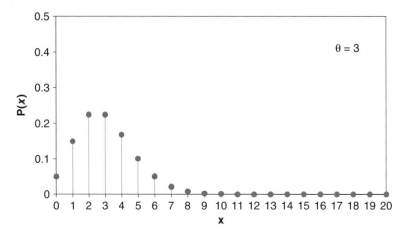

Figure 3.30   Example of Poisson r.v. with $\theta = 3$.

3.3.6.2   Models of continuous random variables

3.3.6.2.1   The Gaussian or Normal model   This is the model of r.v. most widely used in practice and it is generally used for interpreting *errors* (experimental errors, measurement errors, observational errors, etc.).

The Gaussian model is important for a main reason: a probability law called the central limit theorem. This theorem states that if we consider a number n of r.v. identically distributed (according to any model) and we sum up them, we obtain as a result, a r.v. which tends – for increasing n – to follow a Gaussian model.

Therefore, if we imagine that the 'error' – a deviation from what is expected – is the sum (i.e. the result) of many contributing factors, we can reasonably model it as a Gaussian r.v.

The analytical expression of the probability density function of a Gaussian r.v. is:

$$f_X(x) = \frac{1}{\sigma\sqrt{2\pi}} e^{-\frac{(x-\mu)^2}{2\sigma^2}} \qquad (-\infty < x < \infty)$$

A Gaussian r.v. is characterised by two parameters μ and σ. It can be shown that for a Gaussian r.v. X:

$$\Phi_X(t) = E\left\{e^{tX}\right\} = e^{t\mu + t^2\sigma^2/2}$$

$$E\{X\} = \mu$$

$$V\{X\} = \sigma^2$$

The probability density function has a bell shape, is symmetrical about the mean μ, and has two inflection points with oblique tangents at distance $\pm\sigma$ from μ (Figure 3.31).

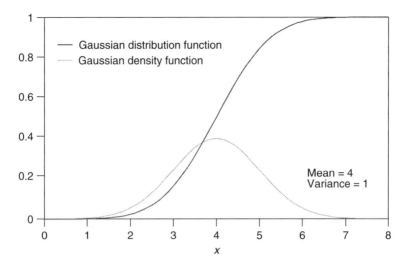

Figure 3.31   Probability density and distribution functions for a Gaussian r.v.

The probability distribution function has no 'closed' analytical expression. It is the integral of the density function:

$$F_X(x) = \frac{1}{\sigma\sqrt{2\pi}} \int_{-\infty}^{x} e^{-\frac{(\xi-\mu)^2}{2\sigma^2}}\, d\xi$$

which is calculated with numerical methods.

3.3.6.2.2   Probability calculations with the Gaussian model. The standard Gaussian
Given a Gaussian r.v. X with parameters μ and σ, through the so-called standardisation:

$$Z = \frac{X - \mu}{\sigma}$$

we get a new r.v. Z, which is still Gaussian with mean 0 and variance 1, that is called a standard Gaussian. Its probability density function is:

$$f_Z(z) = \frac{1}{\sqrt{2\pi}} e^{-\frac{1}{2}z^2}$$

The benefit of the standard Gaussian is to be able to get rid of the values of the parameters and calculate the probabilities (calculation of integrals) only once. These probabilities are usually collected in tables and can be used for a Gaussian r.v. with any parameters, as shown in the following example.

EXAMPLE

Let X be a Gaussian r.v. with mean $\mu = 25$ and variance $\sigma^2 = 9$. Suppose we wish to calculate the following probability: $\Pr\{23 < X < 28\}$. The standardisation leads to:

$$Z = \frac{X - 25}{3}$$

Therefore:

$$\Pr\{23 < X < 28\} = \Pr\left\{\frac{23 - 25}{3} < \frac{X - 25}{3} < \frac{28 - 25}{3}\right\}$$

$$= \Pr\{-0.67 < Z < 1\} = 0.590$$

For the calculation of the last probability we have consulted the standard Gaussian tables. Today with the use of a computer, standardisation and the use of tables are no longer necessary. For example, with Excel® we can use the function NORMDIST that provides the required probabilities for a Gaussian r.v. with any parameters.

3.3.6.2.3 The exponential model   An exponential r.v. is defined on the positive half real line ($x \geq 0$) and usually interprets phenomena of life (e.g. operating life, human life, etc.). One parameter $\lambda > 0$ defines the probability distribution. The probability density function is given by the following expression:

$$f_X(x) = \lambda e^{-\lambda x} \qquad (x \geq 0)$$

The probability distribution function can be analytically calculated:

$$F_X(x) = \int_0^x \lambda e^{-\lambda \xi} d\xi = 1 - e^{-\lambda x}$$

Figure 3.32 shows the probability density and distribution functions of an exponential r.v.

The mean and variance can also be calculated directly:

$$E\{X\} = \int_0^{+\infty} x\lambda e^{-\lambda x} dx = \frac{1}{\lambda}$$

$$V\{X\} = \frac{1}{\lambda^2}$$

3.3.6.2.4 The Weibull model   A r.v. distributed according to a Weibull model also usually interprets phenomena of life. The Weibull model is more flexible than the exponential, as it is characterised by two parameters, α and β, both positive (Figure 3.33).

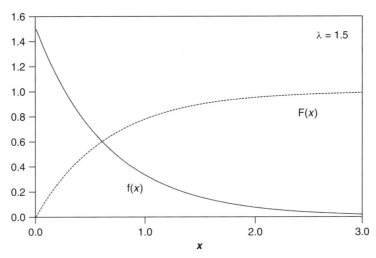

Figure 3.32   Probability distribution function and probability density function for an exponential r.v.

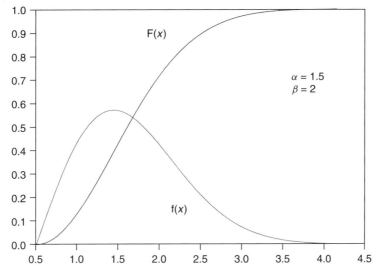

Figure 3.33   Probability distribution function and probability density function for a Weibull r.v.

The probability density function is given by:

$$f_X(x) = \frac{\beta}{\alpha^\beta} x^{\beta-1} \ e^{-(\frac{x}{\alpha})^\beta} \qquad (x > 0)$$

the probability distribution function is:

$$F_X(x) = 1 - e^{-(\frac{x}{\alpha})^\beta} \qquad (x > 0)$$

and the mean and variance are given by the following expressions:

$$E(X) = \alpha\Gamma\left(1 + \frac{1}{\beta}\right)$$

$$V(X) = \alpha^2\Gamma\left(1 + \frac{2}{\beta}\right) - \alpha^2\Gamma\left(1 + \frac{1}{\beta}\right)$$

where the Gamma function is defined as:

$$\Gamma(\vartheta) = \int_0^{+\infty} z^{\vartheta-1}e^{-z}dz$$

3.4   To draw conclusions from observed data

3.4.1   The inferential process

The term inference is synonymous with 'illation': while the former word has a positive connotation and is generally adopted in statistical language, the latter usually has a negative connotation.

In Section 3.1.1 we defined the *population* as the collection of objects (individuals) of interest for a certain study, and defined the *sample* as any subset of the population. We said that a sample is analysed when it is not convenient or even possible to analyse the whole population.

In this last section of the chapter about statistical inference, we try to close the circle which was started with the observation, measurement, collection of data and exploratory analysis. This will be only a hint since statistical inference is an enormous body of methodologies allowing results of an analysis relative to a sample to be brought to the level of the population from which the sample was drawn. Statistical inference is a noteable example of inductive reasoning.

When making inference, the population must be well specified and the sample selected according to good rules. However, the population can also be non-existing at the moment of the inference.

As an example consider: the 'amount of polluting emissions released in a lab test' could be the characteristic of our interest for each individual (specimen) of the population of vehicles under development. The population of vehicles to be produced over the next few years is currently not in existance, but this inference is still relevant today.

It often happens in statistical inference that we denote as population the specific characteristic in which we are interested. Any characteristic of the population can be considered as a r.v. We are used to indicate with X the r.v. interpreting the variability of the characteristic of interest.

3.4.2   Sampling and samples

Sampling is a delicate operation consisting in selecting the right data to use to make a reasonable inference.

Suppose we draw a sample of size n from the population and observe the characteristic X on each individual. Before observing the data, we may think of n r.v. $X_1, X_2, \ldots, X_n$. After the observation we have n values $x_1, x_2, \ldots, x_n$. Such distinction is of utmost importance when discussing the properties of the sample and its possible summaries.

3.4.2.1.1   Random sample   The sample drawn from the population is said to be 'random' if the r.v. $X_1, X_2, \cdots, X_n$ are s-independent and if they are identically distributed with the same probability distribution of the population.

To better understand the concept of a random sample, we use the scheme of repeated extractions from a urn. If the sample unit extracted is observed and then reintroduced in the urn, we can say that the characteristics of the sample do not change, so the sampling is random. Conversely, if the sample unit is not reintroduced in the urn, the characteristics of the population within the urn are every time different when sampling, so the sampling is not random.

3.4.2.1.2   Random sample of a Gaussian population   Hereinafter we will refer to the inferential procedures for the Gaussian model. This is an evident limitation, but coherent with the scope of this book which is not intended to be a 'Statistics' book.

A Gaussian population X is completely determined by the two parameters $\mu = \mathrm{E}\{X\}$ and $\sigma = \sqrt{\mathrm{V}\{X\}}$. Having a random sample, each X_i $(i = 1, \ldots, n)$ is a Gaussian r.v. with parameters equal to those of the population:

$$E\{X_i\} = \mu$$

$$\sqrt{\mathrm{V}\{X_i\}} = \sigma$$

From the observation of the sample, we must be able to determine the population parameters μ and σ. The population parameters are unknown and they will always be unknown, unless we imagine that we draw a sample of infinite size. This is the essence of the inferential process: try to 'guess' the unknown population parameters. We will see that in general the sample size is important: the bigger the sample size, the bigger its informative power.

3.4.2.1.3   Inferential procedures   Three main inferential procedures exist.

- *Point estimation*: we try to guess specific values of the unknown population parameters.

- *Interval estimation*: we try to determine intervals containing the unknown population parameters with a certain confidence.

- *Hypothesis testing*: the sample is used to check the validity of hypotheses regarding the unknown population parameters.

3.4.2.1.4   Sample statistic   In statistical inference we mostly use statistics that provide a special summary of the sample. A statistic is a function of the sample.

$T = \varphi(X_1, X_2, \ldots, X_n)$. It is a function of the r.v. X_i. Therefore it is a r.v.

$t = \varphi(x_1, x_2, \ldots, x_n)$. It is a function of the observed values x_i. Therefore it is a value.

3.4.3   Adopting a probability distribution model by graphical analysis of the sample (probability plot)

Suppose that we have a sample and want to make inferences on the population. The first thing to consider is whether the data refer to a discrete or continuous variable, then we should choose a model of probability distribution to adopt for the population.

Indeed, for any inference we want to do, for example determine a confidence interval for an unknown parameter of the population, we must assume a model of probability distribution for the population and try to determine if the data can be thought of as a random sample from that population.

This problem is called *distribution fitting*. From the outcome of it depends the subsequent application of certain methods of inference.

First and foremost we should assume that the sample is random. This is necessary for the application of many inferential procedures. In order to make that assumption we must refer to the way in which the sample was obtained.

Then, for any given value of the sample, a value of the distribution function is 'estimated'. If we sort the sample data from the minimum to the maximum, we can use the following formula:

$$\hat{F}_i = \hat{F}(x_{(i)}) = \frac{i}{n+1}$$

where $x_{(i)}$ is the value at the i-th position in the sorted sample.

The verification of the assumption of a distribution model passes through the comparison between the empirical distribution function estimated from sample data and a theoretical model of distribution.

For a fast and efficient comparison we may use the graphical technique of *probability plots*. One such method is based on the transformation of the vertical axis of a Cartesian graph in which points of coordinates $(x_{(i)}, \hat{F}_i)$ are plotted so that the curve of the assumed distribution function model would be a straight line.

For example, if we assume an exponential distribution model:

$$F_X(x) = 1 - e^{-\lambda x} \Rightarrow \text{(simplifying the notation)} \Rightarrow \log(1 - F) = \log(e^{-\lambda x}) = -\lambda x$$

So, we obtain

$$-\log(1 - F) = \lambda x$$

If we plot the points with coordinates $(x_{(i)}, -\log(1 - \hat{F}_i))$ on a Cartesian graph and these appear aligned, then we can rightly assume that the sample is drawn from an exponential population.

EXAMPLE

The following 30 results were collected; they refer to the duration, that is life in operation (measured in years), of car halogen lamps:

0.315	0.393	0.419	0.699	0.845	1.055	1.120	0.461	0.836	0.316
1.226	1.513	1.780	2.318	3.887	0.272	1.135	2.301	2.339	1.247
0.017	0.085	0.122	0.202	0.210	0.238	0.265	0.196	0.209	0.084

Table 3.10 shows the sorted data with the corresponding estimates of the distribution function, while the graph of Figure 3.34 shows the probability plot obtained with the above described procedure, having assumed an exponential model for the operating life.

Table 3.10   Sorted data set and preparation of the exponential probability plot.

No.	$x_{(i)}$	$\hat{F}_i$	$-\log(1-\hat{F}_i)$	No.	$x_{(i)}$	$\hat{F}_i$	$-\log(1-\hat{F}_i)$
1	0.017	0.032	0.014	16	0.461	0.516	0.315
2	0.084	0.065	0.029	17	0.699	0.548	0.345
3	0.085	0.097	0.044	18	0.836	0.581	0.377
4	0.122	0.129	0.060	19	0.845	0.613	0.412
5	0.196	0.161	0.076	20	1.055	0.645	0.450
6	0.202	0.194	0.093	21	1.120	0.677	0.491
7	0.209	0.226	0.111	22	1.135	0.710	0.537
8	0.210	0.258	0.130	23	1.226	0.742	0.588
9	0.238	0.290	0.149	24	1.247	0.774	0.646
10	0.265	0.323	0.169	25	1.513	0.806	0.713
11	0.272	0.355	0.190	26	1.780	0.839	0.792
12	0.315	0.387	0.213	27	2.301	0.871	0.889
13	0.316	0.419	0.236	28	2.318	0.903	1.014
14	0.393	0.452	0.261	29	2.339	0.935	1.190
15	0.419	0.484	0.287	30	3.887	0.968	1.491

The points on the graph appear well aligned, which confirms the assumption of the exponential model. The straight line in the graph was obtained as the least-squares line (see Chapter 5) with the constraint of passing through the axes origin. It also allows us to estimate the unknown parameter λ of the distribution (0.43). This method of representing the straight line is actually to be used with caution, of course,

because the presence of outliers in the dataset could lead to incorrect estimates of
the parameter.

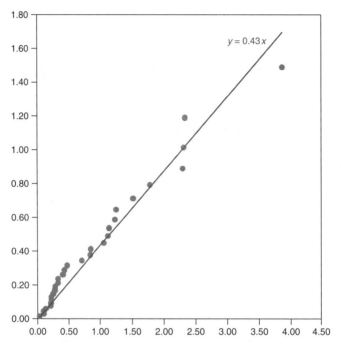

Figure 3.34   Exponential probability plot of halogen light bulb lifetime data.

In the case of a Gaussian distribution, the linearisation of the scale is not so simple
as in the case of an exponential distribution, since we have seen that the probability
distribution function of the Gaussian model has no closed analytical expression.

Therefore, either we resort to ready made paper (Gaussian probability paper) on which
to draw the points of coordinates $(x_{(i)}, \hat{F}_i)$, or we rely on the use of statistical software.
Alternatively, the Gaussian probability plot can be constructed more 'manually' by comparing the sorted sample data with the corresponding quantiles of the assumed distribution
model. The result is a graph called a quantile–quantile plot or just QQ plot. The procedure
will be clarified by the following example.

EXAMPLE

The first column of Table 3.11 shows a set of 30 data already sorted. The second column shows the estimated distribution function. Finally, each cell of the third
column shows the quantile of the standard Gaussian corresponding to a value of
the distribution function equal to the one reported in the cell to its left. If the

Table 3.11   Sorted dataset and preparation of a normal Q-Q plot.

$x_{(i)}$	$\hat{F}_i$	Standard Gaussian quantile
1.231	0.032	−1.849
1.274	0.065	−1.518
1.905	0.097	−1.300
1.963	0.129	−1.131
2.094	0.161	−0.989
2.709	0.194	−0.865
3.188	0.226	−0.753
3.623	0.258	−0.649
4.128	0.290	−0.552
4.523	0.323	−0.460
4.611	0.355	−0.372
4.696	0.387	−0.287
4.960	0.419	−0.204
5.091	0.452	−0.122
5.126	0.484	−0.040
5.270	0.516	0.040
5.871	0.548	0.122
5.909	0.581	0.204
6.065	0.613	0.287
6.067	0.645	0.372
6.173	0.677	0.460
6.218	0.710	0.552
6.724	0.742	0.649
6.918	0.774	0.753
7.043	0.806	0.865
7.428	0.839	0.989
7.452	0.871	1.131
7.693	0.903	1.300
7.863	0.935	1.518
9.111	0.968	1.849

data can be thought of as a random sample drawn from the assumed distribution model, then making the scatter plot of the first and third column of the table we will get a good alignment of points, as observed in Figure 3.35. Here too the least-squares line allows us to estimate the distribution parameters. The mean will be provided by the abscissa of the point of intersection of the line with the horizontal axis and the standard deviation will be given by the difference of the abscissa at the point of the straight line with ordinate value equal to 1, from the previously calculated mean.

From the plot of Figure 3.35 we estimate a value for the mean equal to 5.116 and a standard deviation equal to $7.442 - 5.116 = 2.326$.

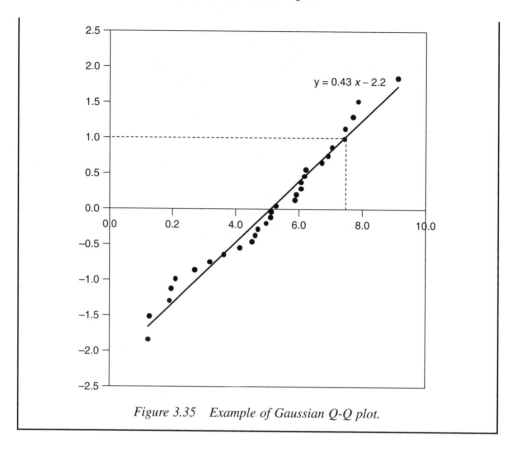

Figure 3.35   Example of Gaussian Q-Q plot.

3.4.4   Point estimation of the parameters of a Gaussian population

3.4.4.1   The sample average

To make inference on the parameter μ of a Gaussian population from a random sample we use the statistic 'average':

$$\overline{X} = \frac{1}{n} \sum_{i=1}^{n} X_i$$

The reason why we use the sample average as 'point estimator' of μ is due to the fact that it has good properties.

- It can be demonstrated that $\overline{X}$ follows a Gaussian distribution.

- The expected value of the sample average $\overline{X}$ is equal to μ. We say that $\overline{X}$ is an unbiased estimator of the population mean. Demonstration hint:

$$E\left\{\overline{X}\right\} = E\left\{\frac{1}{n}\sum_{i=1}^{n} X_i\right\} = \frac{1}{n}E\left\{\sum_{i=1}^{n} X_i\right\} = \frac{1}{n}\sum_{i=1}^{n} E\{X_i\} = \frac{1}{n}\sum_{i=1}^{n} \mu = \mu$$

- The variance of the sample average is equal to the variance of the population divided by n. We say that $\overline{X}$ is a consistent estimator of the population mean. Demonstration hint:

$$V\{\overline{X}\} = V\left\{\frac{1}{n}\sum_{i=1}^{n}X_i\right\} = \frac{1}{n^2}V\left\{\sum_{i=1}^{n}X_i\right\} = \text{being } X_i \text{ s-independent}$$

$$= \frac{1}{n^2}\sum_{i=1}^{n}V\{X_i\} = \frac{1}{n^2}\sum_{i=1}^{n}\sigma_i^2 = \frac{1}{n^2}n\cdot\sigma^2 = \frac{\sigma^2}{n}$$

For n increasing, the variance of $\overline{X}$ decreases, bringing the probability distribution of $\overline{X}$ to concentrate around its mean (which is the population mean) (see Figure 3.36).

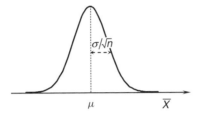

Figure 3.36   The distribution of $\overline{X}$ for a random sample drawn from a Gaussian population.

3.4.4.2   The sample variance

To define a point estimator for the population variance we start from the sample variance as defined in Equation 3.2:

$$\frac{1}{n}\sum_{i=1}^{n}(X_i - \overline{X})^2 \tag{3.8}$$

By taking the expectation of the Equation 3.8 we obtain:

$$E\left\{\frac{1}{n}\sum_{i=1}^{n}(X_i - \overline{X})^2\right\} = \frac{1}{n}\sum_{i=1}^{n}E\left\{(X_i - \overline{X})^2\right\} = \frac{n-1}{n}\sigma^2$$

The previous formula shows that the sample variance as defined in Equation 3.8 is a biased estimator of the population variance since its expectation is not equal to the population variance. Therefore a so-called 'corrected sample variance' is usually defined and used for estimation purposes:

$$S^2 = \frac{n}{n-1}\left[\frac{1}{n}\sum_{i=1}^{n}(X_i - \overline{X})^2\right] = \frac{1}{n-1}\sum_{i=1}^{n}(X_i - \overline{X})^2 \tag{3.9}$$

The expected value of S^2 as defined above is equal to the population variance.

$$E\{S^2\} = \sigma^2$$

In the case of a Gaussian population, the variance of S^2 can also be calculated and it is demonstrated that:

$$V\{S^2\} = 2\frac{\sigma^4}{n-1}$$

Therefore in case of a Gaussian population the corrected sample variance S^2 is an unbiased and consistent estimator of the population variance.

3.4.5   Interval estimation

The interval estimation is an inferential procedure that leads to an interval (confidence interval) which may contain the unknown parameter of the population with a given probability or confidence level. The confidence level is usually assigned by exogenous reasons and it is here denoted by $1 - c$.

If the unknown parameters are more than one, then we talk about confidence regions. That is the most realistic situation, but for simplicity of dealing, it is common practice to assume one parameter at time as unknown.

To construct a confidence interval we start from the random sample $X_1, X_2, \ldots, X_n$ with the aim of reaching a relation of this type:

$$\Pr\{L(X_1, X_2, \ldots, X_n) \leq \vartheta \leq U(X_1, X_2, \ldots, X_n)\} = 1 - c$$

where ϑ is the generic unknown parameter.

$L(X_1, X_2, \ldots, X_n)$ and $U(X_1, X_2, \ldots, X_n)$ are respectively the lower and the upper bounds of the confidence interval.

Accordingly we should find an ancillary function (also called a pivotal quantity), that is a function of the sample and the unknown parameter, say $\psi(X_1, X_2, \ldots, X_n, \vartheta)$, whose probability distribution is completely known (both model and parameters) and does not depend on the unknown parameter ϑ.

Note that the ancillary function is not a statistic. To obtain a confidence interval, the ancillary function must be invertible for ϑ.

3.4.5.1   Interval estimation of the mean of a Gaussian population with known variance

Consider a random sample from a Gaussian population X whose mean μ is unknown, while the variance σ^2 is known. An ancillary function for determining a confidence interval for μ is:

$$Z = \frac{\overline{X} - \mu}{\sigma/\sqrt{n}} \tag{3.10}$$

The r.v. Z follows a standard Gaussian probability distribution (therefore completely known), so we can write:

$$\Pr\left\{-z_{c/2} \leq \frac{\overline{X} - \mu}{\sigma/\sqrt{n}} \leq +z_{c/2}\right\} = 1 - c$$

with $z_{c/2}$ being the quantile that corresponds to an area equal to $c/2$ in the right tail of the distribution (see Figure 3.37):

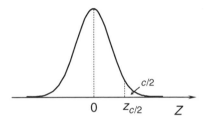

Figure 3.37   The quantile $z_{c/2}$ of the standard Gaussian random variable.

Moreover the quantity in Equation 3.10 is invertible for μ, so we obtain:

$$\Pr\left\{\overline{X} - z_{c/2}\frac{\sigma}{\sqrt{n}} \leq \mu \leq \overline{X} + z_{c/2}\frac{\sigma}{\sqrt{n}}\right\} = 1 - c \tag{3.11}$$

The formula (3.11) is the expression of a confidence interval for the unknown parameter μ. The lower and upper bounds will be determined once the sample is drawn. The logic of the confidence interval is that if repeating the sampling a certain number of times, in the $(1 - c)\%$ of the cases the interval calculated according to Equation 3.11 contains the unknown parameter.

Note: inside the parenthesis μ is not a r.v. The interval bounds are random variables!

3.4.6   Hypothesis testing

As suggested by the name, hypothesis testing is a verification, a check based on sample evidence. As for point estimation and interval estimation, we assume the probability distribution model of the population (Gaussian for this case) and we assume that the sample is random.

The hypotheses may concern one or more parameters of the population, but also here for simplicity we examine the case of only one unknown parameter at time.

A statistical hypothesis on the parameter ϑ can be of two types:

- *simple*: for example $\vartheta = \vartheta_0$;

- *composite*: for example $\vartheta \neq \vartheta_0$ (bilateral) or $\vartheta \geq \vartheta_0$ (unilateral).

The decision following the verification can be to accept or to reject a hypothesis that is conventionally called the 'null hypothesis' and indicated by H_0. It is contrasted with an 'alternative hypothesis' indicated by H_1. The hypotheses H_0 and H_1 shall be incompatible. Theoretically they can be exchangeable; the terminology 'null' and 'alternative' is mostly conventional.

We will check the hypotheses on the basis of a sample; hence the consequent decision (to accept or reject one of the two) has an unavoidable uncertainty related to it.

We define type I risk (r_I) as the probability to make an error of type I. Similarly, we define type II risk (r_{II}) as the probability to make an error of type II (see Table 3.12).

Table 3.12   The general scheme of hypothesis testing.

Decision Hidden truth	Accepting H_0	Rejecting H_0
H_0 is true	Correct decision	Type I error
H_0 is false	Type II error	Correct decision

$1 - r_I$ is the 'significance level': it is the probability to accept H_0 if H_0 is true.
$1 - r_{II}$ is the 'power': it is the probability to reject H_0 if H_0 is false.

For hypothesis testing we make use of statistics that summarise the information contained in the sample (test statistics).

We define a critical region of size r_I as the set of possible values of the test statistic having probability r_I to occur under the null hypothesis H_0 and making us commit a type I error.

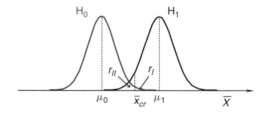

Figure 3.38   Scheme of a hypothesis test for the mean of a Gaussian population.

3.4.6.1 Hypothesis test for the mean of a Gaussian population

The previous reasoning was quite general. Here we consider a Gaussian population X with unknown mean μ and known variance σ^2 and we assume we draw a random sample from it.

As an example, we may formulate the following hypotheses:

$$H_0 : \{\mu = \mu_0\}$$

$$H_1 : \{\mu = \mu_1\} \text{ with } \mu_1 > \mu_0$$

They are both simple hypotheses but are in contrast to each other. To test the hypotheses we adopt the statistic $\overline{X}$. In fact, in the case of a Gaussian population we have seen that $\overline{X}$ follows a Gaussian distribution with parameters μ and σ^2/n.

In the scheme of Figure 3.38, we see how to graphically represent the elements defined above. The two hypotheses delineate two possible scenarios, one of them described by the black curve, the other one described by the grey curve. Once the type I risk r_I is fixed, the critical region is given by the interval $[\bar{x}_{cr}, +\infty[$. Accordingly also the type II risk r_{II} is univocally determined. For the given hypotheses H_0 and H_1 there is only one way to reduce both the type I and type II risks, which is increasing the sample size n. In fact, a bigger sample size would make the two curves in Figure 3.38 more slender.

References

Johnson, R.A. and Bhattacharyya, G.H. (2010) *Statistics: Principles and Methods*, Wiley-VCH Verlag GmbH.

Mood, A.M., Graybill, F.A. and Boes, D.C. (1974) *Introduction to the Theory of Statistics*, McGraw-Hill.

Wonnacott, T.H. and Wonnacott, R.J. (1990) *Introductory Statistics*, John Wiley & Sons, Inc.

4

Advanced managerial techniques

4.1   To describe processes

4.1.1   IDEF0

The Integration Definition for Function Modelling (IDEF0) technique is used to graphically represent processes, supporting the analysis of a complex system (such as an organisational system). The method was developed in the late 1970s by the U.S. Air Force within the project ICAM (integrated computer-aided manufacturing), to describe their productive activities and better manage the integration of these with computer technology. The IDEF0 term identifies a hierarchical set of diagrams that graphically illustrate the processes at increasing levels of detail. The IDEF0 approach is top-down in the sense that the analysis of the process under study proceeds from a more general level to more detailed levels.

At each step, four elements must be identified: inputs, outputs, constraints and resources. The analysis gradually becomes deeper and deeper by breaking down the initial process into sub-processes, activities and operations, up to the level of detail considered most suitable for the purpose of the analysis itself.

In this method, the two concepts 'inputs'' and 'resources', which are generally used as synonyms, here instead assume two different meanings. The inputs are represented by all the materials and information input into the process to undergo a transformation. The resources mostly identify people who work in the process and the machinery used in it. Finally, the constraints relate to anything that affects the performance of the process: information, practices, laws, regulations, and so on.

Statistical and Managerial Techniques for Six Sigma Methodology: Theory and Application, First Edition.
Stefano Barone and Eva Lo Franco.
© 2012 John Wiley & Sons, Ltd. Published 2012 by John Wiley & Sons, Ltd.

EXAMPLE

Figures 4.1 and 4.2 show the first two levels of representation, through IDEF0, of the production process of ceramic tiles. We can see that the process consists of five macro activities, each of which may in turn be represented by a single diagram. For example, the preparation of raw materials usually includes the following activities: dosing, grinding, sieving and atomisation. Compared to suppliers, inputs, process, outputs and customers (SIPOC) (see Section 2.3.1), the IDEF0 technique allows on to reach a level of detail in the process analysis considerably higher while maintaining a well-defined graphical connection between the initial level of analysis and that achieved by two or more phases of detailing.

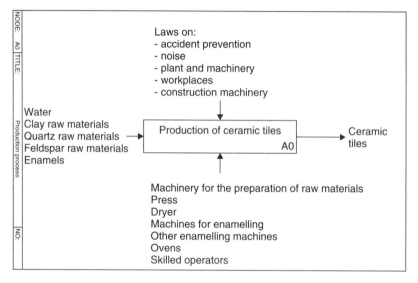

Figure 4.1   IDEF0: production process A0.

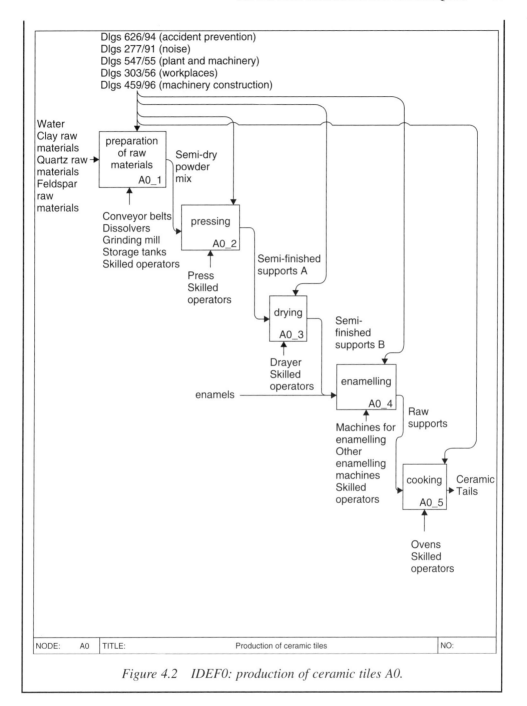

Figure 4.2   IDEF0: production of ceramic tiles A0.

4.2   To manage a project

'Managing a project' means coordinating and directing with the assumption of full responsibility for the planning and control tasks. A project is a complex system of operations, often non-repetitive and with a definite deadline. The management of a project should be directed towards the achievement of a specified goal through a continuous process of planning and controlling different resources, in the presence of interdependent constraints, mainly: time, cost and quality. Usually, the time between the start and completion of a project tends to expand, and with it the necessary financial commitment. In addition, the higher the technological content of a project, the greater the degree of specialisation of resources, hence the importance of project planning and coordination.

4.2.1   Project evaluation and review technique

The program evaluation and review technique (PERT) is a valid alternative to the Gantt chart (Section 2.2.2). Unlike the Gantt chart, which is valid when the phases of the project have a simple sequence, PERT allows one to show the interdependence of activities and is preferable when the project phases are many and have a high level of interdependence.

PERT was initially developed at the time of the Cold War by the U.S. Navy Special Projects Office in 1958, and was subsequently developed by the consulting firm Booz, Allen & Hamilton.

There is another technique that studies the development of a project through a careful planning of its activities: the so-called critical path method (CPM); it will be treated in Section 4.2.2. PERT and CPM are sometimes confused. In fact, PERT and CPM both have some common phases. However, while PERT deals only with the minimisation of time, and durations of different activities may be modelled by random variables, CPM considers cost aspects of the various activities, and uses deterministic activity durations.

The main phases of PERT will be described and illustrated through an example of a daily life project: making pasta carbonara.

4.2.1.1   Phase 1: Identification of project activities

The first phase is to break down the project into activities, trying to keep a homogeneous level of detail. A time should be assigned to each activity. Table 4.1 shows the necessary activities for the preparation of pasta carbonara.

4.2.1.2   Phase 2: Determination of sequence constraints

After identifying the project activities, it is necessary to evaluate the temporal order in which these activities shall be completed. This means to know for each activity, what other activity must have already been completed so that it can start. This operation, called the 'determination of the sequence constraints' can be burdensome, but it is very important.

The constraints represent an AND logic, meaning that an activity can start only after all those preceding it in the sequence have already been completed. In practice, the constraints are determined not only by the technical and logical conditions that prevent the realisation of an activity if the preceding one has/have not been completed, but also

Table 4.1   Description of the activities of the project 'making pasta carbonara'.

Activity	Description	Time (minutes)
A	Put a pot filled of water on the fire	10
B	Put salt in the water	1
C	Cut bacon into small pieces	3
D	Fry the bacon in a pan with oil	5
E	Mix eggs with parmesan cheese and black pepper	5
F	Drop and cook pasta in the pot	8
G	Drain pasta	1
H	Pour the drained pasta to the pan and stir	2
I	Add the egg mixture to the pan and stir with the fire on	4
J	Share the pasta on plates and serve	4

by problems regarding the availability of resources (for example a machine which may not be available for multiple simultaneous tasks).

Table 4.2 reports the sequence constraints for the example project.

4.2.1.3   Phase 3: Construction of the network diagram

Once the project has been broken down into activities and the constraints identified, we can proceed with the construction of the project network diagram (hereinafter just the 'network').

By convention, activities in the network are drawn as continuous oriented arcs, while the circles (nodes) represent the 'start' and 'end' times of each activity.

Each node is an event occurring at the instant of time when **all** the activities entering it are completed. Each activity has a duration, say $t_{i,j}$.

Figure 4.3 shows the network diagram of the example project.

Start and end nodes of the whole project must be clearly visible in the network. When building the network, it is good to keep in mind that:

- all sequence constraints must be respected;

- never create loops;

- do not define two or more activities starting and ending at the same nodes.

Table 4.2   Sequence constraints of the project 'making pasta carbonara'.

Activity	Description	Constraints
A	Put a pot filled of water on the fire	–
B	Put salt in the water	A
C	Cut bacon into small pieces	–
D	Fry the bacon in a pan with oil	C
E	Mix eggs with parmesan cheese and black pepper	–
F	Drop and cook pasta in the pot	B
G	Drain pasta	F
H	Pour the drained pasta to the pan and stir	G
I	Add the egg mixture to the pan and stir with the fire on	E
J	Share the pasta on plates and serve	H, D, I

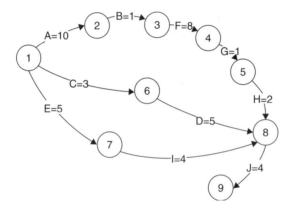

Figure 4.3   Network diagram of the project 'making pasta carbonara'.

The network allows us to calculate the time to complete a project, and the times for all intermediate stages of its realisation.

A dashed arc in the network indicates a 'precedence' constraint between two activities and may be interpreted as a 'dummy' activity of zero duration. For example, in Figure 4.6 the activity F cannot start before the activities B, C and D, but the dashed arc indicates that also the activity E must be completed.

It is also possible to define the network in another way, that is activities represented by nodes and sequence constraints by oriented arcs. This type of notation, rarely used, has the advantage that it is not necessary to use dummy activities.

4.2.1.4   Phase 4: Earliest and latest event times and critical path

The earliest time (ET) for each node of the network is calculated by analysing the network forward, that is from the start to the end nodes.

For the generic node j, if it is reached by several activities (e.g. node 8 in Figure 4.3), its ET will be the maximum sum of the duration of the activity leading to the node and the ET of the previous node:

$$\begin{cases} ET_j = \max \{ET_i + t_{i,j}\} & \forall\ 1 < i < j \\ ET_1 = 0 \end{cases} \tag{4.1}$$

Obviously, if the node is reached by only one activity, its ET will be the sum of the ET of the previous node and the duration of the activity.

In the example project, the ET of node 8 is calculated as:

$$ET_8 = \max \{ET_5 + t_{5,8}; ET_6 + t_{6,8}; ET_7 + t_{7,8}\} = 22$$

The second column of Table 4.3 shows the ETs calculated for each node.

Table 4.3   Earliest times (ET), latest times (LT) and node slacks (NS) for the project 'making pasta carbonara'.

Node	ET	LT	NS
1	0	0	0
2	10	10	0
3	11	11	0
4	19	19	0
5	20	20	0
6	3	17	14
7	5	18	13
8	22	22	0
9	26	26	0

Moreover, we indicate with LT_i the latest time of the node i, intended as the latest time all activities ij starting from node i must be completed. Analysing the network backward from the end of the project, it is calculated as:

$$LT_i = \min \{LT_j - t_{ij}\} \qquad \forall i < j \tag{4.2}$$

Normally we pose the latest time of the end event of the project (the node 9 in the example project) equal to its ET. The third column of Table 4.3 shows the latest times calculated for each node of the example project.

Earliest and latest times calculated for each node allow us to find the so-called *critical path*, that is the sequence of activities whose durations cannot have delays and, summed up, give the minimum duration of the project. To find out the critical path, it is firstly necessary to calculate the so-called slacks for each node and for each activity.

For each node, the slack is calculated as:

$$NS_j = LT_j - ET_j \tag{4.3}$$

If the node slack is zero, the node is said to be critical. In our case only two nodes (6 and 7) are not critical (see Table 4.3).

For each activity, the slack is calculated as:

$$AS_{ij} = LT_j - \left[ET_i + t_{ij}\right] \tag{4.4}$$

The slack expresses the time flexibility of the activity. If the slack is zero it means that no delay is admitted for that activity.

A critical activity is always included between two critical nodes, but it is not said that two critical nodes bound a critical activity. Table 4.4 shows that four activities are not critical: C, D, E and I, with 14, 17, 13 and 13 minutes respectively of admitted delay. All other activities, namely A, B, F, G, H and J are critical and their sequence identifies the critical path. The sum of their durations is the minimum duration of the whole 'project', that is 26 minutes (sorry, this is slow food).

Table 4.4   Activity slacks for the project 'making pasta carbonara'.

Activity	Connecting nodes	Time (minutes)	AS
A	1,2	10	0
B	2,3	1	0
C	1,6	3	14
D	6,8	5	17
E	1,7	5	13
F	3,4	8	0
G	4,5	1	0
H	5,8	2	0
I	7,8	4	13
J	8,9	4	0

4.2.1.5   Phase 5: Probabilistic PERT

We can model the durations of activities as random variables. For each activity it is common to know an optimistic time t_o (best case, if everything works well), a pessimistic time t_p (worst case) and a most probable or 'modal' time t_m.

It is usually assumed that the random variable (r.v.) describing the durations has a probability distribution function of a beta type. This model is adopted because it is delimited by a minimum and a maximum value. The beta distribution is unimodal. In Figure 4.4 an example of a beta distribution function delimited between 1 and 3 is shown.

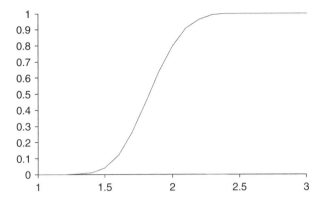

Figure 4.4   Example of beta probability distribution function.

The expected value of T_{ij} distributed according to a beta model is approximated by:

$$\mu_{ij} = \mathrm{E}\left\{T_{ij}\right\} = \frac{1}{3}\left[2t_{m_{ij}} + \frac{1}{2}\left(t_{o_{ij}} + t_{p_{ij}}\right)\right] \tag{4.5}$$

The standard deviation of T_{ij} is:

$$\sigma_{ij} = \sqrt{\mathrm{V}\left\{T_{ij}\right\}} = \frac{1}{6}\left(t_{o_{ij}} + t_{p_{ij}}\right) \tag{4.6}$$

For our example project, Table 4.5 reports the optimistic, pessimistic and modal times of the critical path activities. We have assumed that the previously given t_{ij} are in fact modal values.

Table 4.5   Parameters of the critical path.

Activity I,j	t_o	t_p	t_m	μ	σ^2
1,2	9	15	10	10.7	1
2,3	0.5	1.5	1	1	0.03
3,4	7	12	8	8.5	0.69
4,5	0.5	1.5	1	1	0.03
5,8	1	2	2	1.8	0.03
8,9	2	5	4	3.8	0.25
Total				**26.8**	**2.03**

If we assume that the durations of the critical path activities are r.v. stochastically independent, we can calculate the expected value and the variance of the critical path duration by respectively summing the expected durations of the critical path activities and their variances (see Table 4.5).

If we also assume that the critical path total duration follows a Gaussian distribution, we can easily calculate the probability of exceeding the forecasted project end date. For our example project, with the forecasted end date equal to 26, we have:

$$\Pr\{T_{\text{end}} > 26\} = \Pr\left\{Z > \frac{26 - 26.8}{1.42}\right\} = 0.71$$

4.2.2 Critical path method

The critical path method (CPM) is a method for optimising a project in terms of time and cost. It is based on the identification of links between the project activities, the time and the cost required to carry them out.

CPM was developed in the 1950s by two engineers: M. R. Walker of DuPont and J. E. Kelley of Remington Rand.

The CPM develops through the following phases: (i) identification of project activities; (ii) determination of sequence constraints; (iii) construction of the network diagram and (iv) calculation of parameters. The first three are the same as in PERT. With reference to the calculation of parameters, the CPM uses deterministic activity durations.

In general any activity has a cost–time relationship of the type shown in Figure 4.5. The so-called crash time, t_{L}, indicates a lower limit below which no benefits are expected. The normal time, t_{N}, is the ordinary time that human resources (an individual or a team) take to perform the activity under standard conditions; its cost is denoted by C_{N}. Then t_{U} is the worst time that human resources would take if left without guidance and control.

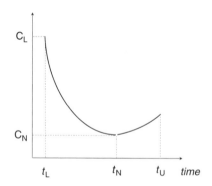

Figure 4.5   Cost–time relationship for a generic activity.

The slope of the curve between t_{L} and t_{N} is denoted the acceleration cost. It can be approximated by:

$$a \cong \frac{C_{\text{L}} - C_{\text{N}}}{t_{\text{N}} - t_{\text{L}}} \tag{4.7}$$

The overall cost structure of a project is complex since it includes:

- direct costs (directly attributable to individual activities), typically decreasing with time;

- indirect costs, typically increasing with time;

- utility costs (such as penalties, bonuses, rentals, etc.) that usually increase stepwise with time.

To determine the project duration corresponding to the minimum cost, it is firstly necessary to calculate the acceleration cost for each activity. The project duration at minimum cost is obtained by a procedure that will be illustrated through the following example.

EXAMPLE

In Table 4.6 the planned activities of the project 'bathroom renovation' consisting of renovating an $8\,m^2$ bathroom, are reported.

Table 4.6   Description of the activities of the project 'bathroom renovation'.

Activity	Description
A	Demolition of floor and wall coverings
B	Installation of drain systems
C	Implementation of plumbing systems
D	Execution of concrete slab on the ground
E	Installation of electrical systems
F	Installation of floor and wall coverings
G	Preparation and painting of walls and ceiling
H	Installation of sanitary ware

As was done in PERT, we build a table (Table 4.7) listing the project activities, the sequence constraints, the nodes, the crash and normal times, the corresponding direct costs (in Euros) and the acceleration costs.

Table 4.7   CPM table for the project 'bathroom renovation'.

Activity	Sequence constraints	Connecting nodes	t_N	t_L	C_N	C_L	a_{ij}
A	–	1,2	3	2	1000	1500	500
B	A	2,3	1	0.5	500	600	200
C	A	2,4	2	1	800	1200	400
D	B	3,4	2	1	500	700	200
E	A	2,5	3	2	900	1200	300
F	C,D,E	4,6	4	3	1000	1500	500
G	F	6,7	3	2	600	800	200
H	G	7,8	1	0.5	500	750	500
Total					**5800**	**8250**	

Through the network diagram (Figure 4.6) we can easily calculate the minimum duration of the project, which, assuming normal times for all activities, is $T_N = 14$

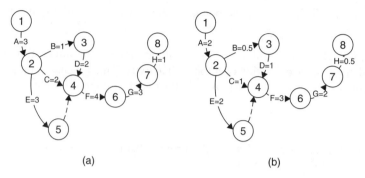

(a)   (b)

Figure 4.6   Network diagram with (a) normal and (b) crash times.

days, while assuming crash times for all activities is $T_L = 9.5$ days. The calculation is based on the concept of earliest time (see formula (4.1)).

With reference to costs, assuming all normal times, the total direct cost is €5800, while assuming all accelerated times it is €8250.

Table 4.8   Earliest times, latest times and node slacks (a) and activity slacks (b), with normal times, for the project 'bathroom renovation'.

(a)

Node	ET	LT	NS
1	0	0	0
2	3	3	0
3	4	4	0
4	6	6	0
5	6	6	0
6	10	10	0
7	13	13	0
8	14	14	0

(b)

Activity	Connecting nodes	Normal times (days)	AS
A	1,2	3	0
B	2,3	1	0
C	2,4	2	1
D	3,4	2	0
E	2,5	3	0
F	4,6	4	0
G	6,7	3	0
H	7,8	1	0

Table 4.8 reports the earliest and latest times calculated for each node on the basis of the network diagram with normal times (Figure 4.6a). The calculation of the activity slacks allows us to identify the critical path. From Table 4.8 it can be noted that only activity C is not critical. The other activities form two critical paths (see Figure 4.6a): 1_2_3_4_6_7_8 and 1_2_5(dummy)_4_6_7_8.

The two identified critical paths have some activities not in common, namely: B, D and E.

Among the common activities – A, F, G, H – the one with the lowest acceleration cost is G (€200). Consequently, if it is assumed to accelerate this activity by one day, the corresponding duration decreases from 3 to 2 (equal to its crash time, so G is saturated), and the corresponding cost increases from €600 to €800. The project direct cost corresponding to a duration 13, will be:

$$C_{13} = C_{14} + a_{67} = 5800 + 200 = €6000$$

Now we update the network of Figure 4.6a, so that it becomes the one in Figure 4.7. The critical paths remain unvaried.

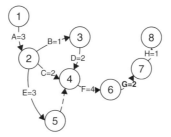

Figure 4.7   Network with total project duration equal to 13.

Now consider the other activities common to the two critical paths: A, F and H. The first two (A and F) have the same acceleration cost as H, but allow the project a time reduction of one day. If we accelerate either A or F the total time would become 12 days and the project direct cost would increase by €500. Moreover, the critical paths remain unvaried in both cases.

Actually, we can consider a third alternative that involves the contextual acceleration of two activities not common to the critical paths, namely D and E. In fact, if we decide to accelerate only one of these two, the project duration would not decrease. However, by accelerating both activities by one day we can reduce the project duration by one day with a higher cost: $a_{34} + a_{25} = 200 + 300 = €500$.

In short, if we proceed with the acceleration of A, F and D conjointly with E, by following one of the possible sequences, in the end we will have a total project duration of 10 days and a direct project cost of $6000 + 500 + 500 + 500 = €7500$.

It must be said that the contextual acceleration of D and E generates a third critical path: 1_2_4_6_7_8. However, the acceleration of C would not bring any project time reduction.

Finally, to get to the minimum project duration (equal to 9.5 days), we must necessarily accelerate the activity H, which is the only activity common to the three critical paths and not yet saturated (having its half-day of 'breath'). The corresponding acceleration cost is equal to 500 (see Table 4.7), so the project direct cost becomes: $C_{9.5} = C_{10} + a_{78} = 7500 + 500 = €8000$.

Figure 4.8 shows the direct costs curves for the example project, as a function of the project duration.

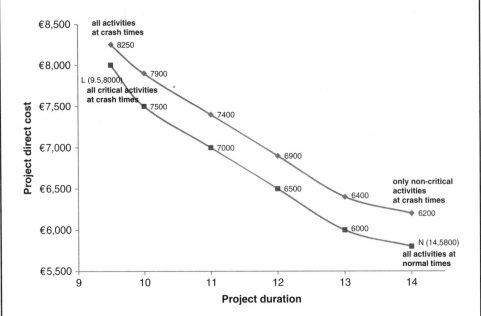

Figure 4.8   Project direct costs curves.

The points on the bottom curve represent the steps of the procedure illustrated above; from the assumption of a project duration of 14 days (normal times for all activities, point N), to the assumption of project duration of 9.5 days (crash times for all critical activities, point L).

The top curve represents the same procedure, but considering also the acceleration of non-critical activities (in our example project only the activity C).

So, corresponding to a project duration 14 days, the project direct cost is equal to the project direct cost for all activities at normal times plus the acceleration cost of activity C: $5800 + 400 = €6200$.

Referring to the last step, reducing the project duration from 10 to 9.5 days, it is not activity H that has to be accelerated, but the activities B and E conjointly, both by half day: $a_{23} + \frac{a_{25}}{2} = 200 + 150$.

The top curve shows that, by accelerating non-critical activities, we have higher project direct costs for the same project duration.

In order to find the optimal project duration, corresponding to the minimum total cost, we should prepare a table reporting not only direct costs but also indirect costs.

Indirect costs are assumed known and distinct in the following categories: penalties, rental fees and rewards. Imagine that the project director and the customer agree that the work will be completed in 12 days. If the duration is exceeded by the limit of one day, the director will have to pay a penalty of €200; a double penalty if the duration will be 14 days.

The equipment needed to carry out the work will be rented; and each day costs €100.

Finally, awards will be given by the customer to the project director in case that the work is completed earlier than the agreed deadline.

In the first row of Table 4.9 the project direct costs are calculated according to the above procedure and are reported (see the points on the bottom curve of Figure 4.8). The other rows of the table report the indirect costs by category; the last row shows the project total costs.

Table 4.9   Project total costs (in Euros) for different durations.

Project duration (days)	14	13	12	11	10	9.5
Direct cost	5800	6000	6500	7000	7500	8000
Penalties	400	200	–	–	–	–
Rental fees	1200	1100	1000	900	800	750
Rewards	–	–	–	(80)	(160)	(200)
Total cost	7400	7300	7500	7820	8140	8450

The optimal project duration, corresponding to a minimum total cost, is 13 days, one day longer than the duration by contract.

4.3   To analyse faults

The concept of reliability expresses the probability that a system (product/service/process) functions as required, and so does not fail, for a predetermined period of time, operating and environmental conditions. Reliability analysis involves the clear identification of the system and the description of a 'functional diagram' showing the subsystems, components and their functional interactions, in addition to any external influences on the system, and vice versa.

Once the system has been identified, depending on the specific case, we may adopt an inductive method (also called bottom-up or hardware), with which we start with a postulated fault at the component level and try to identify the effect on the overall functioning at system level. Alternatively, we can use a deductive method (also top-down or functional) where, in contrast with the former, we start from a postulated fault at system level and try to identify all possible causes, up to the level of various components. Among the most popular methods for the analysis of system reliability are the failure mode and effect analysis (FMEA, bottom-up) and fault tree analysis (FTA, top-down).

4.3.1 Failure mode and effect analysis

FMEA is a bottom-up method defined by specific regulations (US MIL-STD-1629). The methods starts from the analysis of possible failures at component level in order to identify the effects on the operation at system level.

We distinguish:

- design or product FMEA, intended to facilitate the analysis of the product and/or its design;

- process FMEA, whose object is represented by the potential failure modes and their causes that may typically occur during a production process or in general during any process.

The so called 'event', that is the consequence of a failure, can be classified according to the differing degree of severity.

- Catastrophic (level I), when it may cause the death of personnel or the destruction of the system.

- Critical (level II), it can cause serious injuries to personnel and/or serious damage to the system.

- Marginal (level III), it can cause minor injuries to personnel and/or minor damage to the system.

- Not significant (level IV), when it does not cause injury to personnel or damage to the system, but requires maintenance.

FMEA requires one to perform the following activities:

1. determine the system/subsystem/component (project FMEA) or the phase/elementary operation (process FMEA) to analyse;

2. describe the required functions;

3. identify all possible failure modes;

4. identify all possible causes of each failure mode;

5. identify all effects (events) of each failure mode;

6. describe the provided control measures;

7. define responsibilities and timing of the proposed actions;

8. follow the process of implementation of proposed actions;

9. check the success of the proposed actions after they been carried out;

10. keep the FMEA constantly updated in relation to changes to the project/process.

EXAMPLE

Consider the product 'moka express'. It is a device invented in 1933 by Alfonso Bialetti for preparing coffee at home. The moka express revolutionised home coffee brewing and 200 million have since been made. The moka is essentially made of five parts (see Figure 4.9):

1. the boiler or bottom chamber, which contains the water to boil and goes in contact with the heat;

2. the vent valve, which is a spring-type, and ensures that the pressure does not rise excessively;

3. the mechanical filter, which is shaped like a funnel and is filled with ground coffee;

4. the top chamber, which has another filter, a central cannula called the 'chimney' and a cover – this collects the filtered drink;

5. a gasket, which surrounds the top chamber filter and makes firm the top part on the boiler after screwing.

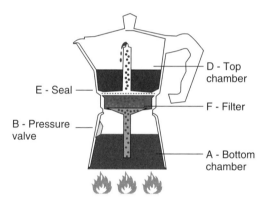

Figure 4.9   Moka express and its components.

The principle of operation is fairly intuitive. First, the boiler is filled with water, the filter is put in its location space and is filled with ground coffee, and the top part is screwed to the lower part. At this point the moka is put on a heat source until the water rises through the filter, transforming into coffee and collecting in the top chamber.

When the water starts boiling the steam that is produced exerts some pressure against the walls of the boiler and against the water surface. Soon the pressure caused by the steam wins against the force of gravity, and the water starts rising through the funnel filter. During this ascent the water passes through the filter containing the coffee, extracting the caffeine and various aromatic substances and becoming the

Table 4.10 FMEA of the moka express.

Activity	Description	Failure mode	Event	Event level	Cause	Control and improvement action
1	Fill the boiler (A) with fresh water being careful to slightly cover the vent valve (B)	Water level does not reach the vent valve (B)	Boiling water struggle to rise through C	IV	Friction pressure is not created	Check the water level during phase 1
		Too much water in A	The coffee flavour is altered	IV	Still cold water bathe the coffee powder	
2	Insert the funnel filter (C) on the top of A	Water appears on the bottom of C	The coffee flavour is altered	IV	Still cold water bathe the coffee powder	Check the water level during phase 1
3	Fill the filter (C) with the coffee powder. The filter must be generously full and the powder not pressed	Excessive pressure of the coffee powder in C	Difficult for coffee to spill out into D	IV	The pressed powder creates a barrier to vapour	Do not press the powder with the spoon
		Small quantity of coffee powder in C	Light consistency of coffee	IV	The ratio between the amount of water passing through the coffee powder and the amount of the latter is incorrect	Adequately dosed quantity of powder

No.	Action	Failure mode	Effect		Cause	Recommendation
4	Screw the top part (D) on the boiler (A) with the filter (C) already inside	Insufficient screwing	Lateral leaking of coffee, damage over time of the moka external surface	III	Closure not strong enough	Dose the torque, control the degree of wear of E
		Too tight screwing	Difficulty of unscrewing and damage of E	III	E worn out; Closure too strong	
5	Put the moka on low heat and leave to simmer	Too high heat	Unpleasant smell of burning, over time damage the walls of moka	III	Coffee starts boiling	Check the heat during brewing
6	Lift the lid when the coffee begins to come out	The lid is not lift	Coffee flavour is altered	IV	The steam condensate falling into D	Monitor the moka during brewing
7	Remove from the heat when all coffee is not yet all out and mix the coffee with a spoon before pouring	Moka is removed from the heat when all coffee has come out	Unpleasant smell of burning, over time damage the walls of moka	III	Coffee starts boiling	Monitor the moka during brewing
		Coffee is not mixed before pouring	Uneven flavour of the poured coffee	IV	The first coffee that comes out, is more rich and dense, tends to stay on the bottom, what comes out at the end is lighter	Mix the coffee with a spoon before pouring

drink 'espresso coffee'. The liquid rises further and causes the freshly formed coffee to collect in the top chamber, passing through the chimney. This rising stops when no water is left in the boiler.

The end result of the application of FMEA, when applied to the process 'use of the moka express', is summarised in Table 4.10.

In this example, the FMEA can be an invaluable help to the producer in order to provide a user manual for the moka and highlight the main points which the user should follow for optimal coffee preparation.

In general, the analysis of failure modes and effects of a product/process can be conducted through the use of inferential methods to estimate the severity and probability of occurrence of failures and their effects. In such cases it is necessary to integrate the sequence of activities provided by the FMEA, using the description of the control measures provided, and then carrying out the following additional activities:

- estimate the probability of occurrence of the effect of the failure mode;

- estimate the severity of the failure mode;

- estimate the probability of preventing the effect of the failure mode, that is how easy is it to detect the cause;

- calculate the risk priority index;

- identify preventive/corrective actions based on the value of the priority risk;

- estimate the new value of the expected priority risk following the activation of the proposed actions.

The analysis when integrated like this, is called Failure Mode, Effects and Criticality Analysis (FMECA).

4.3.2 Fault tree analysis

Fault tree analysis (FTA) is a technique for reliability and risk analysis of complex systems mostly based on a graphical representation called a tree and a top-down analysis. The technique was introduced in the 1960s in the aerospace industry. Today it is widely used in many sectors of business and industry.

The basic idea is the identification and representation of possible failures of a complex system, starting from a 'Top Event', which is normally a failure at system level.

Examples of possible top events are:

- the structural collapse of a bridge;

- the truck engine does not start;

- an unwanted stop of the production line;

- failing an exam (for a student) or missing a flight (for a manager);

- a wrong item is shipped to the customer;

- a patient receives a bad treatment in the hospital.

In FTA the top event is gradually broken down by considering lower levels of the system (subsystems and components), giving rise to a tree graph.

A fault tree is made of 'events' and 'gates'. In Tables 4.11 and 4.12 the most frequently used symbols in FTA are shown.

In Figure 4.10 a simple example of FTA for a paper helicopter is shown (the paper helicopter is a well-known device in quality engineering; it will be further illustrated and adopted in Chapter 5).

FTA is very useful not only for the development of a new system (product, service and production processes), but also when the system is in use. FTA can be simply qualitative or can be complemented with probabilistic methods that aim to calculate the probability of the top event, especially when the probabilities of all basic events are known.

FTA is a flexible tool since it allows us to take into account general types of failures (e.g. due to human factors, natural events, etc.). Some examples of FTA application are the following.

- A team of specialists in a hospital uses FTA to see how incorrect prescriptions may be given to patients. They want to redesign their processes so to prevent such disasters from happening.

- The manufacturer of aircraft parts performs FTA as a standard phase of the design process to identify critical faults that could lead to hazardous failures.

- A quality team in a newspaper press room uses FTA to check potential failures of an improved colour print process.

4.3.2.1 Qualitative FTA

A fault tree cut-set is a set of basic events whose concomitant occurrence determines the top event. A cut-set is said to be 'minimal' if eliminating one of the events from the set means that it is no longer a cut-set.

Table 4.11   Most popular event symbols in fault tree analysis.

Event	Name	Description
	Developed event	An event that is further developed. The Top Event is surely a developed event
	Basic event	A basic event. Examples: failure of a component, a human mistake, a software error
	House event (switch)	It is used for scenario analysis. Such an event can be set to occur or not to occur
	Undeveloped event	An event not developed for several possible reasons: unneeded further details, lack of information or interest
	'Out' triangle	Indicates continuation on a sub-tree (on another sheet)
	'In' triangle	Indicates the start of the sub-tree (from another sheet)

Table 4.12   Most popular gate symbols in fault tree analysis.

Gate	Name	Description
	OR	The output event (above the gate) occurs if <u>at least one</u> of the input events (below the gate) occur
	AND	The output event occurs if <u>all</u> input events occur
	Exclusive OR	The output event occurs if <u>only one</u> input event occurs
	Priority AND	The output event occurs if all input events occur, but according to a given sequence. The sequence is specified within the gate or in an attached oval, otherwise the default sequence is from left to right
r	r out of n	The output event occurs if at least <u>r out of the n</u> input events occur
	Inhibit	The output event occurs if all input events occur and an additional conditional event occurs

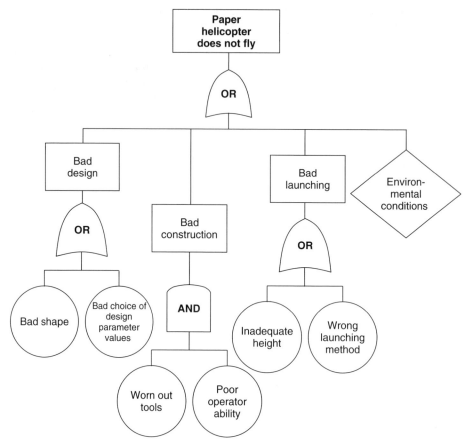

Figure 4.10   A simple illustration of a fault tree analysis.

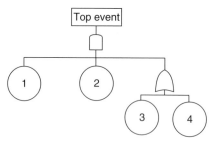

For a complex fault tree it may be burdensome to find the minimal cut-sets. The so-called Mocus algorithm helps find the minimal cut-sets for a fault tree having only AND and OR gates. The algorithm is based on the consideration that an OR gate increases the number of cut-sets, while an AND gate increases the cut-set size.

The procedure in four steps is illustrated by adopting the fault tree shown in Figure 4.12.

1. Enumerate all basic events (1, 2, ...). Some basic events can be found more than once on different branches of the tree.

2. Enumerate all gates (G0, G1, G2, ...).

3. Write the number of the highest gate of the tree in the first row and first column of a table.

4. Follow the tree from the top to the bottom. Replace each OR gate with a vertical displacement of its inputs. Replace each AND gate with a horizontal displacement of its inputs.

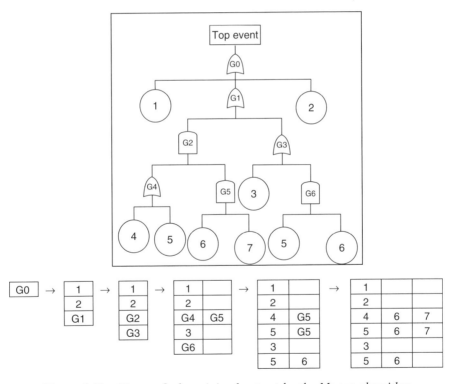

Figure 4.12   How to find a minimal cut-set by the Mocus algorithm.

When all gates are replaced by basic events we read the cut-sets in the rows of the final table. The minimal cut-sets will be obtained by eliminating from the list all cut-sets containing other cut sets (the cut-set {5,6,7} is eliminated since it contains {5,6}). So, for the example fault tree the minimal cut-sets are: {1}, {2}, {4,6,7}, {3} and {5,6}.

Sometimes it is convenient to use a *path-set* instead of a cut-set. A path-set is a set of basic events whose non-occurrence determines the non-occurrence of the top event. A path-set is said to be minimal if eliminating one of the elements from the set means it is no longer a path-set.

It is easy to identify the path-sets by using the *dual* tree of a fault tree. The dual tree can be obtained by:

- replacing each OR gate of the fault tree with an AND gate;

- replacing each AND gate of the fault tree with an OR gate;

- replacing each event of the fault tree with its complementary event.

The cut-sets of the dual tree are the path-sets of the originating fault tree.

4.3.2.2 Quantitative FTA

When the probabilities of the basic events are known, it is possible to perform a quantitative analysis of a fault tree to calculate the probability of the top event.

The procedure is easier if we have identified the minimal cut-sets, which we may denote by:

$$CS_1 = \left\{ B_{1,1}, B_{1,2}, \ldots, B_{1,n_1} \right\} = \left\{ \bigcap_{j=1}^{n_1} B_{1,j} \right\}$$

$$\vdots$$

$$CS_i = \left\{ B_{i,1}, B_{i,2}, \ldots, B_{i,n_1} \right\} = \left\{ \bigcap_{j=1}^{n_1} B_{1,j} \right\} \qquad (4.8)$$

$$\vdots$$

$$CS_m = \left\{ B_{m,1}, B_{m,2}, \ldots, B_{m,n_m} \right\} = \left\{ \bigcap_{j=1}^{n_m} B_{m,j} \right\}$$

where $B_{i,j}$ is the j-th basic event of the i-th cut-set.

In fact, we can reshape the fault tree in terms of minimal cut-sets as shown in Figure 4.13.

From the tree we see that:

$$\text{top event} = \left\{ \bigcup_{i=1}^{m} CS_i \right\} = \left\{ \bigcup_{i=1}^{m} \left(\bigcap_{j=1}^{n_i} B_{i,j} \right) \right\} \qquad (4.9)$$

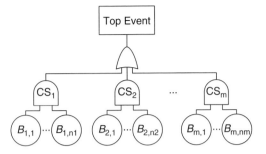

Figure 4.13   Reshaping the fault tree in terms of minimal cut-sets.

Therefore, by extending the formula for calculating the probability of the union of two events (see Section 3.2.3) to the case of $m > 2$ events, the top event probability can be calculated by:

$$\text{Pr}\{\text{top event}\} = \sum_{i=1}^{m} \text{Pr}\{\text{CS}_i\} - \sum_{i=1}^{m-1}\sum_{j=i+1}^{m} \text{Pr}\{\text{CS}_i \cap \text{CS}_j\}$$

$$+ \sum_{i=1}^{m_2}\sum_{j=i+1}^{m_1}\sum_{k=j+1}^{m} \text{Pr}\{\text{CS}_i \cap \text{CS}_j \cap \text{CS}_k\} + \qquad (4.10)$$

$$+ (-1)^{m-1} \text{Pr}\{\text{CS}_1 \cap \text{CS}_2 \cap \ldots \cap \text{CS}_m\}$$

In case the events $B_{i,j}$ are s-independent the calculation is simplified considerably.

EXAMPLE

The fault tree of Figure 4.12 can be reshaped as in Figure 4.14.

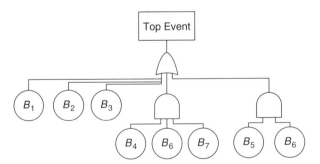

Figure 4.14   The fault tree of Figure 4.12 reshaped in terms of minimal cut-sets.

The top event probability can be calculated as:

$$\text{Pr}\{\text{top event}\} = \text{Pr}\{B_1 \cup B_2 \cup B_3 \cup (B_4 \cap B_6 \cap B_7) \cup (B_5 \cap B_6)\} \qquad (4.11)$$

which can be further developed following the guidelines of Equation 4.10, unless specific assumptions on compatibility/incompatibility and s-dependence/independence of the events can be made which may simplify the calculation. For example, if we assume that B_1, B_2, B_3 are mutually incompatible and incompatible with $(B_4 \cap B_6 \cap B_7)$ and with $(B_5 \cap B_6)$, then Equation 4.11 becomes:

$$\Pr\{\text{top event}\} = \Pr\{B_1\} + \Pr\{B_2\} + \Pr\{B_3\} + \Pr\{(B_4 \cap B_6 \cap B_7) \cup (B_5 \cap B_6)\}$$

$$= \Pr\{B_1\} + \Pr\{B_2\} + \Pr\{B_3\} + \Pr\{B_4 \cap B_6 \cap B_7\} \qquad (4.12)$$

$$+ \Pr\{B_5 \cap B_6\} - \Pr\{B_4 \cap B_5 \cap B_6 \cap B_7\}$$

4.4   To make decisions

4.4.1   Analytic hierarchy process

The analytic hierarchy process (AHP) is a technique that allows us to take into account multiple criteria for decision making. In fact, decision making processes are often complex, characterised by the presence of different stakeholders and multiple evaluation criteria, sometimes conflicting with each other.

The method was developed by Thomas Saaty in the late 1970s. It allows us to make decisions in the presence of several alternatives, by modelling the decision problem by a hierarchical structure, and providing a priority list of alternatives as output.

The AHP leads to rational decisions in the sense that they are based not only on the available objective data, but also on the decision maker's experience, intuition and social-technological background, which are considered equally important as the data to process.

AHP application areas may include:

- the evaluation of projects;

- the allocation and management of resources, particularly human resources;

- evaluation of suppliers;

- marketing strategies;

- cost/benefit analysis;

- financial evaluations;

- risk assessment;

- management of innovation.

The name 'analytic hierarchy process' reflects the approach of the method, which is developed through three main phases.

1. Problem breakdown:

   a. the problem is broken down into its constituent elements;

   b. the problem is structured in hierarchical form.

2. Comparison assessments and their processing:

   a. pairwise comparison judgements are collected;

   b. consistency of assessments is verified;

   c. priorities between elements at each level are determined.

3. Summary of results:

   a. priority list of the alternatives is calculated.

The first phase (problem breakdown) aims to relate the main elements of the problem. A first step is to identify such elements, that is:

- the ultimate goal we intend to reach (for example selecting a Six Sigma project);

- the different possible scenarios (economic, environmental, social, etc.)

- the actors, that is the people directly or indirectly involved in the decision making;

- the possible alternatives to choose from.

The next step is to give a hierarchical structure to the decision problem, pointing out the logical relationships existing among the identified constituent elements.

The hierarchical structure can be created with a top-down approach, identifying the criteria before the alternatives, or a bottom-up approach, starting from the alternatives and then identifying the selection criteria.

In drawing such a hierarchical structure we must ensure that two conditions are met:

- dependence from higher level, that is the elements of a level should be compared in pairs with respect to an element belonging to the upper hierarchical level;

- independence between the elements of each level, that is it should not make sense to compare any two elements of a level with respect to an element of the same or lower level.

The hierarchy in which the problem is structured can be complete or partial (see Figure 4.15). Complete hierarchy occurs when each element of a level depends on each of the elements of the upper level. In this case, all elements of a level must be compared with each other in pairs with respect to each of the elements of the upper level.

Conversely, partial hierarchy occurs when at least one element of a level is independent from at least one element of the higher level.

In the second phase, pairwise comparisons are expressed by the decision maker between elements of the same level in order to determine their relative importance.

The claimed advantage of AHP is that it is easy for the respondent to compare items in pairs since this is an unconscious process of the human mind when making a decision.

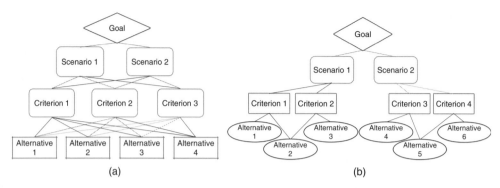

Figure 4.15   (a) Complete and (b) partial hierarchy.

The decision maker expresses his opinion by means of a semantic scale (Table 4.13).

Table 4.13   Semantic scale for the AHP.

Degree of relative importance	Definition
1	Equal importance between the two alternatives
3	Alternative i moderately more important than the alternative j
5	Alternative i noticeably more important than the alternative j
7	Alternative i much more important than the alternative j
9	Alternative i absolutely more important than the alternative j
2, 4, 6, 8	When the judgement lies between two of the previous scores

Based on the assessments of the respondent, each pair of alternatives (i, j) is given a value a_{ij} expressing the degree of importance of the alternative i respect to the alternative j, with reference to a given item of the higher hierarchical level.

Ideally, in AHP the degree of importance is a ratio between absolute weights attributed to two alternatives:

$$a_{ij} = \frac{w_i}{w_j} \tag{4.13}$$

With the a_{ij} we can build a square matrix, say $\underline{\underline{A}}$, with dimensions $(n \times n)$, with n being the number of alternatives. The matrix must be 'reciprocal', since it is assumed that:

$$a_{ji} = \frac{w_j}{w_i} = \frac{1}{a_{ij}} \qquad \forall i \neq j \tag{4.14}$$

It is also obvious that:

$$a_{ii} = 1 \qquad \forall i = 1, \ldots, n \tag{4.15}$$

Still theoretically it should happen that:

$$a_{ij} \cdot a_{jk} = a_{ik} \qquad \forall (i, j, k) \tag{4.16}$$

So the ideal pairwise comparison matrix $\underline{\underline{A}}$ has this form:

$$\underline{\underline{A}} = \begin{bmatrix} 1 & a_{12} & \ldots & a_{1n} \\ 1/a_{12} & 1 & \ldots & a_{2n} \\ \ldots & & & \\ 1/a_{1n} & 1/a_{2n} & \ldots & 1 \end{bmatrix} \tag{4.17}$$

The matrix $\underline{\underline{A}}$ is said to be 'consistent': due to its characteristics, its maximum eigenvalue $\lambda_{\max} = n$, while the other eigenvalues are equal to zero.

In practice, it often happens that the matrix obtained with respondent feedback is not consistent as specified by Equation 4.16. In fact, every time there are more than two or three alternatives, the respondent, in making pairwise comparisons, rarely gives perfectly consistent judgements.

So, the maximum eigenvalue of the pairwise comparison matrix obtained with the respondent feedback is bigger than n, and the other eigenvalues may differ from zero.

For this reason, it is proposed to calculate an index called the 'consistency ratio'.

From the pairwise comparison matrix initially obtained from respondent feedback, say $\underline{\underline{A}}_0$, we should firstly calculate a consistency index (CI) given by:

$$CI = \frac{\lambda_{\max} - n}{n - 1} \tag{4.18}$$

The lower the CI the higher the matrix consistency. The CI is compared to a random index (RI) which expresses an average CI for randomly simulated pairwise comparison matrices. RI values for n up to 15 are reported in Table 4.14.

Table 4.14   Consistency random indexes.

n	2	3	4	5	6	7	8	9	10	11	12	13	14	15
RI	0.00	0.58	0.90	1.12	1.24	1.32	1.41	1.45	1.49	1.51	1.48	1.56	1.57	1.59

The ratio between the CI calculated from $\underline{\underline{A}}_0$ and the corresponding RI is called the consistency ratio:

$$CR = \frac{CI}{RI} \tag{4.19}$$

If $CR \leq 0.01$ we can accept the matrix $\underline{\underline{A}}_0$ as it is, and we can calculate the relative importance weights of the n alternatives through the procedure illustrated in step 2 below, otherwise we 'force' the consistency by the following iterative procedure:

1. determine the matrix $\underline{\underline{A}}_k = \underline{\underline{A}}_{k-1} * \underline{\underline{A}}_{k-1}$ $k = 1, 2, \ldots$ ('*' denotes rows by columns matrix multiplication);

2. calculate the relative importance weights of the alternatives – the elements of each row of the matrix $\underline{\underline{A}}_k$ are summed up and divided by the sum of all matrix elements:

$$\frac{w_i}{\sum\limits_{i=1}^{n} w_i} = \frac{\sum\limits_{j=1}^{n} a_{ij}}{\sum\limits_{i=1}^{n} \sum\limits_{j=1}^{n} a_{ij}} \quad \forall i = 1, \ldots, n \tag{4.20}$$

3. compare the obtained relative importance weights with those obtained in the previous iteration step and stop the procedure when they are stabilised.

The iterative procedure allows us to distribute the inconsistency over the entire matrix. In fact, it can be algebraically demonstrated that by raising the matrix to progressively higher powers, columns closer and closer to proportionality are obtained.

It can be noted that the above procedure is for a single respondent and alternatives on a single hierarchical level. However, for each respondent and for each hierarchical level it is necessary to build a pairwise comparisons matrix. In the example below we show an example with one respondent and two hierarchical levels.

EXAMPLE

The example concerns the evaluation of projects. Specifically, the top manager of a Swedish multinational company needs to decide which of three Six Sigma projects to choose. The projects are already sketched. Also the criteria on which to base the choice have been well identified:

HR – human resources to be committed in the project;

T – time required to complete the project;

I – trans-national impact (i.e. ability to apply the results of the project in other company locations); and

ER – achievable results (lower costs and higher revenues) at one year from the end of the project.

Figure 4.16 shows the hierarchical structure representing this decision problem. It is a complete hierarchy.

The next step consists in the pairwise comparison of the four criteria by the top manager. The matrix of pairwise comparisons is shown in Table 4.15. At the bottom right of the table, the maximum eigenvalue, the consistency index (CI) and the consistency ratio (CR) are indicated. The vector of priorities suggests the following order of importance with regard to the criteria: ER, I, HR and T.

Now, the top manager assesses, still through pairwise comparisons, the three projects respect to the four criteria, one at a time. The four matrixes $\underline{\underline{A}}_0$ are shown in Table 4.16.

Both in the case of pairwise comparisons of Table 4.15 and for the four matrices of Table 4.16, three iterations were necessary to stabilise the weights.

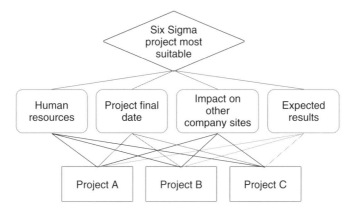

Figure 4.16   Breakdown of the problem into hierarchical levels.

Table 4.15   Pairwise comparison matrix for the criteria.

	HR	T	I	ER	Priority vector
HR	1	3	1/5	1/7	0.102
T	1/3	1	1/3	1/5	0.069
I	5	3	1	1/3	0.269
ER	7	5	3	1	0.558
		$\lambda_{\max} = 4.352$	CI $= 0.117$	CR $= 0.130$	

Finally, it is necessary to determine the composite priorities. The matrix reported in Table 4.17 is purposely constructed. It shows the project priorities with respect to each criterion. By multiplying this matrix by the vector of priorities, we get the final list of priority for the Six Sigma projects.

Project B has to be preferred, according to the top manager.

Table 4.16 Projects pairwise comparison matrices, for each criterion.

Human resources	A	B	C	Priority vector	Time	A	B	C	Priority vector
A	1	3	5	0.651	A	1	1/5	1/7	0.071
B	1/3	1	1/3	0.126	B	5	1	1/3	0.278
C	1/5	3	1	0.222	C	7	3	1	0.650
$\lambda_{max} = 3.287$ CI = 0.144 CR = 0.247					$\lambda_{max} = 3.054$ CI = 0.0268 CR = 0.0463				
Impact	A	B	C	Priority vector	Expected results	A	B	C	Priority vector
A	1	1/5	1/7	0.066	A	1	1/7	1/3	0.080
B	5	1	1/5	0.218	B	7	1	5	0.732
C	7	5	1	0.716	C	3	1/5	1	0.188
$\lambda_{max} = 3.172$ CI = 0.086 CR = 0.149					$\lambda_{max} = 3.054$ CI = 0.027 CR = 0.046				

Table 4.17   Composite priorities matrix.

	HR	T	I	ER	Composite priority
	0.102	0.069	0.269	0.558	
A	0.651	0.071	0.066	0.080	0.134
B	0.126	0.278	0.218	0.732	0.500
C	0.222	0.650	0.716	0.188	0.366

4.4.2   Response latency model

The response latency model allows us to estimate the relative weight of importance of alternatives by using the time a respondent takes in assisted interviews with a computer. In such interviews, the respondent is simply asked to select the alternatives one at a time, from the one considered the most important to the one considered the least important.

This model, developed around 2007 by Barone, Lombardo and Tarantino (2007), is inspired by the 'Preference Uncertainty Theory', developed in the end of nineteenth century, stating that the more a respondent is uncertain about the choice between alternatives, the longer it takes to choose the favourite.

To illustrate the model and the related procedure, firstly suppose that we ask a respondent to choose the most important out of two alternatives (Figure 4.17).

Figure 4.17   Indecision between two alternatives causes response latency.

The relationship between the relative importance weights (according to the respondent) can be thought of as a function of choice time or response latency t_c:

$$\frac{w_1}{w_2} = f(t_c) \tag{4.21}$$

where:

- $0 \leq w_1 \leq 1$ is the relative importance weight of the first chosen alternative (the most important);

- $0 \leq w_2 \leq 1$ is the relative importance weight of the second alternative (the least important); and

- $w_1 + w_2 = 1$.

Theoretically, if the choice time tends to infinity, it means that the respondent is absolutely uncertain about the relative importance between the two alternatives. In this case, the two alternatives have virtually the same importance ($w_1 = w_2 = 0.5$). In formulas:

$$\lim_{t_c \to +\infty} \frac{w_1}{w_2} = 1 \tag{4.22}$$

Conversely, if the choice time tends to zero, this means that the respondent considers the chosen alternative far more important than the second. In formulas:

$$\lim_{t_c \to 0} \frac{w_1}{w_2} = +\infty \tag{4.23}$$

The analytical function that best represents the boundary conditions (Equations 4.22 and 4.23) is:

$$\frac{w_1}{w_2} = 1 + \frac{1}{t_c} \tag{4.24}$$

The right-hand side of Equation (4.24) is not dimensionless. Moreover, we need to consider that different respondents may have different reaction times to the same stimulus. For these reasons, we may define a respondent reference time t^* and modify the relationship (Equation 4.24) as:

$$\frac{w_1}{w_2} = 1 + \frac{t^*}{t_c} \tag{4.25}$$

Being able to measure both respondent choice time and reference time through a computer-assisted interview, the relative weights w_1 and w_2 can be easily calculated by solving the following system of two equations with two unknowns:

$$\begin{cases} \dfrac{w_1}{w_2} = 1 + \dfrac{t^*}{t_c} \\ w_1 + w_2 = 1 \end{cases} \tag{4.26}$$

whose solution is:

$$w_1 = \frac{t_c + t^*}{2t_c + t^*}$$

$$w_2 = \frac{t_c}{2t_c + t^*}$$

(4.27)

We can extend the above described model to a number $n > 2$ alternatives, assuming that the selection basically happens between the most and the second most important alternative:

$$\begin{cases} \dfrac{w_i}{w_{i+1}} = 1 + \dfrac{t^*}{t_{ci}} & i = 1, 2, \ldots, (n-1) \\ \displaystyle\sum_{i=1}^{n} w_i = 1 \end{cases}$$

(4.28)

The solution of the system is found recursively.
The relative importance weights are finally given by:

$$\begin{cases} w_i = w_n \displaystyle\prod_{k=i}^{n-1} a_k & i = 1, \ldots, (n-1) \\ w_n = \dfrac{1}{1 + \displaystyle\sum_{j=1}^{n-1} \left(\displaystyle\prod_{i=j}^{n-1} a_i \right)} \end{cases}$$

(4.29)

where:

$$a_i = 1 + \frac{t^*}{t_{c,i}}$$

By measuring choice and reference times we can determine for each respondent not only the order of importance of alternatives (which is automatically given by the order of choices), but also their relative value.

A software interface has been specifically developed for conducting the interview (see Figure 4.18).

Offline set-up   Interview: ranking task   Interview: reaction time

Figure 4.18   Some steps of the software interface for the response latency method.

Respondents are asked to undergo a brief 'assisted' interview by using a laptop. After providing some input data, the interviewee is invited to read the list of alternatives that appear on the screen and select the favourite one. As soon as the list appears on the screen, a timer starts measuring the time (response latency) until the respondent selects the alternative. After each selection, the selected alternative is removed from the list, which is updated, and reappears on the screen with a new random order. For each choice the timer records the response latency. The procedure continues until the respondent comes to the final choice between the last two attributes.

To estimate t^* a possible expedient is to measure the time taken by the respondent to choose between two coins of €1 and €2 appearing on the screen. This is based on the fact that, from Equation 4.25: $w_1 = 2w_2 \Rightarrow t_c = t^*$.

The results of the assisted interview are stored in a report file containing the ranking of the alternatives, the choice and reference times, and the estimated weights.

EXAMPLE

The response latency technique was applied to investigate young students preferences for the attributes of a cellular phone. Six attributes were preliminary selected for the study (see Table 4.18). Fifty respondents underwent assisted interviews. The sample was balanced in terms of gender (50 % male, 50 % female). The complete data array is reported in Table 4.19.

Table 4.18   Description of the chosen product attributes.

Attribute	Description
A	Integrated antenna
B	Internal memory size
C	Dimensions
D	Bluetooth
E	Digital camera
F	MP3 player

Table 4.19   Dataset: respondent n., gender, reference time (milliseconds), estimated importance weight for each attribute.

R	Gender	t^*	Antenna integrated	Internal memory	Dimension	Bluetooth	Digital camera	MP3 player
1	M	4203	0.22	0.41	0.29	0.06	0.01	0.02
2	M	5000	0.6	0.02	0.22	0.05	0.1	0.01
3	M	4485	0.06	0.03	0.44	0.32	0.01	0.14
4	F	4141	0.03	0.13	0.01	0.06	0.34	0.44
5	F	10 000	0.01	0.3	0.02	0.43	0.07	0.16
6	M	5156	0.27	0.09	0.43	0.16	0.04	0.01
7	F	7047	0.27	0.09	0.57	0.04	0.02	0.01

Table 4.19   *(continued)*

R	Gender	$t*$	Antenna integrated	Internal memory	Dimension	Bluetooth	Digital camera	MP3 player
8	M	3282	0.24	0.16	0.32	0.11	0.1	0.07
9	M	1281	0.21	0.23	0.19	0.15	0.12	0.1
10	M	4984	0.43	0.15	0.29	0.03	0.02	0.08
11	F	1813	0.2	0.27	0.14	0.09	0.25	0.06
12	F	4985	0.07	0.26	0.48	0.03	0.15	0.01
13	F	4984	0.07	0.18	0.42	0.01	0.29	0.03
14	M	5609	0.27	0.07	0.42	0.19	0.03	0.01
15	M	5500	0.29	0.08	0.41	0.16	0.02	0.04
16	M	4734	0.12	0.25	0.19	0.32	0.05	0.09
17	M	5203	0.32	0.06	0.42	0.15	0.03	0.01
18	F	6134	0.32	0.43	0.15	0.01	0.03	0.06
19	F	5503	0.27	0.17	0.41	0.02	0.09	0.05
20	F	4984	0.39	0.09	0.27	0.18	0.05	0.02
21	F	6094	0.45	0.28	0.08	0.15	0.04	0.01
22	M	5047	0.25	0.02	0.56	0.05	0.12	0.01
23	M	4984	0.45	0.03	0.29	0.12	0.06	0.04
24	M	4578	0.05	0.27	0.19	0.35	0.02	0.13
25	M	5219	0.57	0.31	0.00	0.09	0.00	0.03
26	F	4969	0.28	0.14	0.51	0.05	0.02	0.01
27	M	5078	0.01	0.02	0.55	0.04	0.11	0.27
28	F	4984	0.32	0.19	0.28	0.07	0.11	0.33
29	M	11 891	0.00	0.56	0.28	0.03	0.00	0.12
30	M	4203	0.09	0.32	0.23	0.13	0.07	0.16
31	F	4453	0.15	0.27	0.47	0.03	0.07	0.01
32	F	5250	0.08	0.03	0.57	0.01	0.29	0.02
33	F	5078	0.09	0.48	0.25	0.02	0.12	0.04
34	M	1047	0.19	0.25	0.22	0.08	0.15	0.11
35	M	906	0.20	0.23	0.27	0.13	0.09	0.07
36	M	2547	0.26	0.05	0.37	0.16	0.11	0.04
37	F	1814	0.20	0.27	0.14	0.09	0.25	0.06
38	M	6203	0.31	0.17	0.40	0.07	0.01	0.03
39	M	5375	0.27	0.10	0.41	0.05	0.16	0.02
40	F	2531	0.16	0.25	0.31	0.13	0.06	0.09
41	M	4594	0.26	0.38	0.12	0.19	0.04	0.01
42	F	1969	0.19	0.35	0.28	0.10	0.06	0.03
43	F	5047	0.25	0.41	0.09	0.05	0.01	0.19
44	M	4360	0.06	0.36	0.25	0.12	0.03	0.18
45	F	4582	0.39	0.10	0.18	0.27	0.05	0.02
46	F	6503	0.28	0.42	0.17	0.04	0.08	0.01
47	F	1581	0.22	0.24	0.19	0.15	0.12	0.09
48	F	5284	0.45	0.02	0.30	0.12	0.06	0.04
49	F	4955	0.05	0.26	0.36	0.19	0.02	0.12
50	F	3820	0.09	0.33	0.24	0.12	0.06	0.17

Figure 4.19 shows the empirical distributions of the estimated weights distinguished by attribute and gender. No differences appear in terms of gender. Moreover, it is clear that the first three attributes, A, B and C, have been accorded a greater relative importance and the opinion expressed by respondens on these attributes is quite unanimous, while with reference to the other three attributes, D, E and F, there are several outliers.

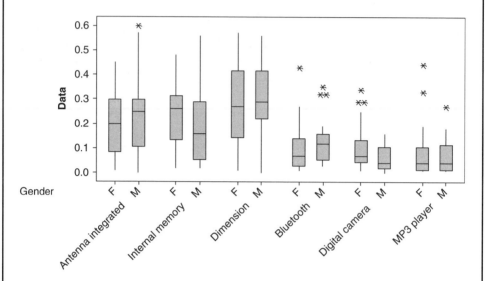

Figure 4.19   Estimated weights by gender and by attribute.

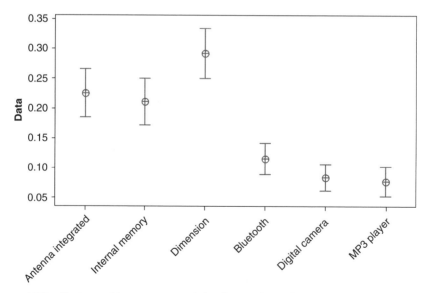

Figure 4.20   Estimated importance weights by attribute (95 % confidence intervals for the means).

The interval plot of Figure 4.20 shows the confidence intervals for the means of the estimated weights (by attribute), the hierarchy of importance of the six selected attributes for the design of a mobile phone, according to the 50 interviewed students, is as follows: (i) dimensions (M = 0.293); (ii) integrated antenna (M = 0.226); (iii) internal memory (M = 0.212); (iv) bluetooth (M = 0.115); (v) digital camera; (vi) MP3 player.

4.4.3   Quality function deployment

Quality Function Deployment (QFD) is a technique that is useful when thinking about the innovation of products and services (hereinafter we will refer only to products, while mean both). It is a systematic way for translating customer requirements into technical features at any stage of development and production. QFD was introduced in the 1960s in Japan in the sector of tyre production and electronics. The first publication was by Yoji Akao, who first formalised the term 'hinshitsu tenkai' (quality deployment).

QFD can help companies to make trade-offs between what customers want and what the company can afford to make.

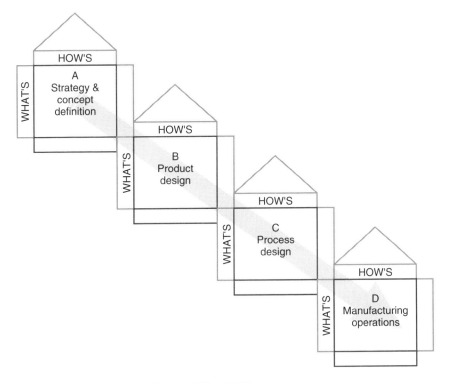

Figure 4.21   QFD matrices.

Usually, to apply QFD the product development process is broken down into four phases. In each phase the customer requirements (What's) serve as an input to establish the engineering characteristics (How's) of the output:

1. strategy and concept definition (customer voice versus measurable objectives);

2. product design (measurable objectives versus design requirements);

3. process design (design requirements versus process requirements);

4. manufacturing operations (process requirements versus production requirements).

The relationships between the inputs and outputs are mapped into matrices whose shape resembles that of a house (Figure 4.21). A generic matrix format used in QFD is called the House of Quality (HOQ). Sometimes, the use of the technique leads to the construction of only one HOQ gathering all the key variables for the development of the product.

If correctly applied, QFD can produce benefits at a general level, such as a deeper understanding of customer requirements, fewer start-up problems and fewer design changes.

EXAMPLE

The example presented here concerns the development of an innovative wheelchair for patients affected by mental retardation. This project was carried out in collaboration with an Italian health research and care institution.

These patients require a posture very different from other disabled people, which is often difficult to set. Moreover, the lack of self-sufficiency of these patients requires constant assistance from paramedical staff and parents/relatives, and consequently the need of easily adjustable setting procedures.

In detail, the new wheelchair should have:

• better performances in terms of weight and manoeuvrability;

• easier manual postural regulation system;

• a diagnostic system able to detect departures from ideal postural settings;

• a pleasurable design.

The study aimed at identifying the specific needs for the innovative chair and translating them into engineering suggestions. It was developed in various phases.

Customer identification

The first and crucial step in achieving customer satisfaction is to clearly identify the customers. In mental disability issues, people around the patient play a central role for therapeutic and rehabilitative functions. The following groups were seen as the main customer categories.

- *Patients*: they are the immediate beneficiary of the improved wheelchair. Due to their disability, these patients are often not able to directly express their needs.

- *Doctors*: they give instructions for the correct posture at different hours of the day.

- *Paramedics*: they follow the indications of doctors for the wheelchair settings.

- *Parents/relatives*: they often act as the intermediary for patients' needs. Moreover, they set the wheelchair at home.

- *National healthcare system*: paying for the assistance of the patients and contributing to the wheelchair purchase.

Benchmarking

After the customer identification, a survey concerning already used wheelchairs was made. Particular attention was given to the types of wheelchair mostly in use at that hospital. The characteristics of 16 models were examined. These characteristics were divided into eight groups: back rest, cushion, lateral push, pelvic waistband, footboard, lumbar push, armrest and headrest. The characteristics inside these groups guarantee postural functionality, stability and comfort. A bar chart was used to show the percentage in which wheelchair characteristics appeared in the examined models. The most frequent characteristics were considered to be basic, while the less frequent were considered to be specific or distinctive for a model and brand.

Customer needs identification

With the aim of collecting both explicit and somewhat technical needs from the customers' and also collecting the implicit and emotional needs, we used different tools with different customers. In particular a structured interview was used for doctors and paramedics. However, a simple questionnaire was prepared for the patients' parents/relatives, based on the Kano model (ref. Chapter 2) and including one item for each attribute. Moreover, the questionnaire was integrated at the beginning by a preliminary set of questions on customers' actual feelings about wheelchairs, and at the end by a set of questions on possible critical incidents (what, when, where and why happened). From the answers to the preliminary questions emerged a poor level of satisfaction for the performance of the currently used wheelchair and a high difficulty to adjust the postural settings.

The complete list of needs coming from doctors/paramedics and parents/relatives are reported in Table 4.20, distinguishing explicit, technical and emotional needs.

Table 4.20   Customer needs.

	Customers	Used method
Explicit/technical needs		
Pathology adaptability	Doctors/paramedics	Structured interview
Armrest adjustability	Doctors/paramedics	Structured interview
Cushion anatomy	Doctors/paramedics	Structured interview
Bodily adaptability	Doctors/paramedics	Structured interview
Pelvis blocking	Doctors/paramedics	Structured interview
Reduction of the sense of weakness	Doctors/paramedics	Structured interview
Transportability	Parents/relative	Critical incident
Lightness	Parents/relative	Kano model
Manoeuvrability	Parents/relative	Critical incident/Kano model
Reducibility	Parents/relative	Critical incident
Setting easiness	Parents/relative	Kano model
Implicit/emotional needs		
Comfort	Doctors/paramedics	Structured interview
Colour and design	Parents/relative	Kano model
Robustness	Doctors/paramedics	Structured interview

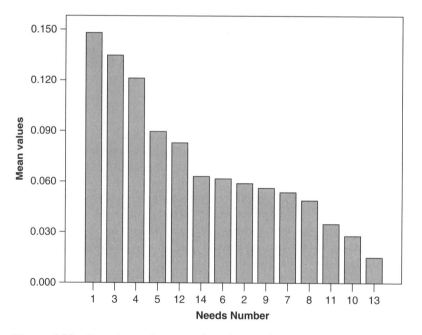

Figure 4.22   Bar chart of mean values for needs weights of importance.

Needs importance assessment

The importance of each identified need was calculated using the response latency method. The average importance weights are shown in Figure 4.22.

Technical response identification

The engineering characteristics responding to the determined needs were identified with the support of medics and paramedics. In particular, the same criteria used for customers' needs was followed: some engineering characteristics are related to explicit/technical needs, while others are related to implicit/emotional needs.

Relationship matrix building

A matrix that assigns a measure of positive correlation to each combination of customer need and engineering characteristic was built. The relationship matrix has a number of rows equal to the number of customer needs and a number of columns equal to the engineering characteristics. Each cell contains an indication of the strength of the link between the i-th customer need and the j-th engineering characteristic. The strength of the link was expressed by using a semantic scale with four levels (0 = no relation; 1 = weak relation; 3 = moderately strong relation; 9 = extremely strong relation) (see Table 4.21).

Planning matrix building

The next step was the compilation of the planning matrix. The matrix contains the following elements:

- importance to customer;
- customer satisfaction performance;
- goals;
- improvement ratio;
- raw weight: this is a summary of the planning matrix – the values of this column are the product of the importance to customer and improvement ratio; the higher the raw weight, the more important the corresponding customer need should be for the development team.

The planning matrix is reported in Table 4.22.

Priorities

At this point it is possible to assign a priority to each engineering characteristic. These priorities will summarise the relative contributions of each engineering characteristic

Table 4.21   Relationship matrix.

	Position indicator	Electronic position system	Lumbar/ridge push system	Cushion regulation system	Balancing roll	Body support	Frame structure	Coupling with endless screw	Frame material	Pelvis push system	Balancing seat	Pelvis waistband	Push system for adduction	Mobile footboard	Modular headrest	Inclinable backrest	Releasable cushion	Length of seat	Washable and transpired fabric	Interchangeable fabric
Pathology adaptability	9	9	9	9	9	9	–	1	–	3	3	1	3	3	–	3	–	1	–	–
Armrest adjustability	9	–	–	–	–	–	–	3	–	–	–	–	–	–	–	–	–	–	–	–
Cushion anatomy	–	–	9	9	–	–	–	–	–	–	–	–	–	1	–	–	–	–	–	–
Adaptability to the body	9	3	9	–	1	9	–	1	–	–	–	–	–	1	–	–	–	9	–	–
Pelvis blocking	9	1	–	–	–	–	–	–	–	9	9	3	3	–	–	–	–	1	–	–
Reduction of sense of weakness	–	–	–	–	–	–	–	–	–	–	–	9	–	–	1	–	–	–	–	–
Transportability	–	–	–	–	–	–	9	–	–	–	–	–	–	–	–	–	9	–	–	3
Lightness	–	–	–	–	–	–	1	–	9	–	–	–	–	–	–	–	–	–	–	–
Manoeuverability	–	–	–	–	–	–	9	3	–	–	–	–	–	–	–	–	–	–	–	–
Reducibility	–	–	–	–	–	–	9	–	–	–	–	–	–	–	–	–	3	–	–	–
Setting easiness	9	9	–	–	–	–	–	3	–	–	–	–	–	–	–	–	–	–	–	–
Comfort	–	–	–	–	9	–	–	–	–	–	–	–	–	–	–	–	–	–	–	–
Colour and design	–	–	–	–	–	–	9	–	3	–	–	–	–	–	–	3	–	–	9	–
Robustness	–	–	–	–	–	–	–	9	–	–	–	–	–	–	–	–	–	–	–	–

to the overall customer satisfaction. The priority for the j-th engineering characteristic is calculated as:

$$p_j = \sum_{i=1}^{I} c_{ij} \times r_i \qquad (4.30)$$

where:

- c_{ij} is the relation value between i-th customer need and j-th engineering characteristic (Table 4.21);
- r_i is the value of the raw weight for the i-th customer need (Table 4.22).

Table 4.22   Planning matrix.

	Importance to customer	Customer satisfaction performance	Goal	Improvement ratio	Raw weights
Pathology adaptability	0.148	2	5	2.5	0.370
Armrest adjustability	0.121	2	2	1.0	0.121
Cushion anatomy	0.09	3	4	1.3	0.119
Bodily adaptability	0.015	2	5	2.5	0.037
Pelvis blocking	0.083	3	5	1.6	0.137
Reduction of sense of weakness	0.056	2	4	2.0	0.112
Transportability	0.062	2	3	1.5	0.093
Lightness	0.135	2	4	2.0	0.270
Manoeuverability	0.063	2	4	2.0	0.126
Reducibility	0.035	3	3	1.0	0.035
Setting easiness	0.054	1	5	5.0	0.270
Comfort	0.059	4	4	1.0	0.059
Colour and design	0.049	3	4	1.3	0.065
Robustness	0.028	1	4	4.0	0.112

The higher the priority, the more influence the engineering characteristic has on customer satisfaction, and therefore the more important it is for the development of the new model of wheelchair. Priorities are reported in Table 4.23.

Three more activities were planned, that is competitive benchmarking, identification of targets and calculation of technical correlations. These were carried out in following phases of product development process, where the product concept created here was compared with competitors, and production constrains forced engineers to solve the potential correlations among the technical characteristics.

The HOQ of the project is shown in Figure 4.23; the numbers indicate the order in which the activities were carried out.

Table 4.23   Priorities of engineering characteristics.

Engineering characteristics	Priorities
Position indicator	8.42
Electronic system for position	6.01
Lumbar/ridge push system	4.73
Cushion regulation system	4.40
Balancing roll	3.89
Body support	3.66
Frame structure	3.14
Coupling with endless screw	2.97
Frame material	2.63
Pelvis push system	2.34
Balancing seat	2.34
Pelvis waistband	1.78
Lateral push system for adduction	1.52
Inclinable backrest	1.30
Mobile footboard	1.27
Modular headrest	1.12
Releasable cushion	0.94
Length of seat	0.84
Washable and transpired fabric	0.59
Interchangeable fabric	0.28

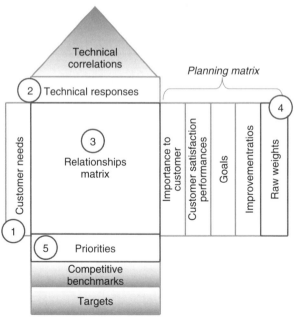

Figure 4.23   The House of Quality for the 'new wheelchair' project.

References

Akao, Y. (1990) *Quality Function Deployment*, Productivity Press, Cambridge, MA.

Barone, S., Lombardo, A. and Tarantino, P. (2007) A weighted logistic regression for conjoint analysis and kansei engineering. *Quality and Reliability Engineering International*, **23**, 689–706.

Barone, S., Lombardo, A. and Tarantino, P. (2008) Analysis of user needs for the re-design of a wheelchair, in *Statistics for Innovation – Statistical Design of 'Continuous' Product Innovation* (ed. P. Erto), Springer, pages 3–25.

Hauser, J.R. and Clausing, D. (1988) The house of quality. *Harvard Business Review*, **66(3)**, 63–73.

Kececioglu, D. (1991) *Reliability Engineering Handbook*, Prentice Hall.

Kerzner, H. (2009) *Project Management*, John Wiley & Sons, Inc.

O'Connor, P.T. (2002) *Practical Reliability Engineering*, John Wiley & Sons, Inc.

Saaty, T.L. (1990) How to make a decision. The analytic hierarchy process, *European Journal of Operational Research*, **48**, 9–26.

5

Advanced statistical techniques

5.1 To study the relationships between variables

5.1.1 Linear regression analysis

Often data are available in which a variable, say Y, can be regarded as 'dependent' on several 'independent' variables, say $X_1, X_2, \ldots, X_r$. This situation raises the opportunity to investigate the relationship between these variables.

EXAMPLE

The amount of carbon monoxide emitted by a vehicle travelling a certain route may depend on several factors related to the engine operating conditions (e.g. speed, gear, type of fuel used) and to environmental conditions (e.g. temperature, humidity). Suppose we set the values of the factors and then repeatedly observe the variable Y. When repeating the observation, the value of Y will not always be identical, although the values of the factors are fixed. So Y is a random variable and can be therefore indicated by Y. The factors are deterministic variables with fixed values $x_1, x_2, \ldots, x_r$. We write the following equation:

$$Y = f(x_1, x_2, \ldots, x_r) + \varepsilon \qquad (5.1)$$

It is an analytical expression that links Y to the $x_1, x_2, \ldots, x_r$. The term ε on the right-hand side justifies the random nature of Y, it is a random variable that we call *error term* or *random error*.

We can imagine that ε only depends on the measurement process. In fact ε may be due to a combination of factors that intervene when observing the data, determining the variation of Y despite $x_1, x_2, \ldots, x_r$ being fixed. The random variable (r.v.) ε represents our rate of ignorance. Usually, it is assumed that ε has zero mean and constant variance:

$$E\{\varepsilon\} = 0 \quad V\{\varepsilon\} = \sigma^2 \tag{5.2}$$

Often it is also assumed that ε follows a Gaussian probability distribution model.

Equation 5.1 is a particular *statistical model*, that is a model that contains a random component in addition to a deterministic component.

Some reflections. Instead of Equation 5.1 we can write:

$$Y = f(X_1, X_2, \ldots, X_r) \tag{5.3}$$

This relationship (Equation 5.3) represents a so-called transformation of random variables. The relationship between the dependent variable and the independent variables in this case is completely known if we specify the function f, but the variability is transmitted from the independent variables (random variables) to the dependent variable.

If, instead of Equation 5.1, we write:

$$Y = f(X_1, X_2, \ldots, X_r) + \varepsilon \tag{5.4}$$

then the situation would be more complex because this model claims that the random variation of Y is not only due to random variation of the known factors, but also to unknown factors that we describe using the error term.

For the remainder of this section we will refer to situations such as those expressed by Equation 5.1.

5.1.1.1   Linear model

To explain the deterministic component of the model (5.1), we resort to the simplest analytical model, imagining a 'linear' or 'additive' behaviour in which each independent variable acts on the Y independently from the others and with its own weight:

$$f(x_1, x_2, \ldots x_r) = \alpha + \beta_1 x_1 + \beta_2 x_2 + \cdots + \beta_r x_r \tag{5.5}$$

The model also has a constant term α representing the fact that the Y can be different from 0 when all factors are 0. This results in the *multiple linear regression* model:

$$Y = \alpha + \beta_1 x_1 + \beta_2 x_2 + \cdots + \beta_r x_r + \varepsilon \tag{5.6}$$

which is reduced to the simple linear regression model if $r = 1$.

5.1.1.2   Simple linear regression model

We present the simple linear regression model not because it is the most widely used in practice, but merely to fix ideas and understand the steps of the methodology of the analysis, which will be extendable to the case of multiple regression.

The simple linear regression model can be written as:

$$Y = \alpha + \beta x + \varepsilon \tag{5.7}$$

Assuming the model (5.7) means, from a deterministic point of view, to argue that if we fix the variable x to a generic value, say x_i, the expected response is equal to $\alpha + \beta x_i$. Unfortunately, due to the random error, the observed value of the dependent variable will be $\alpha + \beta x_i + \varepsilon_i$ where ε_i is the random error that occurs for $x = x_i$ (see Figure 5.1).

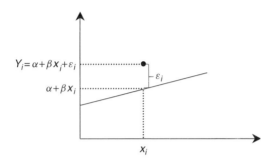

Figure 5.1   Error component at a generic value x_i.

Having a data set of couples (x_i, y_i), with $i = 1, 2, \ldots, n$, the problem is to estimate the two parameters α and β, which means determining the parameters of a straight line.

EXAMPLE

Table 5.1 shows the values of the size of lots produced in a manufacturing company and their production cost. Imagine we want to check that the production cost is linearly dependent on the lot size, and we want to estimate the parameters of this relationship. Figure 5.2 shows the scatter plot of the data (see Chapter 3).

Table 5.1   Production plant data: lot size (x_i) and production cost (y_i).

x_i	y_i
95	214
82	152
90	156
81	129
99	254
100	266
93	210
95	204
93	213
87	150

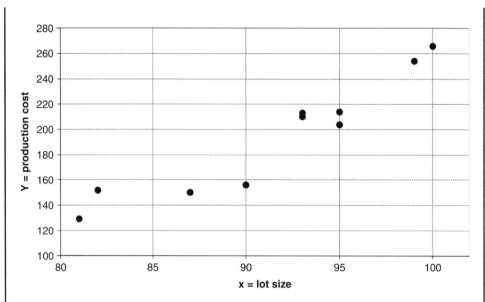

Figure 5.2   Scatter plot of production plant data.

A visual analysis of the scatter plot indicates that the data follow a linear trend quite well, so it makes sense to proceed with the assumption of the linear model (5.7) and the estimation of the parameters α and β. The goal is to estimate the parameters from the available data so to determine the line that best fits them. This line is called the *regression line*.

5.1.1.3   Least-squares line

The renowned scientist Carl Friedrich Gauss (1777–1865), who developed the most popular model of r.v., found that the best regression line is the one for which the sum of the squared distances of points from it is minimum. The method of determining this line (therefore called least-squares line) is the so-called *least-squares method*.

Imagine keeping the values x_i $(i = 1, 2, \ldots, n)$ constant and repeating the observations of the response. Due to the random error, estimates of the parameters α and β will vary randomly.

Note that the two parameters are assumed unknown and will remain unknown, it being only possible to estimate their values. They do not vary with different data sets, but their estimates will vary. Call A and B the r.v. so-called estimators of the parameters which describe the variation of the estimates of α and β.

Given that Y_i is the response that corresponds with x_i, consider the distance (measured along the vertical axis) between the point (x_i, Y_i) and the regression line (see Figure 5.1):

$$Y_i - (A + Bx_i) = Y_i - A - Bx_i$$

The sum of squared distances, denoted by *SSE* (*sum of squares of errors*) is given by:

$$SSE = \sum_i (Y_i - A - Bx_i)^2 \tag{5.8}$$

The goal is to find the formulations of A and B that minimise *SSE*.

To find the minimum of a function of one variable, we usually let the first derivative equal to zero and verify that the second derivative is greater than zero. In this case it can be shown that setting the first partial derivatives of *SSE* equal to zero with respect to A and B is a necessary and sufficient condition to obtain the minimum *SSE*.

Setting the derivatives of *SSE* with respect to A and B equal to zero, we get two equations called *normal equations*:

$$\frac{\partial SSE}{\partial A} = 0 \Rightarrow -2 \sum_i (Y_i - A - Bx_i) = 0 \Rightarrow A = \overline{Y} - B\overline{x} \tag{5.9}$$

$$\frac{\partial SSE}{\partial B} = 0 \Rightarrow -2 \sum_i x_i(Y_i - A - Bx_i) = 0 \Rightarrow B = \frac{\sum_i x_i Y_i - n\overline{x}\,\overline{Y}}{\sum_i x_i^2 - n\overline{x}^2} \tag{5.10}$$

From Equation 5.9 we deduce that the regression line passes through the point with coordinates $(\overline{x}, \overline{Y})$.

To evaluate the optimal values of the estimators A and B, we first calculate their expected values.

Consider the numerator in Equation 5.10 (note that the denominator does not contain random variables):

$$\sum_i x_i Y_i - n\overline{x}\,\overline{Y} = \sum_i x_i Y_i - \overline{x} \sum_i Y_i = \sum_i (x_i - \overline{x})Y_i$$

Starting from the model (5.7) and having assumed $E\{\varepsilon\} = 0$, we know that: $E\{Y_i\} = \alpha + \beta x_i$. Therefore:

$$E\left\{ \sum_i (x_i - \overline{x})Y_i \right\} = \sum_i (x_i - \overline{x})(\alpha + \beta x_i) = \alpha \sum_i (x_i - \overline{x}) + \beta \sum_i (x_i - \overline{x})x_i$$

$$= \alpha \left(\sum_i x_i - n\overline{x} \right) + \beta \left(\sum_i x_i^2 - \overline{x} \sum_i x_i \right) = \beta \left(\sum_i x_i^2 - n\overline{x}^2 \right)$$

Replacing the expression found in the numerator of B, we get:

$$E\{B\} = E\left\{ \frac{\sum_i x_i y_i - n\overline{x}\,\overline{Y}}{\sum_i x_i^2 - n\overline{x}^2} \right\} = \frac{\beta \left(\sum_i x_i^2 - n\overline{x}^2 \right)}{\sum_i x_i^2 - n\overline{x}^2} = \beta$$

A similar argument can be developed for A:

$$E\{A\} = E\{\overline{Y} - B\overline{x}\} = E\{\overline{Y}\} - \overline{x}\beta = E\left\{\frac{1}{n}\sum_i Y_i\right\} - \overline{x}\beta = \frac{1}{n}E\left\{\sum_i Y_i\right\} - \overline{x}\beta$$

$$= \frac{1}{n}\sum_i E\{Y_i\} - \overline{x}\beta = \frac{1}{n}\sum_i E\{\alpha + \beta x_i\} - \overline{x}\beta = \frac{1}{n}[n\alpha + \beta n\overline{x}] - \overline{x}\beta$$

$$= \alpha + \beta\overline{x} - \overline{x}\beta = \alpha$$

We have therefore proved that A and B are unbiased estimators of α and β. In addition, we observe that A and B can be expressed as linear combinations of the responses Y_i:

$$A = \overline{Y} - B\overline{x} = \frac{1}{n}\sum_i Y_i - \frac{\sum_i (x_i - \overline{x})Y_i}{\sum_i x_i^2 - n\overline{x}^2}\overline{x}$$

$$B = \frac{\sum_i x_i y_i - n\overline{x}\overline{Y}}{\sum_i x_i^2 - n\overline{x}^2} = \frac{\sum_i x_i Y_i - \overline{x}\sum_i Y_i}{\sum_i x_i^2 - n\overline{x}^2} = \frac{\sum_i (x_i - \overline{x})Y_i}{\sum_i x_i^2 - n\overline{x}^2}$$

Starting from the model $Y_i = \alpha + \beta x_i + \varepsilon_i$, and assuming that ε_i is Gaussian, Y_i is also Gaussian, hence also A and B have a Gaussian distribution being linear combinations of Y_i.

It can be shown that A and B have the minimum variance over all possible estimators. For all these reasons, they are called *Best Linear Unbiased Estimators* (BLUEs).

However, in the model (5.7) there are three unknown parameters: α, β and σ (equal to the standard deviation of ε).

Hereinafter, to have a more compact notation, we suggest the following:

$$SS_x = \sum_i (x_i - \overline{x})^2 = \sum_i x_i^2 - n\overline{x}^2$$

$$SS_y = \sum_i (Y_i - \overline{Y})^2 = \sum_i Y_i^2 - n\overline{Y}^2 \qquad (5.11)$$

$$SS_{xy} = \sum_i (x_i - \overline{x})(Y_i - \overline{Y}) = \sum_i x_i Y_i - n\overline{x}^2\overline{Y}^2$$

We can rewrite B by replacing its denominator:

$$B = \frac{\sum_i x_i y_i - n\overline{x}\overline{Y}}{\sum_i x_i^2 - n\overline{x}^2} = \frac{\sum_i x_i y_i - n\overline{x}\overline{Y}}{SS_x}$$

At the numerator $\sum_i x_i Y_i - n\bar{x}\bar{Y} = \sum_i (x_i - \bar{x})Y_i$, so we add and subtract the quantity $\sum_i (x_i - \bar{x})\bar{Y}$:

$$\sum_i \left[(x_i - \bar{x})\, Y_i + (x_i - \bar{x})\,\bar{Y} - (x_i - \bar{x})\,\bar{Y}\right] = \sum_i \left[(x_i - \bar{x})\left(y_i - \bar{Y}\right) + \bar{Y}\left(x_i - \bar{x}\right)\right]$$

$$= \sum_i (x_i - \bar{x})\left(Y_i - \bar{Y}\right) = SS_{xy}$$

Being $\sum_i \bar{Y}(x_i - \bar{x}) = \bar{Y}\sum_i (n\bar{x} - n\bar{x}) = 0$.

Therefore: $B = \frac{SS_{xy}}{SS_x}$.

EXAMPLE (*continued*)

For the lot production plant data, we have:

$$\begin{array}{lll} \bar{x} = 91.5 & ss_x = 380.5 & \hat{\alpha} = 6.72 \\ \bar{y} = 194.8 & ss_y = 19263.6 & \hat{\beta} = -419.85 \\ & ss_{xy} = 2556 & \end{array}$$

In Figure 5.3, together with the scatter plot, the regression line estimated by the of least-squares method is drawn. Note that the procedure is very simple with the software Excel®.

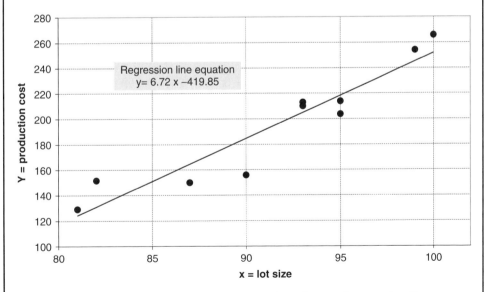

Figure 5.3   Scatter plot of production plant data and regression line.

5.1.1.4 Estimator of the residual variation

We define as residuals each difference between an observed value and the value with the same abscissa on the estimated regression line.

In practice while before observing the data we could write:

$$\varepsilon_i = Y_i - \alpha - \beta x_i$$

after observing the data we write:

$$e_i = y_i - \hat{\alpha} - \hat{\beta} x_i \tag{5.12}$$

and we call the quantity e_i the *residual*.

The sum of squared errors (Equation 5.8) allows us to estimate σ^2: first, using Equation 5.11 we have:

$$SSE = \sum_i (Y_i - A - B x_i)^2 = SS_y - \frac{SS_{xy}^2}{SS_x} \tag{5.13}$$

It can be shown that the ratio $SSE/(n-2)$ is an unbiased estimator of σ^2. To prove it, we need to use the probability distribution model of r.v. known as *chi-square* (χ^2). In fact, it can be shown that:

$$\frac{SSE}{\sigma^2} \sim \chi_{n-2}^2 \tag{5.14}$$

Therefore:

$$E\left\{ \frac{SSE}{n-2} \right\} = \sigma^2 \tag{5.15}$$

EXAMPLE (*continued*)

For the production plant data, the Table 5.1 is completed as follows (Table 5.2).

Table 5.2   Dataset completed with predicted values and residuals.

x_i	y_i	$\hat{y}$	e_i	e_i^2
95	214	218.31	−4.31	18.59
82	152	130.98	21.02	441.67
90	156	184.72	−28.72	825.06
81	129	124.27	4.73	22.41
99	254	245.18	8.82	77.77
100	266	251.90	14.10	198.85
93	210	204.88	5.12	26.25
95	204	218.31	−14.31	204.81
93	213	204.88	8.12	66.00
87	150	164.57	−14.57	212.32

The sum of the values in the last column gives $\widehat{SSE} = 2093.73$, therefore we calculate $\hat{\sigma}^2 = 261.72$.

5.1.1.5   Coefficient of determination

The coefficient of determination is an index of adequacy of the regression line to interpret the data. We start from the following expression:

$$SS_y = \sum_i (Y_i - \overline{Y})^2$$

This expresses the variation of the response. This formula not only reflects the variation of the response caused by the random error, but also considers the range of variation of x.

If we subtract the variation due to the random error from SS_y, using the estimator SSE, and then relate this difference (which therefore measures the variation solely due to the linear relationship) to the overall variation, the obtained ratio is the rate of the overall variation due to the variation of x. It is called the *coefficient of determination*:

$$R^2 = \frac{SS_y - SSE}{SS_y} \qquad 0 \le R^2 \le 1 \tag{5.16}$$

The stronger the linear dependence of Y on x and the smaller the value of SSE, the higher R^2. In practice, the closer the points are to the regression line and the bigger the slope of the regression line, the higher the value of R^2, tending to 1.

If $SSE = 0$ there is no variation around the regression line, so $R^2 = 1$. In this case the regression line fully explains the dependence of Y on x.

If $SS_y = SSE$ it means that the rate of variation due to x is zero and all variation is due to the random error, so the scatter plot will appear as a cloud of points with no discernible linear trend. In such a case $R^2 = 0$.

5.1.1.6   Prediction by the regression line

The regression model allows us to make a prediction of values of the response at non-observed values of x, provided they are within the range of the observed x values.

Suppose we have a data set of n couples of values and have determined the regression line: for a non-observed value of x, say x^*, we can 'predict' the expected response as $\hat{\alpha} + \hat{\beta}x^*$.

5.1.1.7   Analysis of residuals

Analysis of residuals allows us to check if the estimated regression model fits the data well.

Imagine we have estimated the regression line. We construct another scatter plot with the residuals for each point of the dataset. If the regression model is appropriate, the points should be placed around the x-axis without any trend.

EXAMPLE (*continued*)

Figure 5.4 shows the scatter plot of the estimated residuals along with the observed response values for the dataset of Table 5.2. Here the check is successful; the points are placed around the x-axis, without any particular trend.

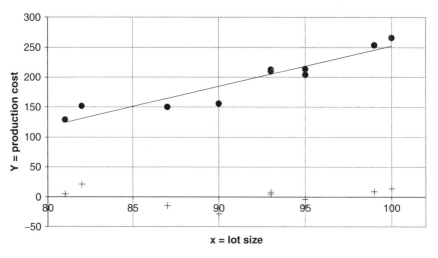

Figure 5.4   Scatter plot of the observed values and estimated residuals.

However, if the linear regression model is not appropriate, we will see some curvilinear or anomalous trend. For example, if with increasing values of x we see an evident increase in the dispersion of points (a so-called funnel shape), then it is not good to assume a constant variance of the error term. In such cases we may use appropriate data transformations.

EXAMPLE

In many situations the actual functional dependence $Y(x)$ is non-linear, for example of the type $Y = \alpha x^{\beta}$ (power law). This relationship can be linearised:

$$\log Y = \log \alpha + \beta \log x$$

If we pose:
$$\log Y = Y' \quad \log \alpha = \alpha' \quad \log x = x'$$

we get:
$$Y' = \alpha' + \beta x'$$

Hence we can try to fit the linear regression model to transformed data.

5.1.1.8 Multiple linear regression

The multiple linear regression model can be conveniently written by adopting matrix notation:

$$\underline{Y} = \underline{\underline{x}} \cdot \underline{\beta} + \underline{\varepsilon} \tag{5.17}$$

where

$$\underline{Y} = \begin{bmatrix} Y_1 \\ Y_2 \\ \vdots \\ Y_n \end{bmatrix} \quad \underline{\underline{x}} = \begin{bmatrix} 1 & x_{11} & x_{12} & \cdots & x_{1k} \\ 1 & x_{21} & x_{22} & \cdots & x_{2k} \\ . & . & . & \cdots & . \\ . & . & . & \cdots & . \\ . & . & . & \cdots & . \\ . & . & . & \cdots & . \\ 1 & x_{n1} & x_{n2} & \cdots & x_{nk} \end{bmatrix} \quad \underline{\beta} = \begin{bmatrix} \beta_0 \\ \beta_1 \\ \vdots \\ \beta_k \end{bmatrix} \quad \underline{\varepsilon} = \begin{bmatrix} \varepsilon_1 \\ \varepsilon_2 \\ \vdots \\ \varepsilon_n \end{bmatrix}$$

Here we have n points and k independent variables or *regressors*.

In the case of multiple regression we can also estimate the parameters by the least-squares method.

$$D = \sum_{i=1}^{n} \varepsilon_i^2 = \sum_{i=1}^{n} \left(Y_i - \beta_0 - \sum_{j=1}^{k} \beta_j x_{ij} \right)^2 \quad \text{min!} \tag{5.18}$$

The normal equations derived from Equation 5.18, written in matrix form, are:

$$\underline{\underline{x}}^{\mathrm{T}} \underline{\underline{x}} \, \underline{\hat{\beta}} = \underline{\underline{x}}^{\mathrm{T}} \underline{y} \tag{5.19}$$

whose solution is:

$$\underline{\hat{\beta}} = \left(\underline{\underline{x}}^{\mathrm{T}} \underline{\underline{x}} \right)^{-1} \underline{\underline{x}}^{\mathrm{T}} \underline{y} \tag{5.20}$$

Once the parameters are estimated, we obtain the predicted model:

$$\underline{\hat{y}} = \underline{\underline{x}} \, \underline{\hat{\beta}} \tag{5.21}$$

Residuals are still calculated as differences between observed and predicted values:

$$\underline{e} = \underline{y} - \underline{\hat{y}} \tag{5.22}$$

The estimate of the error variance is obtained similarly to the case of simple regression:

$$\hat{\sigma}^2 = \frac{\sum_{i=1}^{n} e_i^2}{n - (k+1)} = \frac{SSE}{n - k - 1} \tag{5.23}$$

with $k + 1$ being the number of model parameters.

5.1.2 Logistic regression models

In logistic regression the relationship between the response variable and the regressors is not direct as in linear regression, but it is 'mediated' by probability. In this section we present three cases of logistic regression models at increasing levels of complexity.

5.1.2.1 Case I

Here the response variable is a dichotomous variable taking two possible values: $Y = 0$ and $Y = 1$ (called 'success'). The regressor x is also a dichotomous variable. The model is formulated as follows:

$$\Pr(Y = 1|x) = \pi(x) = \frac{e^{\beta_0 + \beta_1 x}}{1 + e^{\beta_0 + \beta_1 x}} \tag{5.24}$$

The model (5.24) states that the probability of success is a function of x, depending on two parameters β_0 and β_1. The function $\pi(x)$ is defined for any value of x and ranges between 0 and 1.

The ratio between probability of success and probability of failure is called *odds*:

$$\text{odds}(x) = \frac{\pi(x)}{1 - \pi(x)} = e^{\beta_0 + \beta_1 x} \tag{5.25}$$

Taking the logarithm of the odds we get:

$$\log\left[\frac{\pi(x)}{1 - \pi(x)}\right] = \beta_0 + \beta_1 x \tag{5.26}$$

The left-hand side of Equation 5.26 is called *logit*.
The *odds ratio* (OR) is defined as:

$$\text{OR} = \frac{\text{odds}(1)}{\text{odds}(0)} = e^{\beta_1} \tag{5.27}$$

The odds ratio allows us to estimate the effect of the regressor on the propensity of Y to take the value 1, that is the effect of the regressor to have a success. It depends only on the parameter β_1.

5.1.2.2 Case II

Here the response variable is still dichotomous ($Y = 0$, $Y = 1$), while the regressors are k dichotomous variables $x_1, x_2, \ldots, x_k$. The probability of success is modelled as:

$$\Pr(Y = 1|\underline{x}) = \pi(x_1, x_2, \ldots, x_k) = \frac{e^{\beta_0 + \beta_1 x_1 + \cdots + \beta_k x_k}}{1 + e^{\beta_0 + \beta_1 x_1 + \cdots + \beta_k x_k}} \tag{5.28}$$

The odds ratio can be calculated for each regressor:

$$\text{OR}(x_j) = \frac{\text{odds}(x_j = 1)}{\text{odds}(x_j = 0)} = e^{\beta_j}$$

The odds ratio allows us to estimate the effect of the j-th regressor on the propensity of Y to take the value 1. It depends only on β_j so it does not depend on the value of the other regressors.

In case of one polychotomous regressor, the model can be reduced to the case of several dichotomous variables.

5.1.2.3   Case III (ordinal logistic regression)

Here the response variable is ordinal with $m + 1$ categories $0, 1, \ldots, m$ (the category 0 is said to be the 'reference' category). The regressors are k dichotomous variables $x_1, x_2, \ldots, x_k$.

In the *proportional odds* model, the existence of a latent variable Y^* linearly dependent on the regressors is assumed:

$$Y^* = \beta_1 x_1 + \beta_2 x_2 + \cdots + \beta_k x_k + \varepsilon \tag{5.29}$$

The domain of Y^* is divided in $m + 1$ intervals $(-\infty, \tau_1), (\tau_1, \tau_2), \ldots, (\tau_m, +\infty)$ such that:

$$Y^* \in \left(\tau_j, \tau_{j+1}\right) \Leftrightarrow Y = j \qquad j = 1, \ldots, m$$

$$\tau_0 = -\infty \tag{5.30}$$

$$\tau_{m+1} = +\infty$$

τ_{j+1} is said to be the upper threshold of the j-th category.

For each category j of the response variable, the logit between the group of categories higher than j and the group of categories lower than or equal to j (cumulative logit) is given by:

$$c_j(x) = \log\left[\frac{\Pr(Y > j|x)}{\Pr(Y \leq j|x)}\right] = \beta_1 x_1 + \beta_2 x_2 + \cdots + \beta_k x_k - \tau_{j+1} \tag{5.31}$$

We can see that, except for the constant τ_{j+1}, it does not depend on the category of the response variable Y. So the model assumes that the influence of the regressors is the same across the categories of Y.

The procedures for the estimation of model parameters in logistic regression are more complicated than in the case of linear regression. They require the concept of maximum likelihood, which has been intentionally avoided in this book to keep to a desirable level of insight.

However, all estimation methods are well codified, so the interested reader is invited to address the references provided at the end of the chapter. Moreover, the procedures are available on most statistical software packages.

5.1.3   Introduction to multivariate statistics

In observational and experimental investigations it often happens that several variables are observed and measured on each unit of a sample extracted from a reference population. This type of data is said to be multivariate.

EXAMPLE

Table 5.3 shows a multivariate data set. It is related to a survey made on a sample of students attending a course.

Table 5.3   Example of multivariate dataset.

No.	Gender	Age (days)	Height (cm)	English skills	Use of internet
1	M	6580	178	Sufficient	Rare
2	M	6939	178	Sufficient	Very frequent
3	M	6943	172	Sufficient	Frequent
4	M	7187	173	Good	Frequent
5	F	6965	158	Very good	Rare
6	F	7146	160	Very good	Rare
7	F	7157	172	Good	Frequent
8	M	7932	173	Sufficient	Frequent
9	F	6947	160	Good	Frequent
10	F	9028	160	Sufficient	Frequent
11	M	9626	174	Good	Frequent
12	F	7743	164	Good	Frequent
13	M	7664	170	Good	Frequent
14	M	6960	168	Sufficient	Rare
15	M	7475	170	Insufficient	Frequent

5.1.3.1   Bivariate data

When we consider two variables for each statistical unit, this is called bivariate data. For the double variable (X, Y) we call X the first component and Y the second component. Hereinafter, we consider the case in which both first and second components are quantitative variables.

EXAMPLE

The collected height and weight data for a sample of students is an example of a double variable. The data are reported in Table 5.4.

Table 5.4   Example of bivariate data.

Height (cm)	Weight (kg)
173	65
160	52
158	50
165	52
181	80
183	80

Table 5.4   (*continued*).

Height (cm)	Weight (kg)
170	60
167	53
177	65
170	62
165	55
165	70
180	60
171	80
180	82

For bivariate data it is possible to graphically show the *double empirical frequency distribution*. First, we make a double-entry frequency table (Table 5.5) following the same criteria adopted in the case of a single variable.

The frequency distribution is graphically represented by the bivariate histogram of Figure 5.5.

Table 5.5   Tabular representation of a double empirical frequency distribution.

Height	Weight			
	50–58	58–66	66–74	74–82
158.00–164.25	2	0	0	0
164.25–170.50	3	2	1	0
170.50–176.75	0	1	0	1
176.75–183.00	0	2	0	3

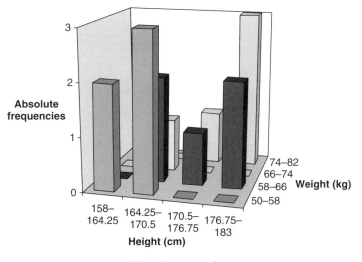

Figure 5.5   Bivariate histogram.

5.1.3.2 Correlation

An issue of interest when studying bivariate data is the presence of *correlation* between the two variables. Two variables X and Y are said to be positively correlated when high values of X typically correspond to high values of Y and low values of X typically correspond to low values of Y.

Conversely, the two variables X and Y are said to be negatively correlated when high values of X typically correspond to low values of Y and vice versa.

In a preliminary analysis, if we want to detect a possible correlation between two variables we can use the scatter plot (Chapter 3).

EXAMPLE

It is supposed that the environmental temperature has an effect on the quality of a certain type of cement. In 20 days we recorded the number of defective specimens and the average daily temperature. We got the following bivariate data matrix (Table 5.6).

Table 5.6   Environmental temperature and number of defective specimens recorded in 20 days.

Day	Temperature	Defects	Day	Temperature	Defects
1	24.2	25	11	24.4	22
2	22.7	31	12	24.8	23
3	30.5	36	13	20.6	20
4	28.6	33	14	25.1	25
5	25.5	19	15	21.4	25
6	32.0	24	16	23.7	25
7	28.6	27	17	23.9	23
8	26.5	25	18	25.2	27
9	25.3	16	19	27.4	30
10	26.0	14	20	28.3	33

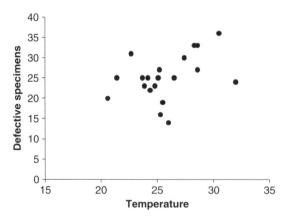

Figure 5.6   Scatter plot of defective specimens versus temperature.

For a preliminary analysis we draw the scatter plot Figure 5.6. The scatter plot shows that the two variables are *somehow* positively correlated.

5.1.3.3   Correlation coefficient

We denote by (x_i, y_i), $(i = 1, \ldots, N)$ the generic couple of a bivariate data set.

Generally, since the two component variables have different dimensions (centimetre and kilogram in the previous example), in addition to different locations and dispersion, we may preferably use *standardised* variables:

$$Z_x = \frac{X - \mu_x}{\sigma_x} \qquad Z_y = \frac{Y - \mu_y}{\sigma_y} \tag{5.32}$$

After observing the value, and estimating the location and dispersion parameters, we have standardised observations:

$$z_{x,i} = \frac{x_i - \hat{\mu}_x}{\hat{\sigma}_x} \qquad z_{y,i} = \frac{y_i - \hat{\mu}_y}{\hat{\sigma}_y}$$

The cloud of points relative to standardised data lies around the origin of the axes in the scatter plot. With the data of the previous example we obtain Figure 5.7.

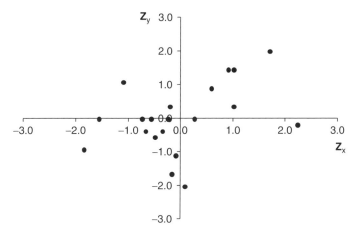

Figure 5.7   Scatter plot of the standardised data.

The products of the point coordinates will be positive if the points lie either in the first or third quadrant, while they will be negative if the points are either in the second or fourth quadrant.

A measure of correlation can be obtained by averaging such products, so obtaining the dimensionless index:

$$\hat{\rho} = \frac{1}{n} \sum_{i=1}^{n} (z_{x,i} \cdot z_{y,i}) \tag{5.33}$$

Referring to the original variables:

$$\hat{\rho} = \frac{\dfrac{1}{n}\displaystyle\sum_{i=1}^{n}(x_i - \hat{\mu}_x)(y_i - \hat{\mu}_y)}{\hat{\sigma}_x\hat{\sigma}_y} = \frac{\hat{\sigma}_{xy}}{\hat{\sigma}_x\hat{\sigma}_y} \tag{5.34}$$

The quantity $\hat{\sigma}_{xy}$ is the estimated *covariance*.
In general we write:

$$\rho = \frac{\text{Cov}\{X, Y\}}{\sqrt{\text{V}\{X\}}\sqrt{\text{V}\{Y\}}} \tag{5.35}$$

The dimensionless index ρ is said to be the correlation coefficient.

As the denominator is always positive, the sign of ρ depends on the sign of the covariance.

5.1.3.4   Properties of the correlation coefficient

If $\rho > 0$ we say that there is 'positive correlation', if $\rho < 0$ we say there is 'negative correlation', otherwise if $\rho \cong 0$ we say there is little or even no correlation.

It is possible to demonstrate that $-1 \le \rho \le +1$.

If there is a linear relationship between the two variables X and Y, $Y = a \pm bX$, then $\rho = \pm 1$.

Note that a high correlation between two variables does not necessarily imply any cause–effect relation.

Figure 5.8 shows three examples of scatter plots of bivariate data. The first two are extreme cases, where there is perfect correlation between the two variables. The third one is also rather extreme, showing a case of very poor correlation, close to zero.

It is also possible to demonstrate that if two variables are s-independent, as a consequence they are uncorrelated. The opposite is not true. A classic example is made with two variables linked by a relationship of the type:

$$(X - a)^2 + (Y - b)^2 = r^2$$

which is the equation of a circumference of a circle with the centre at (a, b) and radius r (see Figure 5.9).

For these reasons ρ is better called the linear correlation coefficient, meaning that it is an indicator of the linear dependence between two variables.

If a linear dependence exists between two variables, but there is an error of observation or measurement to consider, we can write the model as:

$$Y = a + bX + \varepsilon \tag{5.36}$$

where ε is the error term, which we assume has a mean of zero and constant variance σ_ε^2, and is independent from X.

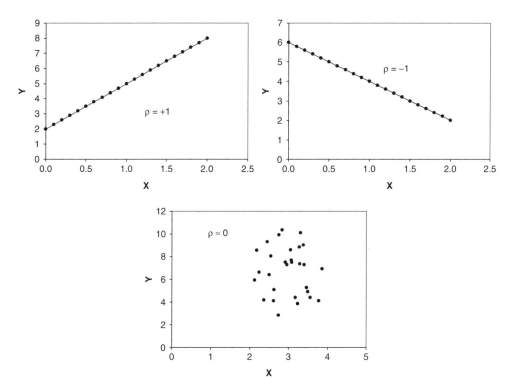

Figure 5.8   Three examples of scatter plots.

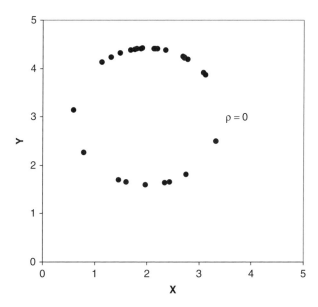

Figure 5.9   Example of two uncorrelated but dependent variables.

Calculating the expected value and the variance from Equation 5.36 we get:

$$\mu_y = a + b\mu_x \tag{5.37}$$

$$V\{Y\} = b^2 V\{X\} + \sigma_\varepsilon^2 \tag{5.38}$$

Subtracting Equation 5.37 from Equation 5.36 we get:

$$Y - \mu_y = b(X - \mu_x) + \varepsilon \tag{5.39}$$

So in the case of an underlying linear relationship between the two variables X and Y we see that the covariance:

$$\text{Cov}\{X, Y\} = E\{(X - \mu_x)[b(X - \mu_x) + \varepsilon]\} = bV\{X\} \tag{5.40}$$

From which, by adopting the definition of ρ previously given in Equation 5.35, we get

$$\rho = \frac{\text{Cov}\{X, Y\}}{\sqrt{V\{X\}}\sqrt{V\{Y\}}} = \frac{bV\{X\}}{\sqrt{V\{X\}\left[b^2 V\{X\} + \sigma_\varepsilon^2\right]}}$$

$$= \frac{1}{\sqrt{1 + \frac{\sigma_\varepsilon^2}{b^2 V\{X\}}}} \tag{5.41}$$

From the expression (5.41) we note that:

- if we replace b with $-b$ we obtain an opposite value of ρ;
- if we pose $\sigma_\varepsilon^2 = 0$ the correlation coefficient is equal to 1.

The correlation coefficient decreases for σ_ε^2 increasing and increases for $b^2 V\{X\}$ increasing (that is the rate of variance of Y due to the linear relation with X).

In summary, the lower the relative weight of the error respect to the linear relation (measured by the ratio $\sigma_\varepsilon^2 / b^2 V\{X\}$) the higher the correlation coefficient (closer to 1).

5.1.3.5   Correlation versus regression

The previous consideration allows us to connect the concept of correlation to the concept of regression (which we could imagine to be strictly correlated!).

A correlation analysis of two variables through a sample of bivariate data can be a preliminary step for the postulation of a regression model.

We can also show that the correlation coefficient ρ is related to the coefficient of determination defined in the linear regression analysis. In fact, in regression analysis we had:

$$R^2 = \frac{\text{total variation of } Y - \text{residual variation}}{\text{total variation of } Y} = \frac{SS_y - SSE}{SS_y}$$

In the regression model we supposed that x was a deterministic variable. If we remove this assumption and also consider X a random variable we get:

$$R^2 = \frac{b^2 V\{X\}}{b^2 V\{X\} + \sigma_\varepsilon^2}$$

From the definition of ρ (Equation 5.41) we get:

$$\rho^2 = \frac{b^2 (V\{X\})^2}{V\{X\} \left[b^2 V\{X\} + \sigma_\varepsilon^2 \right]} = \frac{b^2 V\{X\}}{b^2 V\{X\} + \sigma_\varepsilon^2}$$

So we observe that:

$$\rho^2 = R^2 \tag{5.42}$$

EXAMPLE

Using the height and weight data of Table 5.4 we calculate:

$$\hat{\rho} = \frac{\dfrac{1}{n} \displaystyle\sum_{i=1}^{n} (x_i - \hat{\mu}_x)(y_i - \hat{\mu}_y)}{\hat{\sigma}_x \hat{\sigma}_y} = 0.74$$

Based on the data set, if we assume a linear regression model we get: $\hat{R}^2 = 0.554 = \hat{\rho}^2$ (Figure 5.10).

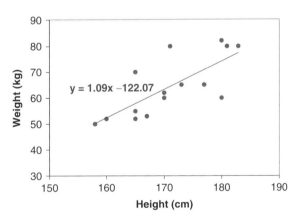

Figure 5.10   Scatter plot and regression line of the bivariate data set of Table 5.4.

5.1.3.6   Bivariate Gaussian distribution

A double or bivariate random variable is characterised by a joint probability distribution function and a joint probability density function.

In particular, if the two components X and Y are s-independent, the joint probability density function is equal to the product of the two *marginal* probability density functions, that is relative to the two components:

$$f_{XY}(x, y) = f_X(x) \cdot f_Y(y) \tag{5.43}$$

If the two components follow a Gaussian distribution we have:

$$f_X(x) = \frac{1}{\sigma_x \sqrt{2\pi}} \exp \left[-\frac{(x - \mu_x)^2}{2\sigma_x^2} \right]$$

$$f_Y(y) = \frac{1}{\sigma_y \sqrt{2\pi}} \exp \left[-\frac{(y - \mu_y)^2}{2\sigma_y^2} \right]$$

If X and Y are s-independent, we can easily *build* the joint density function:

$$f_{XY}(x, y) = \frac{1}{2\pi \cdot \sigma_x \sigma_y} \exp \left[-\frac{1}{2} \left(\frac{(x - \mu_x)^2}{\sigma_x^2} + \frac{(y - \mu_y)^2}{\sigma_y^2} \right) \right] \tag{5.44}$$

representing a bivariate Gaussian model.

Figure 5.11 shows the graph of bivariate Gaussian density function with two s-independent components both with zero mean.

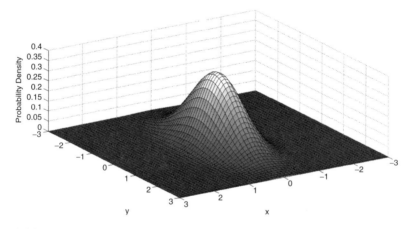

Figure 5.11   Graphical plot of a probability density function of a bivariate Gaussian.

While in the univariate case the representation of the probability density function leads to a bell-shaped curve, in this case the graphical representation of the surface is a 3D bell.

If we imagine we cut the bell with a horizontal plane, by posing $f_{XY}(x, y) = $ constant in Equation 5.44, we obtain:

$$\frac{(x - \mu_x)^2}{\sigma_x^2} + \frac{(y - \mu_y)^2}{\sigma_y^2} = c \tag{5.45}$$

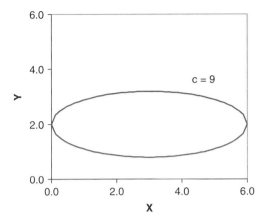

Figure 5.12   Ellipse obtained cutting the bivariate Gaussian density surface with a horizontal plane.

which is the equation of an ellipse centred in the point of coordinates (μ_x, μ_y) (see Figure 5.12 showing a section of a probability density function of a bivariate Gaussian r.v. with parameters $\mu_x = 3, \mu_y = 2, \sigma_x = 1, \sigma_y = 0.4$).

   If the two components are s-dependent, it is necessary to make a reasoning leading to the following formulation of the joint probability density function:

$$f_{XY}(x, y) = \frac{1}{2\pi \cdot \sigma_x \sigma_y \sqrt{1 - \rho^2}} \exp \left[-\frac{1}{2(1 - \rho^2)} \left(\frac{(x - \mu_x)^2}{\sigma_x^2} - \frac{2\rho x y}{\sigma_x \sigma_y} + \frac{(y - \mu_y)^2}{\sigma_y^2} \right) \right]$$

which reduces to the formula (5.44) by setting $\rho = 0$.

   For $\rho \to \pm 1$ the joint probability density function tends to diverge around the line that defines the linear relationship between the two variables.

EXAMPLE

From the height–weight data collected on the sample of students, we could estimate a correlation coefficient $\hat{\rho} = 0.74$. The fairly high positive correlation may be considered as a confirmation of a cause–effect relationship, which we could imagine in this case. If we assume also that height and weight follow a normal distribution over the population of students, then the joint distribution is bivariate normal with s-dependent marginals. The knowledge of the distribution is completed with the estimation of parameters $\mu_x, \mu_y, \sigma_x, \sigma_y$. From data analysis we get $\hat{\mu}_x = 171, \hat{\mu}_y = 64.4$, $\hat{\sigma}_x = 7.58, \hat{\sigma}_y = 11.10$.

   Knowing the joint probability distribution may be useful for certain evaluations. For example, if we wish to design some features of a new campus bus, we may need to estimate the probability that a student is both taller than 185 cm and weighs more than 80 kg. Otherwise, we may be interested to estimate a bivariate *region* weight–height within which there is a high probability (e.g. 0.95) to observe subjects of the population.

5.1.3.7   Covariance and correlation matrix

In the bivariate case, if we have a couple of component variables we can determine a single covariance:

$$\text{Cov}\{X, Y\} = \text{E}\{(X - \mu_x)(Y - \mu_y)\}$$

From such a definition, is evident that:

$$\text{Cov}\{Y, X\} = \text{Cov}\{X, Y\}$$

$$\text{Cov}\{X, X\} = \text{V}\{X\}$$

In the multivariate case we have a vector of random variables $\underline{X}^{\text{T}} = (X_1, X_2, \ldots, X_k)$. If we want to consider covariance aspects in the analysis, we can condense such information into a single matrix called the *covariance matrix*:

$$\underline{\underline{\Sigma}} = \begin{bmatrix} \text{Cov}\{X_1, X_1\} & \text{Cov}\{X_1, X_2\} & \ldots & \text{Cov}\{X_1, X_k\} \\ \text{Cov}\{X_2, X_1\} & \text{Cov}\{X_2, X_2\} & \ldots & \text{Cov}\{X_2, X_k\} \\ \ldots & \ldots & \ldots & \ldots \\ \text{Cov}\{X_k, X_1\} & \text{Cov}\{X_k, X_2\} & \ldots & \text{Cov}\{X_k, X_k\} \end{bmatrix} \tag{5.46}$$

Using the properties listed above we see that:

$$\underline{\underline{\Sigma}} = \begin{bmatrix} \text{V}\{X_1\} & \text{Cov}\{X_1, X_2\} & \ldots & \text{Cov}\{X_1, X_k\} \\ \ldots & \text{V}\{X_2\} & \ldots & \text{Cov}\{X_2, X_k\} \\ \ldots & \ldots & \ldots & \ldots \\ \ldots & \ldots & \ldots & \text{V}\{X_k\} \end{bmatrix} \tag{5.47}$$

Equation 5.47 explains why it is also called a *variance-covariance matrix*. The covariance matrix is a symmetric square matrix having the variances of the components of the multivariate r.v. on the main diagonal. It summarises the variation of the k-varied system of variables.

We can also make a further summary from it.

A summary measure of the covariance matrix is given by its trace (sum of the diagonal elements). It is equal to the sum of eigenvalues of the matrix:

$$\text{tr}\left(\underline{\underline{\Sigma}}\right) = \text{V}\{X_1\} + \text{V}\{X_2\} + \cdots + \text{V}\{X_k\} = \lambda_1 + \lambda_2 + \cdots + \lambda_k \tag{5.48}$$

Another summary measure of the covariance matrix is given by its determinant, which is equal to the product of the eigenvalues:

$$\det\left(\underline{\underline{\Sigma}}\right) = \lambda_1 \lambda_2 \ldots \lambda_k \tag{5.49}$$

From the covariance matrix it is possible to define a correlation matrix:

$$\underline{\underline{P}} = \begin{bmatrix} 1 & \rho_{1,2} & \ldots & \rho_{1,k} \\ \ldots & 1 & \ldots & \rho_{2,k} \\ \ldots & \ldots & \ldots & \ldots \\ \ldots & \ldots & \ldots & 1 \end{bmatrix} \tag{5.50}$$

Also the correlation matrix is a symmetric square matrix of size k, having all '1' on the main diagonal. The trace of the correlation matrix is equal to k.

It can be shown that the determinant of the correlation matrix is an index ranging in $(0,1)$.

The correlation matrix of a set of uncorrelated variables should have k identical eigenvalues (equal to $1/k$). Therefore, the number of eigenvalues exceeding that value is indicative of how many linear relationships are present, which can reduce the dimensionality of the problem (in terms of variables).

5.1.3.8   Autocorrelation in time series and Durbin–Watson test

Correlation has a central role in the study of time series (see Chapter 3).

Usually the study of time series passes through the formulation of statistical models whose validity must then be verified on the basis of available data. In these models there is always a deterministic component and a random component, as previously seen in the regression models.

Suppose we observe a time series at certain monitoring times t_i $(i = 1, 2, \ldots)$.

We define $Y_i = Y(t_i)$ the variable observed at t_i. Suppose that the deterministic component is of the type $a + bt_i$, that is establishing a linear relationship between the observed variable and time. In the observation of the variable there is an error, so the model is written as follows:

$$Y_i = a + bt_i + \varepsilon_i \tag{5.51}$$

The observational error can be decomposed into a pure error component and a component taking into account the dependence on previous observations:

$$Y_i = (a + bt_i) + \rho_1 Y_{i-1} + \rho_2 Y_{i-2} + \cdots + \rho_k Y_{i-k} + v_i \tag{5.52}$$

The coefficients $\rho_1, \rho_2, \ldots, \rho_k$ are called *autocorrelation coefficients* of order 1, 2, $\ldots$, k.

The random variable v_i is usually called *white noise*. There is often a hierarchy of autocorrelation, in the sense that the first-order autocorrelation is generally higher than the second-order autocorrelation, and so on.

We can also define an autocorrelation function and an autocorrelation matrix.

The autocorrelation function defines the trend of the autocorrelation coefficient for varying *lag*, that is the time interval between an observation and the next. Such a trend is generally decreasing.

In the study of a time series the deterministic component is usually estimated at the first stage, so we normally arrive at a situation like this:

$$Z_i = \rho_1 Z_{i-1} + \rho_2 Z_{i-2} + \cdots + \rho_k Z_{i-k} + v_i \tag{5.53}$$

where Z_i are the residuals from the deterministic model.

5.1.3.8.1 Estimation of the first-order autocorrelation If we assume that the auto-correlation of order greater than 1 is negligible, we get a model like this:

$$Z_i = \rho_1 Z_{i-1} + v_i$$

Based on the data, we can verify the presence of first-order autocorrelation using the Durbin–Watson test.

The statistic for the test is provided by the following expression:

$$DW = \frac{\sum_{i=2}^{n} (z_i - z_{i-1})^2}{\sum_{i=1}^{n} z_i^2} \tag{5.54}$$

The critical values of the test vary with the number of parameters estimated in the deterministic component of the model and with the number of observations. The critical values are reported in tables as lower threshold D_L and higher threshold D_U.

Table 5.7 provides the threshold values for a number of parameters of the deterministic model equal to two (a straight line), for two common values of the Type I risk and a number of observations n from 15 to 200.

Table 5.7   Critical values D_L and D_U for the Durbin-Watson test.

n	$\alpha = 0.01$		$\alpha = 0.05$	
	D_L	D_U	D_L	D_U
15	0.81	1.07	1.08	1.36
20	0.95	1.15	1.20	1.41
25	1.05	1.21	1.29	1.45
30	1.13	1.26	1.35	1.49
40	1.25	1.34	1.44	1.54
50	1.32	1.40	1.50	1.59
70	1.43	1.49	1.58	1.64
100	1.52	1.56	1.65	1.69
150	1.61	1.64	1.72	1.75
200	1.66	1.68	1.76	1.78

Operationally, we proceed as follows:

1. calculate the statistic DW

2. calculate $D = \min(DW, 4 - DW)$

3. compare D with the threshold values of Table 5.7, and:

   a. If $D < D_L$ it is possible to argue that there is positive first-order autocorrelation

   b. If $D > D_U$ there is no first-order autocorrelation

   c. If $D_L \leq D \leq D_U$ we cannot say anything because the test is not conclusive.

5.2 To monitor and keep processes under control

Monitoring a process means observing one or more variables of interest related to the process over time. Imagine for simplicity that it is a single variable. Assume that the variable is affected by variability, so it will be denoted by $X(t)$. It is a *random function* (of time).

The monitoring intervals may be constant or variable. If a random function is observed at discrete times we get a time series (see Chapter 3). In this case the deterministic variable t takes on a discrete number (finite or infinite) of values $t_0, t_1, \ldots, t_n$.

Often we must ensure that the variable X satisfies some specifications over time, for example that it does not exceed prefixed limits (*specification limits*).

EXAMPLE

Imagine the production process for a soft drink. For the production process, certain ingredients (inputs) are used in certain quantities. The final product (output) is the soft drink. Some characteristics of the drink, such as the calories, should be kept within certain *specification limits*. These characteristics should therefore be monitored (Figure 5.13).

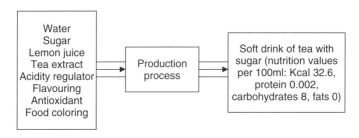

Figure 5.13   Production process of a soft drink of tea.

Let us turn our attention to a single variable. Depending on the criticality of the variable we decide how frequent the monitoring should be.

If the manufacturing plant like the one in the example always works under identical conditions, we could imagine that the monitored variable keeps constant over time. However in reality, even if the conditions in which the process takes place are controlled, this variable will oscillate around a mean value. In this case we would say that the process is in ideal conditions and *stationary*.

A random function that describes the evolution of the monitored variable when the process is under ideal conditions is called *stationary*. A *section* of the random function at a given time t^* is a r.v. $X(t^*)$. If the random function is stationary, the sections at different monitoring times will be identically distributed.

If the monitored variable can be described as a stationary random function, then it is possible to adopt the control chart technique. This will be the subject of Section 5.2.2.

5.2.1 Process capability

Process capability is the attitude of a process under study to ensure uniform output over time. This attitude depends on the natural variability of the process. In the event that the process is stationary, and its section can be modelled as a Gaussian r.v. with parameters μ and σ, the probability of observing values of the variable within the interval ($\mu - 3\sigma$, $\mu + 3\sigma$) is equal to 0.997. This interval will be hereinafter called the *natural variability interval*. By extension, even if the monitored variable is not Gaussian, the interval:

(expected value $-$ 3 · standard deviation; expected value $+3$ · standard deviation)

$$(5.55)$$

will be called the natural variability interval.

5.2.1.1 Process capability indices

Process capability indices provide a summary on how much the process is able to produce outputs that meet the specification limits. The process capability assessment should be done before the process is monitored with control charts.

If we assume that the process is stationary and that the descriptive section is a Gaussian r.v. with parameters (μ, σ), the simplest proposal of the process capability index is as follows:

$$C_p = \frac{USL - LSL}{6\sigma} \qquad (5.56)$$

This index is simply the ratio between the widths of the specification interval and the natural variability interval. The idea is shown in Figure 5.14.

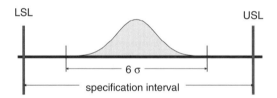

Figure 5.14   Visualisation of the C_p index rationale.

Traditionally, if $C_p \geq 1.33$ we say that the process has a satisfactory capability; if $1 \leq C_p < 1.33$ we say that the capability is sufficient, and finally if $C_p < 1$ the process capability is insufficient, since inevitably many produced units do not comply with the prescribed specification limits. However, in accordance with the Six Sigma principles, we will have a fully satisfactory capability only if $C_p \geq 2$.

The C_p index is a simple indicator, but it has a flaw because it takes into account only the natural variability of the process, without considering its 'centring'. In fact, as can be seen from Equation 5.56 and Figure 5.14, it does not consider whether or not the process mean is centred on the target, which often coincides with the midpoint of the specification interval.

If we want to take into account the centring of the process, it is preferable to use the index defined as follows:

$$C_{pk} = \min\left(C_{pu}, C_{pl}\right) \tag{5.57}$$

where:

$$C_{pu} = \frac{USL - \mu}{3\sigma} \qquad C_{pl} = \frac{\mu - LSL}{3\sigma} \tag{5.58}$$

The C_{pk} index therefore considers both dispersion and centring of the process, assuming of course that the ideal situation occurs when the process is centred on the target ($\mu = ob$, see Figure 5.15).

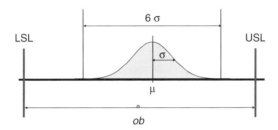

Figure 5.15   Visualisation of the C_{pk} index rationale.

For a Six Sigma process, admitting a shift of 1.5σ of the process expected value (see Section 1.2.3), the $C_{pk} \geq 1.5$.

Another index that takes into account both dispersion and centring is as follows:

$$C_{pm} = \frac{USL - LSL}{6\sqrt{\sigma^2 + (\mu - ob)^2}} \tag{5.59}$$

In the C_{pm} formula (5.59) the quantity $\sigma^2 + (\mu - ob)^2$ appears in the denominator. We will see later (Section 5.3.1), that it is linked to the concept of *expected loss* due to variation.

Yet another index, resulting from the combination of C_{pk} and C_{pm}, is as follows:

$$C_{pmk} = \frac{\min(USL - \mu; \mu - LSL)}{3\sqrt{\sigma^2 + (\mu - ob)^2}} \tag{5.60}$$

For a Six Sigma process, we deduce that $C_{pm} \geq 1.11$ and $C_{pmk} \geq 0.83$.

In general, the capability indices give good summary indications about the status of a process, with the advantage of being dimensionless and easily interpretable.

5.2.2   Online process control and main control charts

The statistical process control and the well-known *control charts* were introduced around the 1930s by W.A. Shewhart. Control charts provide a graphical representation of the course of a process (its status and its trend), and are very useful to those who must ensure consistency of a process variable over time.

Control charts are Cartesian graphs that show the monitoring times on the horizontal axis and the values of the monitored variable on the vertical axis. In addition to the specification limits (which can also be not represented on the chart), three horizontal lines are usually represented on the chart, from top to bottom (see Figure 5.16):

- an upper control limit (UCL) at $+3$ times the standard deviation of the monitored variable from its expected value;

- a central line (CL), at the expected value of the monitored variable;

- a lower control limit (LCL) at -3 times the standard deviation of the monitored variable from its expected value.

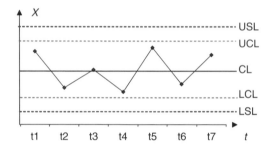

Figure 5.16   Specification limits and control limits.

Control charts are divided into two main categories.

1. *Control charts for continuous variables*: here we report only the control chart for the average ($\overline{X}$ chart) and the chart for the standard deviation (S chart).

2. *Control charts for discrete variables*: here we report only the chart for the fraction non-conforming (p chart), the chart for the number of defects (c chart) and the chart for the average number of defects per unit (u chart).

5.2.2.1   Common causes and special causes of variation

We will assume that the monitored variable X is stationary. Suppose also that lower and upper specification limits are prescribed for X (e.g. in agreement with regulations, or imposed by the customer). These limits should never be exceeded.

At one time an event may occur (e.g. a sudden failure of the production line) such that the variable overcomes the specification limits.

We define as *special causes* the phenomena that determine an abrupt *shift* of the expected value of X, or a sudden change (usually an increase) of its dispersion.

Then we define as *common causes* those phenomena that lead to an inevitable but *natural* variation of the observed value of X around its expected value with a nearly constant dispersion. Only common causes act in a stationary process.

Thus a need arises to monitor the variable in order to avoid exceeding the specification limits as a result of the occurrence of special causes.

So, to avoid the risk of exceeding the specification limits it is necessary to set limits more restrictive than the specification limits: these are so-called *control limits*: the UCL and the LCL. They, as mentioned above, together with the CL, characterise a control chart (Figure 5.16).

Before fixing the control limits it is necessary to examine the natural variation of the process, determined by the common causes. In fact, it might be the case that natural variation is such that it is very likely to exceed the specification limits and therefore it is completely useless to carefully set control limits. In this case, we should try, if possible, to reduce the process variation or (if possible) to review the specification limits.

Even before the comparison with the control limits, the plot of time series of the monitored variable can provide useful information such as a drift of process in average or dispersion (Figure 5.17), that is a slow and progressive deterioration, which may be due for example to wear.

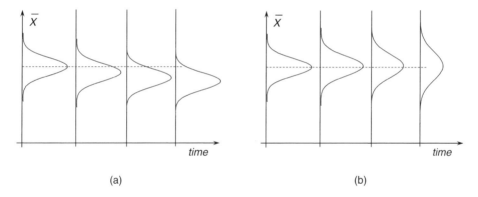

(a)   (b)

Figure 5.17   Drift of the process in mean (a) and dispersion (b).

5.2.2.2   Control chart for the average, or '$\overline{X}$ chart'

Suppose that the monitored process is continuous and stationary. Suppose we can assume a Gaussian model for the $X(t)$.

Suppose that at various monitoring times $t_1, t_2 \ldots, t_i$ a random sample of the variable X is drawn. In statistical process control language such samples are called *rational subgroups*. Suppose that the sample size is constant and equal to n.

For every monitoring time we consider the sample average:

$$\overline{X}_1 = \frac{1}{n}\sum_{j=1}^{n}X_{1j} \quad \overline{X}_2 = \frac{1}{n}\sum_{j=1}^{n}X_{2j} \ldots \overline{X}_i = \frac{1}{n}\sum_{j=1}^{n}X_{ij}$$

If $X(t)$ is stationary and Gaussian, the sample average will also be stationary and Gaussian (Figure 5.18):

$$\overline{X}_i \sim Gaussian\left(\mu, \frac{\sigma}{\sqrt{n}}\right).$$

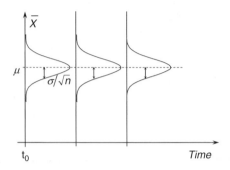

Figure 5.18   The stationary process of the sample average.

Setting the UCL and LCL according to the logic (Equation 5.55) means properly considering the process as 'in control' in 99.7 % of cases. However, there is always a 0.003 probability (0.3 % of cases) to observe values outside the control limits even though there has been no change in parameters (μ, σ).

Obviously the control limits can be relaxed (e.g. ± 4 sigma), but on the other hand if we do so, the control limits could overcome the specification limits (this is a very rare event for a Six Sigma process), and most of all we could not be able to notice the occurrence of special causes.

In the '$\overline{X}$ chart' we plot the observed sample averages.

The CL is placed at the expected value of the sample average (μ), the LCL equal to $\mu - \dfrac{3}{\sqrt{n}}\sigma$ and the UCL equal to $\mu + \dfrac{3}{\sqrt{n}}\sigma$.

Every time we plot a new point on the chart, it will be compared with the control limits. If the new point is within the limits, the process will be said to be 'in control' and no human intervention is required. Conversely, if the new point is outside the limits, the process will be said to be 'out of control' and an investigation should follow to trace the possible causes. The test may reveal that this is a *false alarm* as a value outside the control limits is conceptually compatible with a stationary process. It is unlikely, but possible.

In fact, the control chart is a hypothesis test repeated at the various monitoring times, in which the null hypothesis represents the status of stationary process.

If the parameters μ and σ are assumed known, it will be a *control chart with pre-scriptions*. However, if the parameters are not known it will be a control chart *without prescriptions*. In this case the CL and the control limits are estimated using the procedure described below.

5.2.2.3   Average run length of the $\overline{X}$ chart

Suppose that at some moment the process mean steps from the value μ to a value $\mu + \delta$, that is it is affected by a shift equal to δ (see Figure 5.19). The following problem arises:

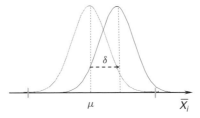

Figure 5.19   Abrupt shift of the mean due to a special cause.

if we use the control chart with limits set according to the previously seen formulas, how many inspections we will have to do before discovering the shift?

Since the shift, $\overline{X}_i$ will no longer have mean μ, but $\mu + \delta$ so the probability of observing a point within the control limits previously set is:

$$\Pr\left\{\text{LCL} \leq \overline{X}_i \leq \text{UCL}\right\} = \Pr\left\{\frac{\text{LCL} - (\mu + \delta)}{\sigma/\sqrt{n}} \leq \frac{\overline{X}_i - (\mu + \delta)}{\sigma/\sqrt{n}} \leq \frac{\text{UCL} - (\mu + \delta)}{\sigma/\sqrt{n}}\right\}$$

$$= \Pr\left\{-3 - \frac{\delta}{\sigma/\sqrt{n}} \leq Z \leq +3 - \frac{\delta}{\sigma/\sqrt{n}}\right\} = \pi \qquad (5.61)$$

where Z is a Standard Gaussian r.v.

We denote by Y the r.v. that counts the number of inspections carried out before we detect the shift. The probability that Y is equal to a specific value y is given by the product of y times the probability π that, despite the shift, the observed value falls within the control limits, multiplied by the probability $(1 - \pi)$ that the $(y + 1)$-th time we detect the shift:

$$\Pr\{Y = y\} = \pi^y (1 - \pi) \qquad (5.62)$$

The expression (5.62) coincides with the probability mass function of a model of discrete r.v. called *Geometric*. It can be shown that the expected value of Y is:

$$\text{E}\{Y\} = \frac{1}{1 - \pi} \qquad (5.63)$$

This is therefore the expected number of inspections to be carried out to detect the shift.

This value is a performance indicator of the control chart and is called the *average run length* (ARL). As we can notice from the Equations 5.63 and 5.61, it depends on the size of the shift, on the dispersion of the process and on the rational subgroup size.

5.2.2.4   Control chart for the standard deviation or 'S chart'

It may happen that despite the expected value of the monitored variable keeping constant over time (i.e. it is not affected by any shift), the dispersion around the mean increases, as shown in Figure 5.17b. It is therefore necessary to avoid the possibility that the presence of points outside the control limits of the $\overline{X}$ chart is attributed to a shift of the expected value μ, when it is in fact due to a change in the dispersion.

For this reason it is good to plot near the $\overline{X}$ chart (usually above or below), a second chart that allows us to monitor the dispersion of the process.

If the size n of the rational subgroup is ≥ 10 we should definitely use the chart for the standard deviation or S chart. Alternatively, for very small values of n we may use the chart for the range.

We call S_i the sample standard deviation of the rational subgroup at the monitoring time t_i:

$$S_i = \sqrt{\frac{1}{n-1} \sum_{j=1}^{n} (X_{ij} - \overline{X}_i)} \tag{5.64}$$

The control chart is constructed plotting a CL at the expected value of sample standard deviation S and the control limits at ± 3 sigma (here sigma means the standard deviation of S).

It can be shown that the expected value of S is equal to:

$$E\{S\} = c_4 \cdot \sigma \tag{5.65}$$

being:

$$c_4 = c_4(n) = \frac{\Gamma\left(\frac{n}{2}\right)\sqrt{2}}{\Gamma\left(\frac{n-1}{2}\right)\sqrt{n-1}}$$

while the standard deviation of S is equal to:

$$\sqrt{V\{S\}} = \sigma \sqrt{1 - c_4^2} \tag{5.66}$$

For common values of n the coefficient c_4 is reported in tables.

Thus, the CL is drawn at $c_4 \cdot \sigma$.

The LCL and UCL, derived in accordance with Equation 5.55, are:

$$LCL = \sigma \left(c_4 - 3\sqrt{1 - c_4^2} \right)$$

$$UCL = \sigma \left(c_4 + 3\sqrt{1 - c_4^2} \right)$$

5.2.2.5 Construction of the $\overline{X}$ and S charts without prescriptions

In general the values μ and σ are not known when we start our control activity. In this case we must 'initialise' the chart through an estimation of μ and σ.

The estimation can be made on the basis of some preliminary sampling. This initial set-up is sometimes referred to as *Phase I* to distinguish it from *Phase II*, the actual control phase.

Suppose that the Phase I consists of k samples. We start by initialising the S chart.

If S_i, given by Equation 5.64, is the sample standard deviation in the i-th rational subgroup, the average of the standard deviations of the samples in Phase I will be:

$$\overline{S} = \frac{S_1 + S_2 + \cdots + S_k}{k}$$

It provides an estimate of $E\{S\}$ and it is the value at which we draw the CL of the chart. According to Equation 5.65 we can estimate the unknown parameter σ:

$$\hat{\sigma} = \frac{\overline{S}}{c_4} \tag{5.67}$$

The UCL and LCL of the S chart are:

$$\text{UCL} = \overline{S} + 3\frac{\overline{S}}{c_4}\sqrt{1 - c_4^2}$$

$$\text{LCL} = \overline{S} - 3\frac{\overline{S}}{c(n)}\sqrt{1 - c_4^2}$$

To build the $\overline{X}$ chart without prescriptions we must first estimate the parameter μ on the basis of Phase I samples:

$$\hat{\mu} = \overline{\overline{X}} = \frac{\overline{X}_1 + \overline{X}_2 + \cdots + \overline{X}_k}{k} \tag{5.68}$$

Equation 5.68 will be adopted for the calculation of the CL of the chart.

Having estimated the parameter σ by Equation 5.67, the UCL and LCL of the $\overline{X}$ chart will be calculated as:

$$\text{UCL} = \overline{\overline{X}} + 3\frac{\overline{S}}{c_4\sqrt{n}}$$

$$\text{LCL} = \overline{\overline{X}} - 3\frac{\overline{S}}{c_4\sqrt{n}}$$

5.2.2.6 Control chart for the fraction non-conforming or 'p chart'

The p chart allows us to monitor the fraction of defective items characterising a process (which could be a production process or service).

At every monitoring time, we examine a rational subgroup of n units and verify for each unit if a certain condition is respected or not, qualifying the unit as conforming (good) or non-conforming (no good).

Suppose, as usual, that the process is stationary. In this case the probability of observing a non-conforming unit is constant and equal to p.

If the sampling is random, the number of defective units in the rational subgroup is a r.v. X distributed according to a binomial model:

$$\Pr\{X = x\} = \binom{n}{x} p^x (1 - p)^{n-x} \qquad x = 0, 1, \ldots, n \tag{5.69}$$

The fraction non-conforming X/n is a r.v. which follows the so-called *Bernoulli proportion* model (a model of r.v., immediately derived from the binomial, which assumes a discrete number of values between 0 and 1).

From the formulations of the binomial model (see Chapter 3) it follows that:

$$E\{X/n\} = p \tag{5.70}$$

$$\sqrt{V\{X/n\}} = \sqrt{p(1-p)/n} \tag{5.71}$$

Still adopting the criterion (5.55), the parameters of the control chart are:

$$UCL = p + 3\sqrt{p(1-p)/n}$$

$$CL = p$$

$$LCL = p - 3\sqrt{p(1-p)/n}$$

At every monitoring time we will take a rational subgroup of size n, count the number of defective units in the rational subgroup, divide it by n and post this value onto the control chart, checking that it is within the control limits and examining the trend of the time series.

If the parameter p is not known, then we can make an estimation in Phase I. The estimate of p is equal to the average fractions of non-conforming units in the k preliminary samples.

When the rational subgroup size is not constant, we can adopt variable control limits:

$$p \pm 3\sqrt{p(1-p)/n_i}$$

where n_i is the rational subgroup size at the monitoring time t_i.

5.2.2.7   Control chart for the number of defects or 'c chart'

If at each monitoring time we count the number of defects present in a given unit of material (e.g. the number of defects in a welded joint), then it is possible to construct a control chart for the number of defects.

Suppose that the number of defects is rather small, such that we may assume a Poisson model for the discrete r.v. X which counts the number of defects.

If we call c the expected number of defects then we know that:

$$E\{X\} = c; \quad V\{X\} = c; \quad \sqrt{V\{X\}} = \sqrt{c} \tag{5.72}$$

Still adopting the criterion (5.55), the parameters of the control chart for the number of defects are:

$$UCL = c + 3\sqrt{c}$$

$$CL = c$$

$$LCL = c - 3\sqrt{c}$$

If the parameter c is not known, then we may estimate it in Phase I before putting the process under control. The estimate of c will be given by the average number of defects found in the preliminary samples.

At each monitoring time we will count the number of defects in the analysed material and we will plot this value onto the control chart, compare it to the limits and examine the trend of the series over time.

5.2.2.8   Control chart for the average number of defects per unit or 'u chart'

Suppose that the defects are still rare events, but in this case, unlike the previous case, we have multiple units of material, say n, on which we count the defects. If, as in the previous case, we choose c to be the total number of defects, the average number of defects per unit of material is:

$$u = \frac{c}{n}$$

From which, adopting still the criterion (5.55), the parameters of the control chart for fraction that is defective will be:

$$UCL = u + 3\frac{\sqrt{u}}{\sqrt{n}}$$

$$CL = u$$

$$LCL = u - 3\frac{\sqrt{u}}{\sqrt{n}}$$

If the parameter u is not known, then we can estimate it in Phase I, before putting the process under control:

$$\hat{u} = \frac{1}{k}\sum_{i=1}^{k}(d_i/n)$$

where k is the number of preliminary samples of Phase I.

5.2.2.9   Other control charts

In addition to the control charts presented so far, there are many others, which differ according to the characteristic of the monitored variable.

Here we provide an example of a control scheme which can be useful when it is not possible to have a rational subgroup at each monitoring time, but only one value of the monitored variable is given.

EXAMPLE

Some years ago we collaborated on a project aimed at the implementation of a self-assessment model at 'Oasi Maria SS', a health care organisation devoted to health research and care. One of the main activities of the project consisted in the definition of a control system for two key processes: diagnosis and rehabilitation. For this purpose a set of indicators was defined. Subsequently, data concerning two years (2005 and 2006) were collected and analysed through appropriate control charts.

As an example, we report a summary of the results obtained through the analysis of the indicator called 'D1.1BA', which is calculated as the difference between

(i) the number of actual admissions (of patients) over the month and (ii) the number of planned admissions for the same month, and it measures the management effectiveness of the waiting list.

We did some preliminary considerations: the two components (i) and (ii) can be thought of as being random variables distributed according to a Poisson model; the high values of the observed frequencies allow us to treat the data as coming from a Gaussian distribution.

After having examined the autocorrelation of each component and the correlation between the two components, and checked the Gaussian assumption, we built individual and moving range charts (see Figure 5.20).

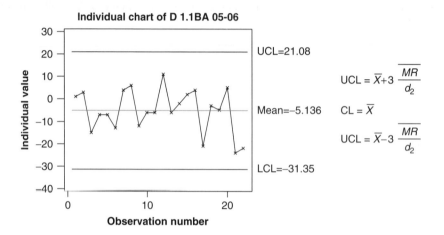

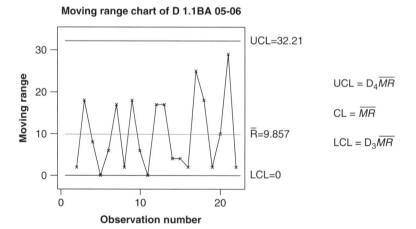

Figure 5.20   Example of individual X – moving range charts.

These control charts for variables are useful when the available data are individual data, in our case one number per month was provided. The individual X chart monitors the position of the process over time, based on a single observation. The moving range chart monitors the variation between consecutive observations. When constructing both charts it is necessary to calculate the moving range. In our case it is based on two successive observations:

$$MR_i = |x_i - x_{i-1}|$$

The formulas for calculating the CL and the control limits are also shown in Figure 5.20. The constants d_2, D_3 and D_4 depend on the sample size.

When the moving range is based on two observations then $d_2 = 1.128$, $D_3 = 0$ and $D_4 = 3.269$.

In Table 5.8 values of d_2, D_3 and D_4 for sample sizes from 2 to 10 are provided.

Table 5.8   Constants for the 'individual X – moving range' charts.

Sample size	d_2	D_3	D_4
2	1.128	0	3.269
3	1.693	0	2.574
4	2.059	0	2.282
5	2.326	0	2.114
6	2.534	0	2.004
7	2.704	0.076	1.924
8	2.847	0.136	1.864
9	2.97	0.184	1.816
10	3.078	0.223	1.777

5.2.3   Offline process control

The statistical control of a process can be made either with the process running or 'offline', having in this case the so-called acceptance control. In theory, the acceptance control may be carried out on 100 % of the process output, but more usually it is made either on a random sample of produced lots, or continuously, depending on the type of process (Figure 5.21).

In the past the sample inspection of incoming lots from a supplier was very popular in industry. In recent years it has become more natural for customers and suppliers to cooperate to improve or at least control the quality of the traded goods and services, thus reducing the use of inspection for that purpose.

On the other hand, the techniques that we discuss in this section are general and can also be applied in different contexts from the inspection of products by a customer.

Acceptance control is a set of techniques based on which a lot of production (defined as a quantity of a good produced from a single supplier under conditions that are presumed

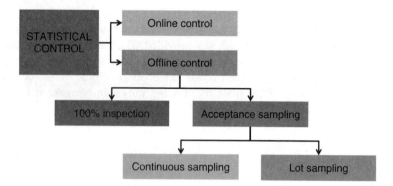

Figure 5.21   Classification of statistical control.

to be homogeneous) is accepted or rejected on the basis of the outcome of an inspection of a sample of the lot, in particular the proportion of defective units it contains.

As for the online process control, in the offline control we distinguish between control for variables and control for attributes.

5.2.3.1   Sampling plans in acceptance control for attributes

A sampling plan defines the procedures for sample extraction and decision rules for acceptance or rejection of the lot based on the outcome of the sample check. The elements that characterise a sampling plan are:

- N = lot size

- n = sample size ($n < $ N)

- n_a = acceptance number ($n_a < n$): it is the maximum acceptable number of defective units detected in the sample for which a lot is surely accepted;

- n_r = rejection number ($n_a < n_r < n$): it is the minimum number of defective units detected in the sample for which the lot is surely rejected.

Sampling plans can be classified as follows:

- *simple* when they are based on the extraction of a single sample;

- *double* if they foresee the extraction of two samples;

- *multiples* if they foresee the possible extraction of more than two samples.

Figures 5.22 and 5.23 depict the decisional process schemes in the case of single and double sampling plans, respectively.

So-called *sequential* sampling plans also exist, in which the units are taken one at a time from the lot and, after the inspection of each unit, a decision is made on the acceptance/rejection of the lot or the extraction of a further unit.

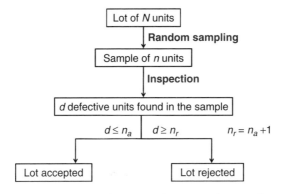

Figure 5.22   Decisional process for a simple sampling plan.

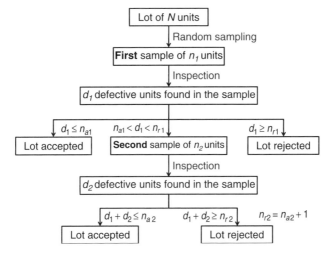

Figure 5.23   Decisional process for a double sampling plan.

5.2.3.2   The operating characteristic curve

The operating characteristic curve characterises a sampling plan. It graphically shows the probability that the lot is accepted (P_a) as a function of the fraction of defective (p) in the same lot. The operating characteristic curve expresses the discriminatory ability of a sampling plan.

5.2.3.2.1   Ideal operating characteristic curve   Theoretically, in a plan that perfectly discriminates between good and bad lots, we surely accept lots with a defect rate $\leq p_0$, and surely reject lots with defect rate $> p_0$ (p_0 is agreed between the customer and the supplier).

The so-called ideal operating characteristic curve represents the above conditions. However, to obtain it, the inspection needs to be conducted on 100 % of produced units.

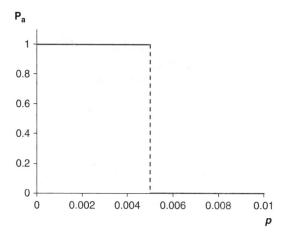

Figure 5.24   Ideal operating characteristic curve ($p_0 = 0.005$).

This will become clear later. Figure 5.24 shows an ideal operating characteristic curve for a p_0 equal to 0.005.

5.2.3.2.2   Construction of the OC curve for a single-sampling plan   Remember that the characteristics of the sampling plan are: N, n, n_a.

Suppose we draw a sample of size n from a finite population of size N (the lot) containing D defective units. The probability of drawing a sample with d defective units is provided by the so-called *hypergeometric* formula:

$$\Pr\{d \text{ defective units in the sample}\} = \frac{\binom{D}{d}\binom{N-D}{n-d}}{\binom{N}{n}} \tag{5.73}$$

This probability is the ratio between the number (favourable cases) of all possible samples of n units that contain d defective, and the number (possible cases) of all samples of n units that can be extracted from the lot of N units.

The probability that the lot is accepted is given by:

$$P_a = \Pr\{d \le n_a\} = \sum_{d=0}^{n_a} \frac{\binom{D}{d}\binom{N-D}{n-d}}{\binom{N}{n}} \tag{5.74}$$

By Equation 5.74, for different values of $p = D/N$, we construct the so-called operating characteristic curve of type A. Figure 5.25 shows four operating characteristic curves of type A for different values of N and n and for a fixed common value $n_a = 0$ (if we find one or more defective units in the sample, the lot is rejected). The figure shows that for the same sample size n, the lot size N has a marginal effect on P_a, while for the same

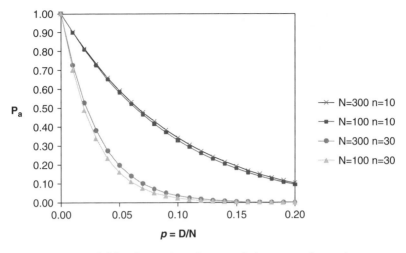

Figure 5.25   Operating characteristic curve of type A.

lot size, the sample size has a noticeable effect. The figure also shows that even for the same ratio n/N (curve with squares and curve with circles), there is a big difference in terms of P_a which, however, depends only on the sample size n.

The curve of type A has a limitation. In fact, it assumes that each lot has the same number D of defective units, which in general is not the case. In fact, the number of defective units in the lot depends on the defect rate of the production process (say p). Therefore it is more correct to refer to p in the calculation of P_a. It can be shown that if we refer to p we get:

$$P_a(p) = \sum_{d=0}^{n_a} \binom{n}{d} p^d (1-p)^{n-d} \tag{5.75}$$

The Equation 5.75 shows that P_a is a cumulative binomial probability distribution (see Chapter 3) with parameters n and p. For different values of p it provides the so-called operating characteristic curve of type B (Figure 5.26). We can see that it does not depend on lot size N.

In Figure 5.27 is shown a comparison between the curves of type A and type B for a fixed and common value $n_a = 1$ (if there are two or more defective units in the sample, the lot will be rejected). Thus we see that in both cases for $n > 0$ the curve has a change of curvature, which will be necessary to understand the subsequent considerations.

5.2.3.3   The agreement between supplier and customer

Normally, the supplier requires from a sampling plan, with his commitment to provide lots of defect rate p not exceeding a certain threshold (p_0), that the P_a is not lower than a certain predetermined value, say $1 - \alpha$.

The maximum value of the probability of rejection is called the supplier's risk and is indicated by α. In practice, values adopted for α range from 0.05 to 0.01.

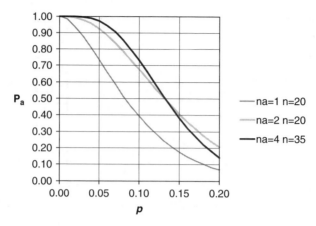

Figure 5.26   Operating characteristic curve of type B for different values of n *and* n_a.

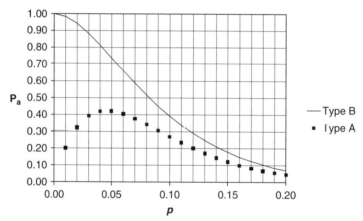

Figure 5.27   Comparison between an operating characteristic curve of type A (N = 100, n = 20, n_a = 1) *and one of type B (* n_a = 1).

Fixing α for a given value of p_0 means forcing the operating characteristic curve to go through the point with coordinates $(p_0, 1 - \alpha)$. For fixed α we can have different sampling plans that have different degrees of discrimination for $p > p_0$.

Figure 5.28 shows the operating characteristic curves of three sampling plans, with all three passing through $(p_0 = 0.01; 1 - \alpha = 0.80)$. This means that if a supplier producing with a defect rate of 1 % wants to have a probability of acceptance at least 80 % there are different possible sampling plans meeting such a requirement. Obviously, the supplier will prefer the sampling plan corresponding to the blue curve of the graph ($n = 82$, $n_a = 1$), which implies a higher probability of acceptance if p becomes higher than p_0.

So, for the unambiguous definition of a sampling plan, there should be another constraint. This constraint is given by the so-called *customer's risk*, β. This is the maximum value of P_a for a given value $p_1 > p_0$. Values usually adopted for β range from 0.05 to 0.10.

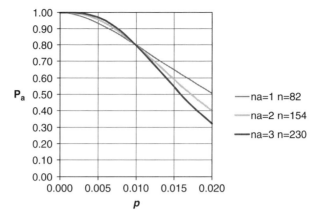

Figure 5.28   Different sampling plans for fixed p_0 *and* $1 - \alpha$.

In this way, the operating characteristic curve passes through two points: the point of coordinates $(p_0, 1 - \alpha)$, where the maximum defect rate guaranteed by the supplier is also known as the acceptable quality level (AQL), and the point of coordinates (p_1, β), where the minimum defect rate that guarantees the customer (p_1) is called the tolerable quality level (TQL).

Once these constraints have been set, then n, n_a and the operating characteristic curve of type B are obtained by solving the following system of equations:

$$1 - \alpha = \sum_{d=0}^{n_a} \binom{n}{d} p_0^d (1 - p_0)^{n-d}$$

$$\beta = \sum_{d=0}^{n_a} \binom{n}{d} p_1^d (1 - p_1)^{n-d}$$

Figure 5.29 shows an operating characteristic curve for a sampling plan in which supplier and customer agree on the following constraints: $(p_0 = 0.01; 1 - \alpha = 0.85)$ and $(p_1 = 0.015; \beta = 0.45)$. We see that to obtain these requirements, the plan requires a sample of 600 units and an acceptance number equal to 8. Obviously, the example allows us to reflect on the fact that if we want both the supplier's and customer's risks to be low, with a steep sloping curve between p_0 and p_1, then the sample size must increase dramatically. At the limit, to obtain the ideal curve (Figure 5.23), the sample size must be infinite (i.e. we must inspect 100 % of the production).

5.3   To improve products, services and production processes

5.3.1   Robustness thinking

Conceiving and developing robust products and processes, including services (hereinafter we will use the term 'systems' to indicate all of them), is more than a technique. It is

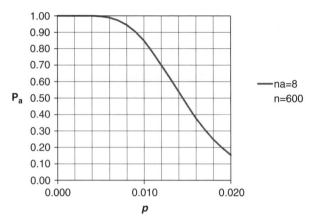

Figure 5.29   Operating characteristic curve with specifications AQL = 0.01, supplier's risk = 0.85, TQL = 0.015 and customer's risk = 0.45.

a way of thinking about the basis of a continuous search for quality and excellence, by focusing on unwanted variation of system performances and trying to reduce it as much as possible.

According to a well-known scheme, we generally distinguish three phases in the development of a new system: *system design*, *parameter design* and *tolerance design* (Figure 5.30).

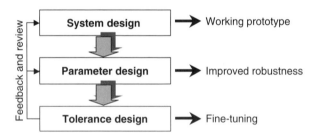

Figure 5.30   The three phases for developing a new 'robust' system.

In system design the aim is to develop a working prototype of the system under study. It can either be physically built one or a virtual one, that is simulated by means of models and computer algorithms. The system design is essentially developed by exploiting technical knowledge and experience.

Parameter design is the first phase where attention is paid to improving system robustness, that is insensitivity to sources of variation, by looking at some selected *key system characteristics* (KSC) and by acting on system factors purposely chosen, called *design parameters*.

KSC are characteristics requiring special attention because their variation may have a negative impact on system functionality, on compliance with regulations, on safety or more generally, on the quality of the system. The concept of KSC is quite well known in business and industry. The KSC can also be the result of a quality function deployment (QFD) study (Section 4.4.3).

Design parameters are factors whose values can be modified in this phase without large expense. Experts select design parameters that could significantly improve the KSC. The effect of changing design parameter values is usually evaluated through experiments aimed at finding the best design solution or *design setting*, that is the best combination of design parameter values.

In this phase, it is also necessary to identify sources of variation – the *noise factors* (NFs) – and understand their impact on the KSC, especially when the system will work in realistic conditions. Once identified, the NFs can be opportunely introduced in the experimental arrangement.

A parameter design phase is generally schematised using a *P-diagram* (Figure 5.31).

Figure 5.31   P-diagram of a parameter design phase.

Usually design parameters are depicted as an input of the P-diagram, but since they are often part of the system itself, we prefer to place them into the system box.

Signal factors can be also considered in a parameter design phase when there is an express interest in the relationship between the KSC and any of them. The signal factors can be varied either in development phases or in usage of the system to achieve an intended value of KSC. Despite a good parameter design phase, it can happen that KSC variation is still too high. In such case, a tolerance design phase is called for.

In general 'tolerance' means an admitted range of variation. Basically, a tolerance range can be equivalent to the distance between lower and upper specification limits (tolerance range = USL – LSL). In a tolerance design phase, tolerances are set or reviewed in order to meet an acceptable level of KSC variation. This phase can call for more expensive measures, as manufacturing cost can increase when tolerances tighten (more accurate manufacturing work, more advanced technologies, more expensive components, etc.).

5.3.1.1   Specific objectives of robustness

It is usual to make the following distinctions, depending on the specific objective to be achieved in relation to the target for the KSC:

- if the target is a finite value, the problem is called *nominal-the-best;*

- if the target is 'the least possible value', the problem is called *the-smaller-the-better;*

- if the target is 'the highest possible value', the problem is called *the-larger-the-better;*

- if we consider systems whose KSC assumes different values depending on the values of a signal factor, and the interest lies on the signal-response relationship, then we refer to *signal-response systems*.

5.3.1.2   An illustrative example throughout this section

The paper helicopter (Figure 5.32) is an *educational toy*, well known in the literature. In fact it is very useful: it has a simple design; it is easy, fast and cheap to make specimens and experiments; its visible performance (free-fall or free-flight) is appealing although the underlying aerodynamic phenomenon is not trivial.

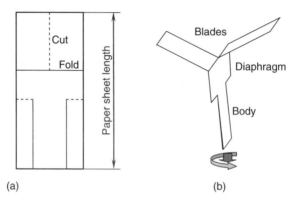

Figure 5.32   The paper helicopter: (a) construction drawing and (b) in free-fall.

The system design phase for the paper helicopter is when an original idea and technical experience led to a working prototype, whose basic design is presented in Figure 5.32a.

Using it as an example, several authors have demonstrated in different ways how, by means of a parameter design phase, it is possible to achieve better performances.

Considering a paper helicopter as a new system in a development phase, we will take it as an example and follow it throughout the process.

Imagine that the experts of the 'paper helicopter company' want to improve the robustness of their product. So they firstly consider all system characteristics which could be of interest for the customer (Table 5.9). They decide to initially focus on the minimisation of the vertical velocity. The flight time from a certain release height is easily measurable, hence they take it as the main KSC. A parameter design phase is carried out, experiments are planned and performed and finally a good design setting is obtained.

5.3.1.2.1   Key system characteristics variation   Experts of the paper helicopter factory, after they have identified a good design setting via a parameter design phase, are not fully satisfied. They believe that the variation of the KSC flight time is still too high and should be further reduced.

5.3.1.2.2   Classification of sources of variation   In general, KSC variation can be due to many causes. A 'classical' and widely accepted classification distinguishes between

Table 5.9   Characteristics of the 'paper helicopter' system.

System characteristics	Measurability	Measurable variables	Priority
Flight regularity (stability)	Qualitative (subjective)	Duration of the transient phase; duration of the transient phase; duration of the steady motion phase height H1 at steady rotation	3
Vertical velocity	Quantitative	Vertical velocity in the steady motion phase; flight time (given H_0) duration of steady rotation	1
Predictability of landing point	Quantitative	Precision; accuracy	2
Simplicity to build	Qualitative	Time and resources necessary to build a prototype (# of lines to cut/fold)	4
Cost (actual cost of the paper)	Quantitative	Dimensions of the prototype; paper type	5

sources manifesting and acting during system production from sources acting during system usage. The sources of variation of the first type determine differences between manufactured specimens (*unit-to-unit variation*). Sources of variation of the second type result in different behaviour of a single specimen when repeatedly used (*in-use variation*). Sources of variation acting during usage can be further distinguished between *external sources* and *internal sources*. External sources could be environmental conditions changing from time to time when the system is used. Internal sources could be the conditions of the system itself, changing from time to time when it is used, due to the action of irreversible and unavoidable physical processes causing wear or degradation.

A summary Ishikawa diagram (see Section 2.1.1) of such a cause-and-effect relationship is presented in Figure 5.33.

Experts took several helicopter specimens manufactured with the same design and the same procedure, and released them only once under identical conditions. They observed different flight times due to the unit-to-unit variation, caused by some factors acting during the production process. Subsequently, in another experiment, they took only one new helicopter specimen and released it several times, in a controlled environment (a purpose-made lab). They observed variation in flight time which could be partially due to the measurement system and partially due to factors acting during the usage and having influence on the flight time, for example uncontrolled air flows. The combined effect in terms of variation of these last two factors is sometimes called *experimental error*.

Another classification of sources of variation is also possible (see the scheme in Figure 5.34). In fact, sources of variation can be known or unknown. If the sources of

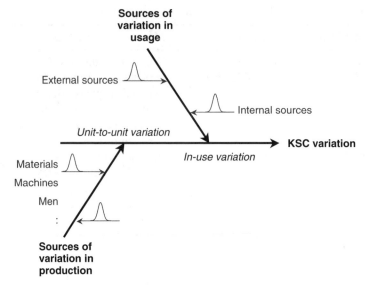

Figure 5.33   Sources of KSC variation.

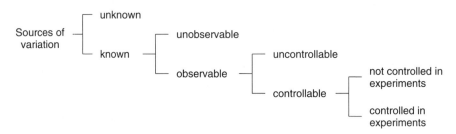

Figure 5.34   Another classification of sources of KSC variation.

variation are known they can be observable, either in experimental situations or during the system usage, if suitable equipment is available to measure their value; otherwise they are unobservable, if their observation requires unavailable or too costly equipment. Notice that the concept of 'observability' is here used in the same way as 'measurability'.

If the sources of variation are observable, they can be either controllable or uncontrollable in experimental situations. By controllability we mean that it is possible to set the factor at any specific and intended value before each experimental run. Finally if a source of variation is controllable, it can either be included as a controlled factor in planned experiments or not.

5.3.1.2.3   Experimental implications of parameter and tolerance design   Here we discuss the experimental implications of parameter design and tolerance design aimed at choosing the best setting of nominal values and tolerance values.

Consider the workflow of Figure 5.30 and the effect of external and internal sources of variation (Figure 5.33).

For illustration purposes let us consider a KSC influenced by only one design parameter and by only one external source of variation.

We assume the design parameter to be a deterministic variable denoted by x. We denote by W the external source of variation (it is a random variable). As a consequence of the random variation of W the KSC is also a random variable, denoted by Y.

The response surface $Y = \varphi(x, W)$ is illustrated in Figure 5.35. The function of one variable $Y = \varphi(x_0, W)$ corresponding to an initial setting x_0 for x, is also represented in the figure (it is the intersection of the surface with a vertical plane passing through x_0).

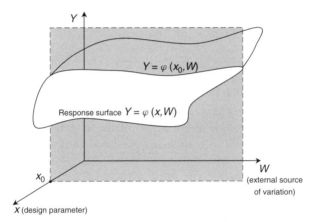

Figure 5.35   Response surface of a KSC, function of a design parameter and an external source of variation.

In parameter design we assume that the design parameter is perfectly controllable. Suppose that two values x_1 and x_2, different from the initial setting, are chosen for x. The functions $Y = \varphi(x_1, W)$ and $Y = \varphi(x_2, W)$ are shown in Figure 5.36a,b, respectively.

When the KSC is repeatedly observed then, due to the random variation of W (note the Gaussian curve on the W axis), variation of Y is observed as a result.

The best design setting (in this case the best design parameter value) is the one making the system most robust, that is the most insensitive to the action of the external source of variation W. From Figure 5.36 it is evident that in correspondence of x_2 less variation of the KSC can be eventually gained. It is also clear that the choice between the two values x_1 and x_2 (or three if we consider also the initial value) is obvious if the aim is to minimise the KSC (the-lower-the-better objective). Otherwise (nominal-the-best, higher-is-better cases), a consideration jointly of the location and dispersion of Y is needed.

As previously mentioned, the search for the best design setting is made via a parameter design phase. When more than one design parameter is involved (as it is often the case) suitable experimental plans are used that accommodate both design parameters and external sources of variations – whenever it is possible (look back at Figure 5.34).

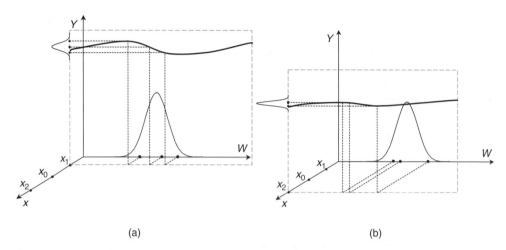

(a)   (b)

Figure 5.36   Sections of the response surface for two settings of the design parameter.

So, in the parameter design phase it is assumed that the design parameter value is fixed and the KSC variation is only due to the action of the external source. In the tolerance design phase, however, the KSC variation is considered as if it was caused by sources of variation acting during the production process and by internal sources of variation acting during usage (Figure 5.33). The direct effect of these is the variation of design parameter values.

Let us now consider the situation in Figure 5.37.

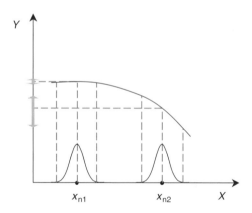

Figure 5.37   Effect of design parameter variation and choice of nominal setting.

From the figure it is evident that if the nominal value of the design parameter is set at x_{n2}, the high sensitivity of the KSC with respect to the design parameter value determines a big variation of the KSC (the function $Y = \varphi(X)$ has high slope in correspondence of $x = x_{n2}$). However, if the nominal value of the design parameter is set at x_{n1}, a lower variation of the KSC is obtainable even with the same tolerance on the design parameter.

The real situation considering KSC variation due to both kind of sources – external sources and those determining design parameter variation – has to be taken into consideration (Figure 5.38).

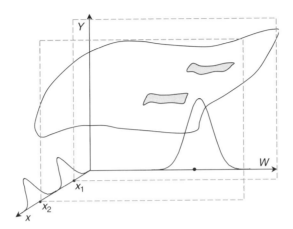

Figure 5.38   KSC variation due to internal and external sources.

The KSC is a function of two random variables, $Y = \varphi(X, W)$. The level of difficulty of the problem is higher. A parameter design phase followed by a traditional tolerance design phase is sub-optimal, as demonstrated by several authors in the literature.

5.3.1.3   Robustness indicators

Once we have understood the aims of robust design, its phases and experimental implications, we should now define the robustness indicators, merit criteria to evaluate the robustness of a system, with particular reference to one or more KSC.

As an example, look at Figure 5.39. Y is a random variable that models a KSC for which a target of 3 and a specification interval (2.6, 3.4) were set. Also assume that Y can be described with two alternative models, a Gaussian distribution of parameters $\mu = 3$; $\sigma = 0.1$ and a uniform distribution in the interval (2.6, 3.4). Both are centred on the target.

If the distribution of Y were the Gaussian, a high percentage of products will fall within specifications, while in the second case we would get 100 % compliance. Which should we prefer? The main difference is that, with the Gaussian model, we get a higher percentage of products with performance close to the target.

Based on reasoning started around the 1960s, initially by G. Taguchi and then widely accepted by the scientific community, we can say that:

- if the value of KSC is different from the target, a loss occurs;

- the loss increases moving away from the target;

- the loss at the target is minimal and we can say equal to zero.

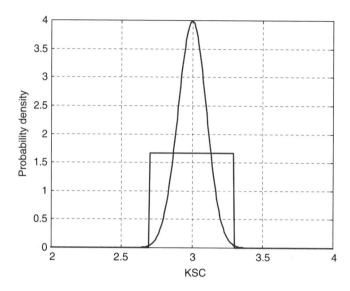

Figure 5.39   Two alternative probability distribution models for the KSC.

The loss, indicated by L, is thus depending on the value assumed by the KSC:

$$L = L(Y) \tag{5.76}$$

In general it is difficult to obtain an exact expression of the functional relationship (5.76). Therefore we resort to a Taylor series expansion of the function in the neighbourhood of the target:

$$L(Y) \cong L(ob) + L'(ob)(Y - ob) + \frac{L''(ob)}{2}(Y - ob)^2$$

Based on the previous assumptions:

$$L(ob) = 0, \, L'(ob) = 0$$

So we get:

$$L(Y) \cong k(Y - ob)^2 \tag{5.77}$$

The constant k in Equation 5.77 can be calculated on the basis of economic criteria. For example, say that $ob \pm \delta$ are the specification limits of the KSC and say that c is the monetary cost at those limits (symmetric case), from Equation 5.77 we obtain:

$$k = \frac{c}{\delta^2}$$

For *the-smaller-the-better* and *the-larger-the-better* problems, or in the case of the deviation from the target producing different losses depending on its sign (asymmetric loss), we make some changes to Equation 5.77.

A quadratic-type loss function is a reasonable and useful tool for assessing the robustness of the system.

However, given that Y is a random variable, we deduce also that the loss L is a random variable, for which we can calculate the expected value. Let:

$$E\{Y\} = \eta \qquad V\{Y\} = \sigma^2$$

From Equation 5.77 we obtain:

$$E\{L(Y)\} = k[(\eta - ob)^2 + \sigma^2] \tag{5.78}$$

Equation 5.78 allows us to observe that the expected loss can be minimised both by taking the average value of the KSC on the target ($\eta \to ob$), and minimising the variance of the KSC ($\sigma^2 \to 0$).

For practical purposes, however, Equation 5.78 is not sufficient to discriminate between design configurations of the system under study characterised by KSC with different means and variances but getting the same values of the expected loss.

Sometimes, in order to get the expected value of KSC on target, we can act on the signal factors. In fact, if there is a signal factor linearly related to the KSC that does not interact with other design parameters (the concept of interaction will be clarified in Section 5.3.4), it can be simply shown that by modulating the input signal, the expected value of the KSC is made equal to the target. It is therefore possible to define a *corrected* KSC:

$$Y_C = \frac{ob}{\eta} Y$$

whose parameters are:

$$E\{Y_C\} = ob \qquad V\{Y_C\} = \frac{ob^2}{\eta^2}\sigma^2$$

The corresponding quadratic loss has the expected value:

$$E\{L(Y_C)\} = k \qquad ob^2(\sigma/\eta)^2 \tag{5.79}$$

In contrast to Equation 5.78, Equation 5.79 is operational since it allows us to choose between different possible design configurations of the system characterised by different values of σ and η, with the assurance that the performance will be centred on the target via the signal factor. Obviously, it is not always possible or easy to identify these special signal factors.

The objective of searching for a design setting which minimises the corrected expected loss (Equation 5.79) is equivalent to searching for the design setting that maximises the dimensionless quantity defined as:

$$S/N = -10\log_{10}(\sigma/\eta)^2 \tag{5.80}$$

which is called the 'signal/noise ratio', based on an analogy with acoustics.

In the experimental phase, in order to estimate the S/N ratio, because we don't have the parameters η and σ, we use their estimators $\overline{Y}$ and S, based on a sample of measurements of the random variable Y, so that the theoretical formulation (5.80) is replaced by the estimator:

$$\text{S/N} = -10\log_{10}\left(S/\overline{Y}\right)^2$$

In *the-smaller-the-better* and *the-larger-the-better* problems, the formulations of the loss function and the S/N ratio are obtained from the previous formulations by adopting suitable adjustments.

In *the-smaller-the-better* problems, using the identity:

$$\frac{1}{n}\sum_i Y_i^2 = \frac{1}{n}\sum_i \left(Y_i - \overline{Y}\right)^2 + \overline{Y}^2 = \frac{n-1}{n}S^2 + \overline{Y}^2$$

we obtain the following expression for the S/N:

$$\text{S/N} = -10\log_{10}\left(\frac{1}{n}\sum_i Y_i^2\right) \tag{5.81}$$

In *the-larger-the-better* problems the formulation of the S/N ratio is obtained from Equation 5.81 and, instead of Y_i, we consider the reciprocals:

$$\text{S/N} = -10\log_{10}\left[\frac{1}{n}\sum_i \left(\frac{1}{Y_i}\right)^2\right] \tag{5.82}$$

For *signal–response systems* where the relation signal–response is linear this can be taken as a robustness indicator to maximise the ratio defined as follows:

$$\text{S/N} = \log\left(\frac{\hat{\beta}_1^2}{s^2}\right) \tag{5.83}$$

where:

$\hat{\beta}_1$ is the estimate of the slope of the regression line of the signal–response relationship (see Section 5.1.1);

s^2 is the estimate of the variance of the error term of the simple linear regression model.

5.3.2   Variation mode and effect analysis

Variation mode and effect analysis (VMEA) is a technique inspired by the increased attention to robustness thinking and by a wide use of failure mode and effect analysis (FMEA) (Section 4.3.1) in business and industry.

However, while FMEA develops around the concept of failure, VMEA develops around the concept of variation. In fact, failures are often caused by variation, as illustrated in Figure 5.40 with the well-known load-strength scheme.

If the strength is higher than the load (as in the nominal case, $s_0 > l_0$) no failures occur. However, either load or strength or both are generally affected by random variation about

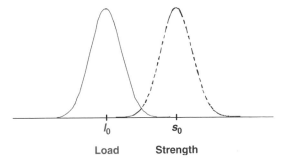

Load   Strength

Figure 5.40   Load-strength scheme.

their nominal values. Hence, failures might occur when a particularly high load 'meets' a particularly low strength.

This scheme, traditionally adopted in reliability engineering, is in fact very general. Here are some examples:

- the number of phone calls (load) reaching a call centre with a fixed number of operators (strength);

- the orders (load) arriving at a production plant of a fixed capacity (strength);

- the number of simultaneously running tasks (load) on a computer with a limited available memory (strength).

Since failures are mostly caused by unwanted variation, it is advisable to focus on variation as early as possible when developing a new system.

The VMEA technique can be applied once KSCs are identified. Each selected KSC is then broken down into a number of sub-KSCs. These are known and controllable factors influencing the KSC. For each sub-KSC, one or more noise factors (NFs) are identified that affect the sub-KSC.

The process described above is a typical KSC causal breakdown. It is a basis for making sensitivity and variation size assessments, which in turn allows us to calculate a variation risk priority number (VRPN), a summary index which can be provided for each NF and for each sub-KSC.

Depending on the depth of the analysis to be achieved, experts can carry out VMEA at three possible levels. The procedure is the same for all three levels, but the assessment methods differ somewhat.

The level I VMEA is easy. Sensitivity and variation size assessments are usually made on a 1–10 Likert scale.

With more information and data available on the system under study, it is possible for experts to carry out a level II VMEA, using more accurate sensitivity and variation size assessment tools.

Finally, with all data and models of the system at hand, experts can apply a level III VMEA. It uses the most accurate estimation of sensitivities and variation sizes, thus resulting in the most accurate VRPNs.

5.3.2.1   VMEA procedure

The proposed four-step procedure for carrying out a VMEA is the same for all three levels. It is the following.

1. *KSC breakdown*: the KSC is broken down into sub-KSCs and NFs.

2. *Sensitivity assessment*: experts assess the sensitivity of the KSC to the action of each sub-KSC and the sensitivity of the sub-KSCs to the action of each NF.

3. *Variation size assessment*: experts identify NFs and assess their variation size.

4. *Variation risk priority*: a VRPN is calculated for each NF based on the assessments made in the previous two steps:

$$\text{VRPN}_{\text{NF}_{ij}} = \alpha_i^2 \cdot \alpha_{ij}^2 \cdot \sigma_{ij}^2 \tag{5.84}$$

where:

α_i is the sensitivity of the KSC to the i-th sub-KSC ($i = 1, 2, \ldots, m$),
α_{ij} is the sensitivity of the i-th sub-KSC to the action of the j-th NF,
and σ_{ij} is the variation in the size of the NF.
If a sub-KSC is affected by several NFs, it is possible to calculate its VRPN as the sum of the VRPNs of the NFs acting on it:

$$\text{VRPN}_{\text{Sub - KPC}_i} = \sum_j \text{VRPN}_{\text{NF}_{ij}} \tag{5.85}$$

A Pareto chart of the calculated VRPNs is used for the prioritisation of NFs and sub-KSCs. This is the final step in identifying the 'areas' of the system where more attention is needed.

5.3.2.2   The 'method of moments' as the foundation of VMEA

What is the justification of the VMEA method? To show that, we consider a typical KSC breakdown. The formulations can be easily generalised.

We denote the KSC by Y. A generic sub-KSC affecting Y is denoted by X_i ($i = 1, 2, \ldots, m$), and a generic NF affecting X_i is denoted by N_{ij} ($i = 1, 2, \ldots, m$ and $j = 1, 2, \ldots, n_i$). The randomness of the NFs is transmitted to the sub-KSCs and, through them, finally to the KSC. The KSC breakdown is graphically illustrated in Figure 5.41.

We can write:

$$Y = \varphi(X_1, \ldots, X_m)$$

We use the following notation:

$$\mu_y = \text{E}\{Y\}; \sigma_y^2 = \text{V}\{Y\}; \mu_i = \text{E}\{X_i\}; \sigma_i^2 = \text{V}\{X_i\}; \sigma_{ij}^2 = \text{V}\{N_{ij}\}.$$

It is assumed that $\text{E}\{N_{ij}\} = 0$.

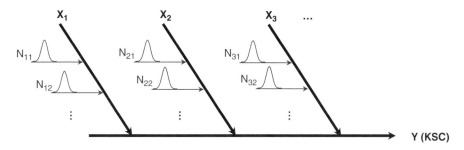

Figure 5.41   Typical KSC breakdown.

Furthermore, it is assumed that the sub-KSCs X_i $(i = 1, 2, \ldots, m)$ are mutually stochastically independent and the NFs N_{ij} are mutually stochastically independent ($\forall i = 1, 2, \ldots, m$ and $j = 1, 2, \ldots, n_i$).

The sensitivity of Y to X_i is a partial derivative which is defined as:

$$\alpha_i = \left. \frac{\partial Y}{\partial X_i} \right|_{\mu} \tag{5.86}$$

where $\mu = (\mu_1, \mu_2, \ldots \mu_m)$.

In a similar way, the sensitivity of X_i to N_{ij} is defined as:

$$\alpha_{ij} = \left. \frac{\partial X_i}{\partial N_{ij}} \right|_{\underline{0}}$$

where $\underline{0}$ is the null vector of dimensions n_i.

Consider the second-order Taylor expansion of Y:

$$Y \cong \varphi\left(\underline{\mu}\right) + \sum_{i=1}^{m} \left. \frac{\partial \varphi}{\partial X_i} \right|_{\underline{\mu}} (X_i - \mu_i) + \sum_{i=1}^{m} \left. \frac{\partial^2 \varphi}{\partial X_i^2} \right|_{\underline{\mu}} \frac{(X_i - \mu_i)^2}{2}$$

$$+ \sum_{i=1}^{m} \sum_{\substack{k=2 \\ k > i}}^{m} \left. \frac{\partial^2 \varphi}{\partial X_i \partial X_k} \right|_{\underline{\mu}} (X_i - \mu_i)(X_k - \mu_k) \tag{5.87}$$

By using the previous formula, it is possible to calculate the first two moments (expected value and variance) of Y. For the expected value, the full second-order approximation is generally used. For the calculation of the variance, only the first-order terms are considered in order to get a simpler formulation:

$$\sigma_y^2 = \sum_{i=1}^{m} \alpha_i^2 \sigma_i^2 \tag{5.88}$$

By following the same reasoning for the sub-KSC X_i, its variance, σ_i^2, can be further decomposed as follows:

$$\sigma_i^2 = \sum_{i=1}^{n_i} \alpha_{ij}^2 \sigma_{ij}^2 \tag{5.89}$$

Therefore, by combining Equations 5.88 and 5.89, we obtain:

$$\sigma_y^2 = \sum_{i=1}^{m} \alpha_i^2 \left(\sum_{i=1}^{n_i} \alpha_{ij}^2 \sigma_{ij}^2 \right) \tag{5.90}$$

which is – for the chosen KSC breakdown – the expression of the so-called method of moments, a well-established result in statistics.

The model described above does not exclude the possibility that an NF directly affects the KSC (Figure 5.42). In such case, it is possible to introduce a 'dummy' sub-KSC (with a sensitivity equal to 1) connecting the NF to the KSC.

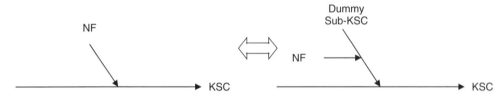

Figure 5.42   Introduction of a dummy sub-KSC.

5.3.2.3 A simple illustrative example

A company produces rectangular metal plates. The plates are manufactured by cutting a big metal sheet into four pieces (see Figure 5.43).

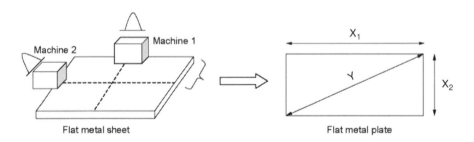

Figure 5.43   Manufacturing of a rectangular metal plate.

The cuttings are made using two machines of the same type. Recently the company is having customer complaints about the quality of the product since the machines are not precise; both are affected by sources of variation. Hence, deviations are observed around the nominal values of the plate length and width.

The company is particularly interested in the diagonal of the plate. This is considered as the main KSC. It is evident that the variation of the plate length and width leads to variation of the diagonal.

5.3.2.3.1 Level I VMEA   Since no data were initially available regarding the variation of the metal cutting process, experts carried out a level I VMEA. They broke down the KSC (diagonal Y) into sub-KSCs. The causal breakdown is illustrated in Figure 5.44.

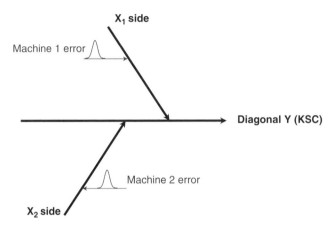

Figure 5.44   Breakdown of the KSC (plate diagonal).

In the second step, experts assessed the sensitivity of the diagonal Y (KSC) to the length X_1 (sub-KSC$_1$) and width X_2 (sub-KSC$_2$) on a $1-10$ scale (see Table 5.10).

On the basis of their past experience, they agreed that the sensitivity of Y to X_1 was higher than the sensitivity of Y to X_2. They also agreed that both sensitivities were quite low. Therefore, they placed weights 4 and 3, respectively. Next, they assessed the sensitivity of X_1 to the NF 'Machine 1 error' and the sensitivity of X_2 to the NF 'Machine 2 error'. Experts thought that the impact of the NFs on both sub-KSCs was equal because the manufacturing process is the same and the machines are of the same type. They also believed that the weight was moderate. Therefore, they placed an equal weight 5 on both of them.

In the third step, experts assessed the variation size of the two NFs. They knew that the cutting machine acting on X_1 was older than the cutting machine acting on X_2. Hence, they accepted that the size of the variation of 'Machine 1 error' was slightly larger than that of 'Machine 2 error'. Furthermore, they believed that this variation was relatively high and thus placed weights 8 and 7, respectively. The assessment scale is provided in Table 5.11.

VRPN are calculated according to Equations 5.84 and 5.85. The summary results of the level I VMEA are given in Table 5.12.

Experts could judge that the portion of variation channelled by X_1 is 2.32 times greater than the portion of variation contributed by X_2.

Table 5.10   Sensitivity assessment scale.

Criteria for sensitivity assessment	Score
Very low sensitivity. *The variation of NF (sub-KPC) is (almost) not at all transmitted to sub-KPC (KPC).*	1–2
Low sensitivity. *The variation of NF (sub-KPC) is transmitted to sub-KPC (KPC) to small degree.*	3–4
Moderate sensitivity. *The variation of NF (sub-KPC) is transmitted to sub-KPC (KPC) to moderate degree.*	5–6
High sensitivity. *The variation of NF (sub-KPC) is transmitted to sub-KPC (KPC) to high degree.*	7–8
Very high sensitivity. *The variation of NF (sub-KPC) is transmitted to sub-KPC (KPC) to very high degree.*	9–10

Table 5.11   NF variation assessment criteria.

Criteria for assessing the variation of NF	Score
Very low variation of NF. *NF is considered to be almost constant.*	1–2
Low variation of NF. *NF exhibits small fluctuations.*	3–4
Moderate variation of NF. *NF exhibits visible but moderate fluctuations.*	5–6
High variation of NF. *NF exhibits visible and high fluctuations.*	7–8
Very high variation of NF. *NF exhibits very high fluctuations.*	9–10

Table 5.12   Summary results of level I VMEA.

KSC	Sub-KSC	KSC sens. to sub-KSC	Noise Factor	Sub-KSC sens. to NF	NF variation size	VRPN (NF)	VRPN (sub-KSC)
Diagonal, Y	Length X_1	4	N_{11}	5	8	25 600	25 600
	Width X_2	3	N_{21}	5	7	11 025	11 025

5.3.2.3.2 Level II VMEA Some weeks later, experts wanted to carry out a level II VMEA. They met again and made new sensitivity and variation size assessments. They used the *sensitivity fan* (Figure 5.45) for the sensitivity assessment.

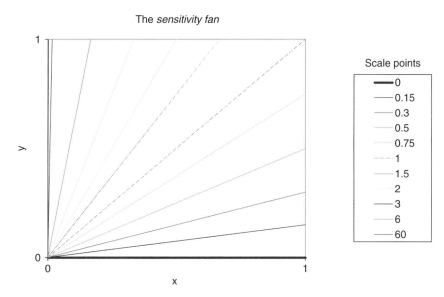

The *sensitivity fan*

Scale points

- 0
- 0.15
- 0.3
- 0.5
- 0.75
- 1
- 1.5
- 2
- 3
- 6
- 60

Figure 5.45 The 'sensitivity fan' for sensitivity assessment in level II VMEA.

They also used the *range method* for the variation size assessment. The range method can be used when it is possible to have estimates of the NF range. In this case one can assess the size of the NF variation by the following formula:

$$\hat{\sigma} = \hat{R}/6$$

In fact, the company experts had the necessary data at hand, so they completed the procedure, and the summary results are shown in Table 5.13.

Table 5.13   Summary results of the level II VMEA.

KSC	Sub-KSC	KSC sens. to Sub-KSC	Noise factor	Sub-KSC sens. to NF	NF variation size	VRPN (NF)	VRPN (sub-KSC)
Diagonal, Y	Length X_1	0.75	N_{11}	1	0.75	0.316	0.316
	Width X_2	0.50	N_{21}	1	0.60	0.090	0.090

Now the experts were more confident to say that the estimated portion of variation contributed by X_1 was 3.52 times greater than the portion of variation contributed by X_2.

5.3.2.3.3 Level III VMEA Some time later, one of the experts, wishing to deepen the problem and having new data available, developed a level III VMEA. He realised that the mathematical relationship between the length, width and the diagonal of the plate could be written as:

$$Y = \sqrt{X_1^2 + X_2^2}$$

Using the Equation 5.86 he got:

$$\alpha_1 = \frac{\mu_1}{\sqrt{\mu_1^2 + \mu_2^2}} \qquad \alpha_2 = \frac{\mu_2}{\sqrt{\mu_1^2 + \mu_2^2}} \tag{5.91}$$

Then he thought that the variation of length and width can be modelled by:

$$X_1 = \mu_1 + N_{11} \text{ and } X_2 = \mu_2 + N_{21}$$

where N_{11} and N_{21} are NFs causing variation in the cutting processes. As the cutting process involves two different machines, he assumed that N_{11} and N_{21} are independent random variables with:

$$E\{N_{11}\} = 0, E\{N_{21}\} = 0, V\{N_{11}\} = \sigma_{11}^2, V\{N_{21}\} = \sigma_{21}^2$$

By applying Equation 5.90 he finally obtained:

$$\sigma_y^2 = \frac{\mu_1^2}{\mu_1^2 + \mu_2^2}\sigma_{11}^2 + \frac{\mu_2^2}{\mu_1^2 + \mu_2^2}\sigma_{21}^2 = \frac{\mu_1^2\sigma_{11}^2 + \mu_2^2\sigma_{21}^2}{\mu_1^2 + \mu_2^2} \tag{5.92}$$

He assumed that the variances of the two cutting processes were equal (since the machines were of the same model and they had the same maintenance schedule) and he denoted their common value by σ_ε^2. Therefore from Equation 5.92 he got the non-intuitive result $\sigma_y^2 = \sigma_\varepsilon^2$.

To apply the level III VMEA, the expert needed data collected previously. The nominal length was $\mu_1 = 500$ mm and the nominal width was $\mu_2 = 400$ mm. He estimated that $\sigma_{11} = 0.70$ and $\sigma_{21} = 0.50$. Based on Equation 5.91 he calculated: $\alpha_1 = 0.78$ and $\alpha_2 = 0.62$.

Table 5.14 provides the summary results of the level III VMEA.

Table 5.14   Summary results of the level III VMEA.

KSC	Sub-KSC	Partial derivative	Noise factor	Sub-KSC sens. to NF	Sigma	VRPN (NF)	VRPN (sub-KSC)
Diagonal, Y	Length X_1	0.78	N_{11}	1	0.70	0.299	0.299
	Width X_2	0.62	N_{21}	1	0.50	0.098	0.098

Now, from the new calculations, the experts could finally judge that the estimated portion of variation channelled by X_1 was 3.06 times greater than that by X_2.

5.3.2.4   Industrial applications of VMEA

VMEA can find large applicability in business and industry. So far we have looked at several applications of the technique and we have seen many more done by others.

The technique was developed during a European project called EURobust, in the fifth Framework Programme of the European Commission. A software procedure was also developed to facilitate and foster the application of the technique. It was posted on the internet as a freeware. It is still there (http://eurobust.ivf.se/vmea/) and so is available for the interested reader.

5.3.2.5   Final considerations on VMEA

At early phases of development of a new system, we rarely know the relationships between the KSCs and the factors affecting them. Thus, it is not possible to apply quantitative methods to model these relationships. Nevertheless, it is important to anticipate the critical areas from a variation standpoint. VMEA is very useful and its use is indicated in these situations.

However, VMEA is not only useful in development phases, but also for any improvement related activity, as shown in the illustrative example.

In VMEA it is assumed that sub-KSCs are mutually s-independent. This is often reasonable, especially if we choose the right sub-KSCs and NFs. In fact, if there is a correlation between sub-KSCs, some of them can be excluded from the model.

Another assumption is the negligibility of interaction effects. These effects are analytically expressed by the second-order mixed derivatives of Equation 5.87. This assumption gives rise to a simple and manageable formulation and easy applicability of the VMEA technique. It is motivated by the natural principle of effects hierarchy (see Section 5.3.4). However, when it is believed – by *a priori* knowledge – that interaction effects are relevant, it is possible to test the assumption by making some experiments. If the interaction effects appear to be significant, VMEA should not be utilised.

5.3.3   Systemic robust design

Robust design problems usually concern specific components or functions of the system. This is a serious limitation that is mainly due to a lack of methodologies supporting experts with a systemic view of variation. The approach called *variation risk management* was one of the attempts to provide such a framework.

When developing the VMEA technique we created a simple approach that we called systemic robust design. After a complete breakdown of all KSCs, one can look at the robust design problem in a systemic way, that is looking at the whole system instead of looking at specific single subsystems or components.

The flow-tree-block (FTB) diagram of Figure 5.46 illustrates the idea.

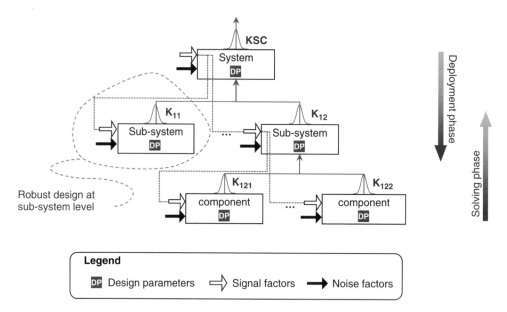

Figure 5.46   The idea of a systemic robust design.

We can look at three phases: (i) deployment phase, (ii) selection phase and (iii) solving phase.

5.3.3.1   Deployment phase

In the *deployment phase* we make explicit the relationships between all sub-KSCs and the corresponding subsystems (components) generating them.

The robust design problem for the whole system (RD problem at system level) and for a chosen KSC deploys into a series of RD problems concerning subsystems or components (RD problem at subsystem/component level). The process can be pushed down at the wanted depth level, which is normally the same level chosen for the KSC breakdown.

The FTB diagram of Figure 5.46 is operational since it separates the performance domain, that is the domain of the characteristics (KSC and sub-KSCs), from the physical domain, that is the domain of the subsystems generating them.

At the various levels of breakdown, the subsystems can be part of the whole system or even the system itself (the example below will clarify this aspect).

Each block of the FTB diagram represents a robust design problem concerning a subsystem (or the whole system), characterised by a specific set of design parameters (DP) and NFs. Signal factors are input to the system (i.e. achievement of a target for the KSC) which transfers to the subsystems and eventually to the components.

Note that the deployment phase is a deductive process (from the general to the particular). It is a very general process covering other well-known engineering processes. If the KSC is a binary variable related to the functionality of the system (its values can be: 'fault'/'no fault') the deployment phase produces a fault tree analysis (FTA). If the problem is related to examining the variation of the KSC (e.g. a geometric dimension of a mechanical assembly) and to reviewing (tightening or relaxing) the allowed variation of sub-KSCs at subsystem/component levels, then the deployment phase produces a tolerance allocation process.

5.3.3.2   Selection phase

After the deployment phase has taken place, or during it, experts can evaluate if specific sub-KSCs exhibit very limited or uninfluential variation. As a consequence, for these sub-KSCs it can be decided to neglect the corresponding RD problem and the further development at deeper levels of detail. This *selection phase* is therefore essentially based on past experience.

5.3.3.3   Solving phase

The third phase is a *solving phase* where the selected robust design problems are faced mostly through planned experiments. The solving phase is a bottom-up process that should start from the lowest level (component) up to the whole system level. The rationale of this bottom-up flow is that the sets of DP at higher levels generally contain sets of DP at lower levels; so that optimising and fixing the DP at lower levels simplifies problems at higher levels.

5.3.3.4   VMEA and systemic robust design for the paper helicopter

Let us still consider the paper helicopter case introduced in Section 5.3.1. After an initial brainstorming, five system characteristics were identified (see Table 5.9). The system characteristics were prioritsed according to their impact on quality.

The first KSC to be considered is the flight time (from a given release height H_0), to which the top priority was assigned. The flight time is a larger-the-better (LB) KSC, since helicopters falling slower were considered better.

It was necessary to identify possible noises, that is agents whose variation directly cause variation in flight time. The first level of breakdown is illustrated in Figure 5.47a. It shows two noises (environmental conditions and release modalities) and one sub-KSC (aerodynamic force). The noises identified at this level are usually called 'external noises'. They can determine variation of the KSC even on a single helicopter specimen repeatedly used several times.

Once the noises at the first level of breakdown have been identified, the sub-KSCs to be identified are those agents primarily influencing the mean value of the KSC. In this case, the identified sub-KSC is called 'aerodynamic force'. It represents the aerodynamic characteristic of the helicopter. It has a direct effect on the mean value of flight time but it can also channel variation to the KSC from lower levels, as can be observed from the second level of breakdown, illustrated in Figure 5.47b. The 'aerodynamic force' is further broken down into two second-level sub-KSCs: 'lift' and 'drag'. This is a typical breakdown of the aerodynamic force acting on the paper helicopter during its flight. The lift is the vertical force counteracting the weight force. The higher the lift, the slower the vertical motion of the helicopter. The drag is the resisting force due to the viscosity

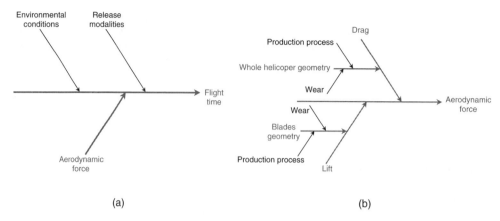

Figure 5.47   KSC breakdown for the paper helicopter: (a) simple and (b) deeper.

of the fluid (air). The higher the drag, the slower will be the rotational motion of the helicopter and consequently the faster will be the vertical descent. In summary, it is desired to have high lift and low drag.

At a lower level of breakdown it can be observed that the lift is produced by the motion of the blades, the body of the helicopter not contributing to it. Then lift is affected by blades geometry only. Conversely, both blades and body (whole helicopter geometry) contribute to the drag.

At the final chosen level of breakdown it is possible to identify two other important noises: wear and production process.

Wear is the natural, unavoidable and irreversible process affecting even a single helicopter specimen repeatedly used for several times, for example reducing paper stiffness.

Production process can instead induce variation among different produced helicopter specimens.

From the scheme in Figure 5.47 it is also possible to observe the difference between the two ways of conceiving engineering design: the traditional one – old fashioned – making use only of the deterministic knowledge, and the new one – enhanced engineering design – which takes into full account the effect of variation since the first phases of system development.

The deployment of the robust design problem is depicted in Figure 5.48.

5.3.4   Design of experiments

The scientific discoveries, technological innovations or simply desirable improvements from 'Six Sigma' projects and related activities frequently arise from experiments and not just from theoretical speculation.

For all the experimental activities that take place in an organisational context, the planning of experiments prior to data collection and analysis is crucial.

An *experiment* is a test or a set of tests in which changes are imposed on one or more inputs in order to observe and measure changes of a *response*; the response can be

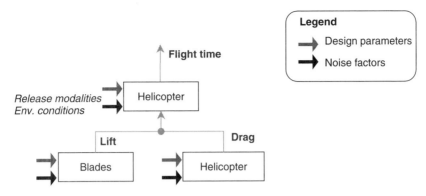

Figure 5.48   Robust design deployment for the paper helicopter case.

a performance of a product or a specific output of a process we are interested in. The response may be a single variable, as we will see in most of the examples, or more than one (multi-response problems). The inputs are named *control factors*. The *experimental unit* is the system on which the measurements are made.

For example:

1. in case of agricultural experiments the experimental unit could be the lot of soil treated with some fertiliser;

2. in case of environmental experiments the experimental unit could be a sample of waste water treated with a certain chemical agent;

3. in experimental pharmacology the experimental unit could be a patient treated with some antibiotics;

4. in technological experimentation, the experimental unit could be the metal specimen on which to make a certain machining.

An experimental *treatment* is a combination of test conditions, that is control factor *levels*, that are imposed on an experimental unit.

Referring to the examples above:

1. the experimental treatment could be a specific combination of: type of fertiliser, amount and type of insecticide and type of seeds;

2. the experimental treatment could be: type of chemical catalyst and temperature of the mud;

3. the experimental treatment could be: type of antibiotic and frequency of dispensing;

4. the experimental treatment could be the type of surface treatment, the alloy composition and processing temperature.

The aims of the experiments are always a starting point. Only after clarifying these we can talk about responses and control factors.

The design of experiments (DOEs) means planning, making a project of the complete sequence of operations by which we can obtain appropriate data, in order to perform objective analyses and valid inferences.

Despite all nice and good theory, in practice, in the real world we see that experimentation is carried out with different approaches.

The poorest approach is the so-called 'best-guess', that is to select the combinations of control factors only on the basis of intuition and experience.

The second approach, a bit better than the previous one, is the so-called *one-factor-at-time* (OFAT) experimentation. It involves the iterative search for the best combination of control factors by varying one factor at a time and keeping the other factors at fixed values. This approach, though still widespread, has been shown to be far less effective and efficient in comparison with the following approach.

The third possible approach is the so-called factorial experimentation, where we pursue the best combination of control factor levels by adopting a scientific method and using carefully studied experimental designs.

The objectives of the experiment may be varied:

- to improve the knowledge of the system (product/process/service) under study;

- to determine which of the control factors are more influential on the response;

- to set the significant factors at levels that guarantee a response close to a target value;

- to set the significant factors at levels that ensure minimal variation of the response;

- to set the significant factors at levels that ensure the minimum sensitivity of the response to the variation of factors outside our control;

- to obtain data on which to rely to solve quality problems, such as the identification of special causes of variation;

- to use data from a small number of experiments, carried out efficiently, that maximise the acquisition of knowledge at minimum cost.

5.3.4.1 Experiments with a single control factor

We introduce this topic with an example. Mr. Johansson, a truck driver, realised that to move from the warehouse where the goods were generally loaded, to the company where he usually unloaded the goods, he had five possible routes to follow (Figure 5.49).

One day he decided to measure the travel time for each route. He began on Monday with route 1 and ended on Friday with the route 5. He put the data together and created the following Table 5.15.

At the end of the week he was convinced that the route 5 was the best. During the weekend he had a doubt: '...is it possible that the travel time of route 5 came out first just because it was on Friday, a weekday usually less busy compared to the others? Maybe I should do another test...'. Then for the next week he worked out another plan

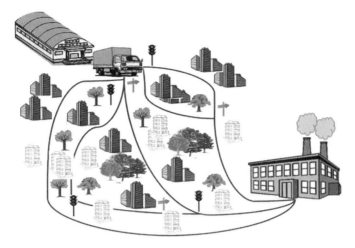

Figure 5.49   Five possible routes for Mr. Johansson.

Table 5.15   Data of the first 'experimental campaign'.

Route	Travel time (minutes)
P1	44.0
P2	40.4
P3	30.8
P4	37.8
P5	27.7

(Table 5.16), by randomly choosing the route to follow for each day of the week. From the experiments of the second week, Mr. Johansson realised some things:

- the best travel time this time was P3;

- on the same route he observed variation of travel time;

- the randomisation of the order of routes within the week was important, but what about the difference between weeks? Is it possible that in week II the traffic was changed due to the approach of the holiday season?

'Maybe I should make a programme of experiments', he said, 'in which I decide how many times to take each single route and then randomly choose the route for each day'.

These ideas represent a concept well-known to statisticians, which is the complete randomisation of the order of testing.

So, Mr. Johansson decided to follow each route four times and do the experimentation for the whole of the next month (four weeks). He randomly chose the sequence of tests:

Table 5.16   Data after the second week of experiments.

Week	Day	Route	Travel time (minutes)
I	Monday	P1	44.0
I	Tuesday	P2	40.4
I	Wednesday	P3	30.8
I	Thursday	P4	37.8
I	Friday	P5	27.7
II	Monday	P3	18.6
II	Tuesday	P1	37.5
II	Wednesday	P5	33.5
II	Thursday	P4	51.6
II	Friday	P2	32.6

he prepared four notes on which he wrote P1, four with P2, and so on, up to four notes with P5. Then he put the notes in a large bowl, mixed them and took them out one by one at random, not putting them back in the urn, and noting the sequence. At this point he was ready and in the following month he carried out the experiments. He collected the raw data in a table like Table 5.17.

Table 5.17   Data of the planned experiments of Mr. Johansson.

Day	Route	Travel time (minutes)
1	P1	49.4
2	P5	35.7
3	P4	42.6
4	P5	23.9
5	P1	52.9
6	P3	31.5
7	P3	43.3
8	P1	35.1
9	P1	45.2
10	P2	36.0
11	P3	24.6
12	P3	24.6
13	P5	26.0
14	P2	36.9
15	P2	41.1
16	P4	47.3
17	P4	37.4
18	P5	24.0
19	P2	34.1
20	P4	22.8

Table 5.18   Rearranged data of the third experimental campaign.

					Average travel time
P1	49.4	52.9	35.1	45.2	45.7
P2	36.0	36.9	41.1	34.1	37.0
P3	31.5	43.3	24.6	24.6	31.0
P4	42.6	47.3	37.4	22.8	37.5
P5	35.7	23.9	26.0	24.0	27.4

He then condensed the data in Table 5.18.

At this point Mr. Johansson realised, having studied mathematics at school, that it was not enough to calculate the average time, as he had done (see last column of Table 5.18), but to make a final decision he should have done the appropriate analysis. But how?

The above example provides the basics for understanding some key concepts of experimental design.

The *replications* are repetitions of the test made in 'the same' experimental conditions (the quotation marks are necessary since we understand that we will rarely have exactly 'the same' experimental conditions).

The experimental plan of the example is *balanced*, that is for each experimental situation (in the following we will say *experimental treatment*), we perform the same number of replications (see Table 5.18).

The opportunity to make replications arises from the need, as we understand by the example, to examine the variation – often natural and unavoidable – inherent in the system under study. An exception to this rule is for experiments executed by computer where the outcome is the result of a deterministic computation and exactly repeatable. In the case of so-called *computer experiments* other problems arise, which are not covered by this book.

The replications make it possible to measure the uncertainty of the estimates that will be based on experimental data.

From the example we immediately understand how important it is to randomise the order of execution of tests, to avoid confusing effects due to the factor or factors that we want to study, with effects due to uncontrolled variables, which we often call noise factor (NFs). Complete randomisation is often boring, but it is the most reliable. In some cases we cannot make a complete randomisation, but partial one, and, even then, it is good practice to take that into account in data analysis.

5.3.4.1.1   The statistical model for data analysis   The model to which we refer to analyse the experimental results is as follows:

$$Y_{jk} = \eta_j + \varepsilon_{jk} \quad j = 1, \ldots, a \quad k = 1, \ldots, b \tag{5.93}$$

where

Y_{jk} is the random variable modelling the data related to the j-th treatment and the k-th replication;

η_j is the parameter 'expected value' of the response to the j-th treatment;

ε_{jk} is the error term, which models the variability of the observed phenomenon;

a is the number of experimental treatments;

b is the number of replications, which for simplicity we assume constant and identical for each experimental treatment.

A general average can be defined as the average of the treatment means:

$$\eta = \frac{1}{a} \sum_{j=1}^{a} \eta_j$$

So we may pose:

$$\eta_j = \eta + \alpha_j$$

where α_j is defined as the effect of the factor at the j-th level.

It is clear that by construction we have:

$$\sum_{j=1}^{a} \alpha_j = 0$$

The model (5.93) can be rewritten as:

$$Y_{jk} = \eta + \alpha_j + \varepsilon_{jk} \tag{5.94}$$

It is called a *fixed effects model* because the levels of the factor are constant values chosen and set by the experimenter. In the case where the levels of the factor were also random variables, then we would talk about a *random effects model*.

5.3.4.1.2 The hypothesis testing

With reference to the example above, a doubt of Mr. Johansson was: are the routes not different or there is a *significant* difference between them? In other words, is the difference between the calculated arithmetic means significant when considering the variation of results obtained with each treatment?

Answers to this type of question are given by the test of hypotheses developed in case where more than two treatments are tested, the so-called *one-way analysis of variance* (ANOVA). The first thing to do is specify the null hypothesis:

$$H_0 : \{a_j = 0 \quad \forall j = 1, \ldots, a\}$$

The null hypothesis is an assumption of no difference between treatments. In contrast to this hypothesis, an alternative hypothesis can be very general such as:

$$H_1 : \{H_0 \text{ not true}\}, \text{ or at least a value } j \text{ exists such that } \alpha_j \neq 0$$

5.3.4.1.3 Test statistics

For the hypothesis test, as seen above (Chapter 3), we must use statistics. First we write the full experimental plan in general terms:

Factor level	Observations	Averages
1	$Y_{11}, \ldots, Y_{1k}, \ldots, Y_{1b}$	$\overline{Y}_1$
j	$Y_{j1}, \ldots, Y_{jk}, \ldots, Y_{jb}$	$\overline{Y}_j$
a	$Y_{a1}, \ldots, Y_{ak}, \ldots, Y_{ab}$	$\overline{Y}_a$

The statistics to use for a balanced experimental plan with a levels of the factor (treatments) and b replications are as follows.

- *The treatment average*:

$$\overline{Y}_j = \frac{1}{b} \sum_{k=1}^{b} Y_{jk}$$

- *The general average*:

$$\overline{Y} = \frac{1}{ab} \sum_{j=1}^{a} \sum_{k=1}^{b} Y_{jk}$$

For the test statistics, we use the decomposition of the *total deviance* (sum of squared differences between all the observed responses and their overall average).

$$SS_{\text{tot}} = \sum_{j=1}^{a} \sum_{k=1}^{b} (Y_{jk} - \overline{Y})^2$$

Then SS_{tot}, which measures the overall variation, can be decomposed into two addends: (i) the sum of the squares of differences between treatment averages and overall average, which measures the variation due to different treatments (different levels in this case, being only one factor), and (ii) the sum of squared differences between observed responses for each treatment and their treatment average, which measures the variation due to the experimental error within each treatment:

$$SS_{\text{tot}} = b \sum_{j=1}^{a} (\overline{Y}_j - \overline{Y})^2 + \sum_{j=1}^{a} \sum_{k=1}^{b} (Y_{jk} - \overline{Y}_j)^2 = SS_{\text{treatment}} + SS_{\text{error}}$$

5.3.4.1.4 Assumptions on the error terms   The error terms are assumed independent and identically distributed according to a Gaussian distribution with zero mean and constant variance:

$$\varepsilon_{jk} \sim \text{IIDN}(0, \sigma^2)$$

Made these assumptions, under the null hypothesis H_0, we have that:

$SS_{\text{treatment}}$ follows a chi-square distribution with $(a - 1)$ degrees of freedom;

SS_{error} follows a chi-square distribution with $(n - a)$ degrees of freedom; being $n = ab$ the total number of observations.

In addition, we have that:

$$\frac{SS_{\text{treatment}}}{a - 1} \text{ and } \frac{SS_{\text{error}}}{n - a} \text{ are s-independent} \tag{5.95}$$

The ratios (Equation 5.95) are called *mean squares*. Therefore under the null hypothesis the ratio between the two mean squares, called the F ratio, follows a *Fisher distribution* with $(a - 1)$ and $(n - a)$ degrees of freedom.

However, being still under the null hypothesis:

$$E\{SS_{\text{treatment}}\} = (a - 1)\sigma^2 \tag{5.96}$$

$$E\{SS_{\text{error}}\} = (n - a)\sigma^2 \tag{5.97}$$

The F ratio should be around the value 1.

Instead, under the alternative hypothesis, we have:

$$E\{SS_{\text{treatment}}\} = (a - 1)\sigma^2 + b \sum_{j=1}^{a} \alpha_j^2 \tag{5.98}$$

with Equation 5.97 being still valid.

Therefore, under the alternative hypothesis, the numerator of the F ratio is inflated and can lead, if the effect of the experimental factor on the response is significant, to values of F much greater than 1.

The procedure for the Fisher's test is summarised in the so-called ANOVA table (see Table 5.19).

Table 5.19 ANOVA table.

Source of variation	Sum of squares	Degrees of freedom	Mean squares	F	p
Treatment	$SS_{\text{treatment}}$	$(a - 1)$	$\dfrac{SS_{\text{treatment}}}{a - 1}$	$\dfrac{SS_{\text{treatment}}}{a - 1} \bigg/ \dfrac{SS_{\text{error}}}{n - a}$	–
Error	SS_{error}	$(n - a)$	$\dfrac{SS_{\text{error}}}{n - a}$	–	–
Total	SS_{tot}	$(n - 1)$	–	–	–

The p value in the last column represents the conclusion of the Fisher test and will be explained by an example.

As stated above, the F ratio is modelled by a Fisher r.v. under the null hypothesis. Figure 5.50 shows the Fisher probability density function with 4 and 15 degrees of freedom, for the example above. As we can see, the distribution is not symmetric, on the

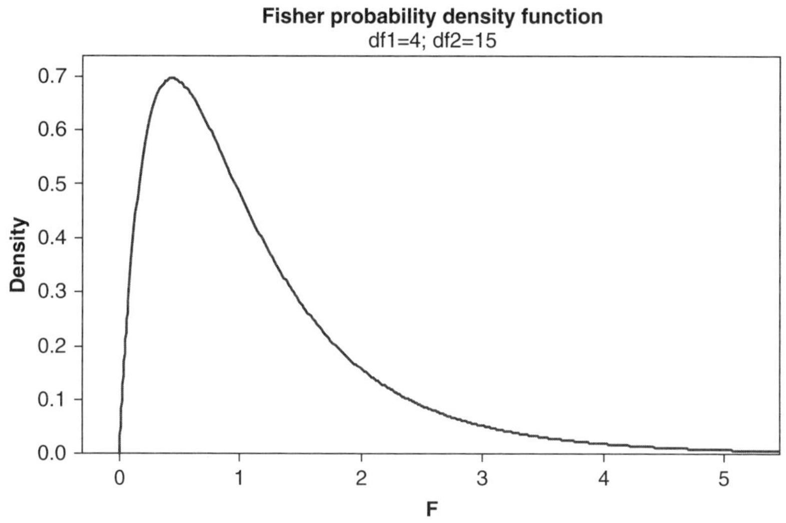

Fisher probability density function
df1=4; df2=15

Figure 5.50   Fisher probability density (4 and 15 degrees of freedom).

positive half-line (the r.v. is defined as a ratio of positive quantities), and is concentrated around the value 1.

The hypothesis test is 'one-tail' (right tail) since, as said, under the alternative hypothesis the F ratio is always higher than 1.

With the example data we get the following ANOVA table (Table 5.20).

Table 5.20   ANOVA table for the example data.

Source of variation	Sum of squares	Degrees of freedom	Mean squares	F	p
Treatment	781.0	4	195.2	3.35	0.038
Error	874.0	15	58.3	–	–
Total	1655.0	19	–	–	–

The p value corresponds to the area in the right tail of the distribution that is formed from the calculated value of the F ratio (Figure 5.51). The smaller the p value the more we say the treatment is significantly influential on the response. The p value is compared with the usual values of type I risk (0.05, 0.01).

In our example, we get a p value small enough that supports the claim that the followed routes are not the same, but that there is a significant difference between them.

5.3.4.1.5   The analysis of residuals   The randomisation of the order of the experiments tends to ensure the independence of errors. However, to validate the results of the ANOVA it is necessary *a posteriori* (i.e. after acquired the data) to perform some checks.

The analysis of residuals is the ultimate test.

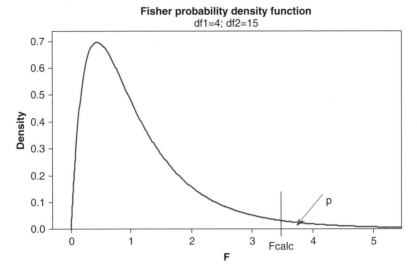

Fisher probability density function
df1=4; df2=15

Figure 5.51   The p value is the probability in the right tail from the calculated F ratio.

The residuals are the differences between the observed data and values predicted by the model for the same experimental treatment, estimated using the procedures previously seen.

$$r_{jk} = \hat{\varepsilon}_{jk} = y_{jk} - \hat{\eta}_j = y_{jk} - \overline{y}_j \qquad (5.99)$$

The calculation of the residuals is very simple in this case of a single factor. Once we have calculated the residuals we can construct some diagnostic graphs. For the example data, we report in Figure 5.52 a couple of very important diagnostic graphs.

The first, a Gaussian probability plot of residuals, allows us to verify if the Gaussian assumption is correct, while the second (residuals versus fits) allows us to test the hypothesis of constant variance (in this case, independent from the average value of the response).

There are other diagnostic graphs, such as the plot of the residuals in the timeline (to see if there is a trend in the data due to uncontrolled variables), in which case we should record the order of data acquisition.

5.3.4.2   Simple factorial plans

A *factorial plan* (or *factorial design*) is a programme of experimental tests that involves testing two or more factors, each of them with two or more levels.

A factorial plan is said to be *complete* if it contemplates testing of all possible combinations of factor levels. Otherwise, we talk about *reduced* or *fractional* plans, since they involve a fraction of the treatments foreseen by the full factorial plan.

The advantages of using a factorial plan with respect to the OFAT approach are:

- greater efficiency;

- greater effectiveness and accuracy;

- possible to estimate interaction effects;

- creating a wider inductive basis.

A factorial plan is said to be *symmetric* if for each factor we adopt the same number of levels.

In general, l^k indicates a symmetric full factorial plan with k factors, each with l levels. The experimental treatments (combinations of factor levels) in the plan are equal

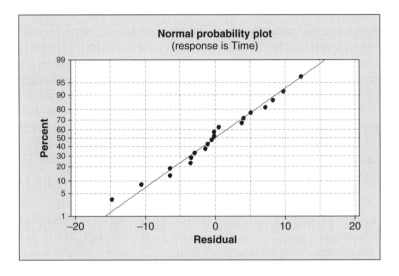

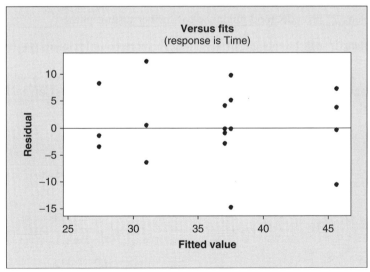

Figure 5.52   Diagnostic plots for the residuals.

in number to l^k and the factorial effects which can be estimated are equal in number to $l^k - 1$.

The factorial effects are classified into *main effects* and *interaction effects*. A main effect of a factor is the mean effect on the response due to the change in the level of the factor, whereas an interaction effect between two or more factors is the mean effect on the response due to the conjoint change of levels of the factors considered for the interaction. This concept, which is crucial in DOE, will be clarified in the course of this section through the given examples.

5.3.4.3   Full factorial plans symmetric with two levels (2^k)

When all factors are tested with two levels and all possible treatments are tested, we have 2^k plans. The two levels will be hereinafter always indicated with '-1' and '$+1$'. This is a very convenient coding for the algebraic operations that we will show in the following.

If the factor is a categorical variable, the two levels -1 and $+1$ are simply codes, but if the factor is a numerical variable, the two adopted levels can always be normalised so that the low level is equal to -1 and the higher level is equal to $+1$. However, it is good to keep in mind this normalisation from the original levels when making conclusions from the analysis of experimental data.

For a simpler notation, the two levels are often marked with the symbols '$-$' and '$+$'. The 2^k plans have some advantages.

- The resulting number of experimental treatments is relatively small. For example, if we have to experiment with three factors, using a factorial plan (2^3) we will test only 8 treatments, but if we consider three levels for each factor, the number of experimental treatments immediately jumps to 27, almost 20 more, often meaning a considerable additional time and cost for the experimentation.

- The plans 2^k provide good guidance for further testing phases.

- The plans 2^k can be thought of as *blocks* of more complex plans.

Figure 5.53 shows the tabular representation and the so-called 'control domain' (convenient graphical representation of the experimental treatments) of a 2^2 plan. The experimental treatments are listed in the so-called *standard order*.

Treatment	Factor A	Factor B
1	−1	−1
2	+1	−1
3	−1	+1
4	+1	+1

Figure 5.53   Tabular representation and control domain for a 2^2 plan.

Next to each experimental treatment will be reported the observed response values. Here for simplicity we assume to make only one replication per treatment. If we carry out a randomisation of treatment order then another column will appear that will show the order of execution. The table is completed as follows (Table 5.21).

Table 5.21   Complete tabular representation of the factorial plan.

Treatment	Factor A	Factor B	Run order	Response
1	−1	−1	−	Y_1
2	+1	−1	−	Y_2
3	−1	+1	−	Y_3
4	+1	+1	−	Y_4

The effects that can be estimated are $(4 - 1) = 3$. These are the two main effects of factors A and B and their interaction effect on the response. Below are the formulas for estimating these effects:

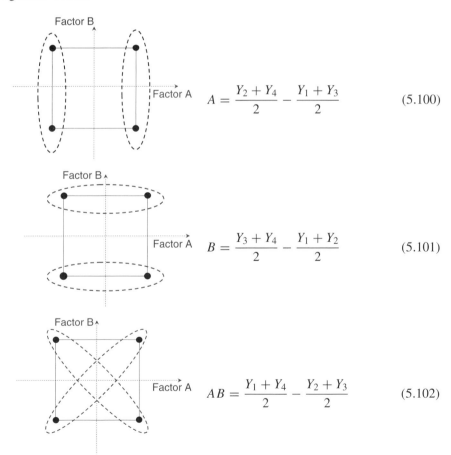

$$A = \frac{Y_2 + Y_4}{2} - \frac{Y_1 + Y_3}{2} \qquad (5.100)$$

$$B = \frac{Y_3 + Y_4}{2} - \frac{Y_1 + Y_2}{2} \qquad (5.101)$$

$$AB = \frac{Y_1 + Y_4}{2} - \frac{Y_2 + Y_3}{2} \qquad (5.102)$$

To facilitate the calculation of the interaction effect it is convenient to add to the experimental plan table a column AB (Table 5.22). It is not needed during the experimental phases, but only for the analysis of the results.

Table 5.22   A column for the estimation of the interaction AB is added to the tabular representation of the plan.

Treatment	Factor A	Factor B	Run order	AB	Response
1	−1	−1		+1	Y_1
2	+1	−1		−1	Y_2
3	−1	+1		−1	Y_3
4	+1	+1		+1	Y_4

We may notice that the values in the column AB are obtained as a product of the values in the two columns A and B. This algebraic artifice justifies the use of the coded levels.
Below is a tabular representation and the 'control domain' for a plan 2^3 (Figure 5.54).

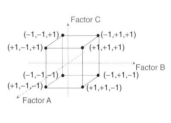

Treat.	A	B	C	Run order	AB	AC	BC	ABC	Response
1	−1	−1	−1		+1	+1	+1	−1	Y_1
2	+1	−1	−1		−1	−1	+1	+1	Y_2
3	−1	+1	−1		−1	+1	−1	+1	Y_3
4	+1	+1	−1		+1	−1	−1	−1	Y_4
5	−1	−1	+1		+1	−1	−1	+1	Y_5
6	+1	−1	+1		−1	+1	−1	−1	Y_6
7	−1	+1	+1		−1	−1	+1	−1	Y_7
8	+1	+1	+1		+1	+1	+1	+1	Y_8

Figure 5.54   Tabular representation and control domain for a 2^3 plan.

The estimable effects are in this case $8 - 1 = 7$: three main effects, three interaction effects between two factors and an interaction effect between the three factors.
Here below are some of the formulas used to calculate these effects:

$$A = \frac{Y_2 + Y_4 + Y_6 + Y_8}{4} - \frac{Y_1 + Y_3 + Y_5 + Y_7}{4} \qquad (5.103)$$

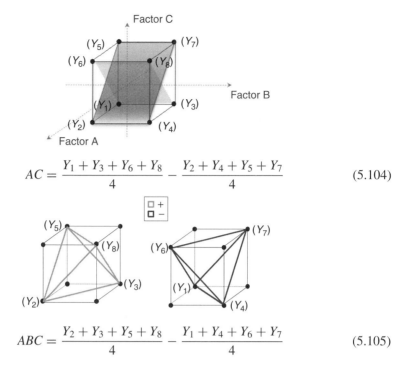

$$AC = \frac{Y_1 + Y_3 + Y_6 + Y_8}{4} - \frac{Y_2 + Y_4 + Y_5 + Y_7}{4} \qquad (5.104)$$

$$ABC = \frac{Y_2 + Y_3 + Y_5 + Y_8}{4} - \frac{Y_1 + Y_4 + Y_6 + Y_7}{4} \qquad (5.105)$$

Let us make a first reflection on the uncertainty associated with the estimates of the factorial effects. The previous formulas, both for the case of two factors and for the case of three factors, emphasise that a factorial effect, whether a main effect or an interaction effect, can be calculated as a difference between two means.

In all previous cases we have assumed that there was only one replication for experimental treatment. If we remember the model (5.93) written for the case in which each treatment corresponded to the level of a single factor, then we realise that even for the case of several factors we may think that the observation at a generic treatment has an uncertainty that is due to the experimental error. If we consider also here the assumptions on the error terms made before, then we can say that the variance of a factorial effect is given by:

$$\sigma_{\text{effect}}^2 = \frac{\sigma_\varepsilon^2}{T/2} + \frac{\sigma_\varepsilon^2}{T/2} = 4\frac{\sigma_\varepsilon^2}{T} \qquad (5.106)$$

where T is the number of experimental treatments.

If instead of a single replication per treatment, we experiment say r replications per treatment (balanced case), then for the estimation of factorial effects in the above formulas we use the treatment average instead of the individual observations. For example, instead of Equation 5.100 we will write:

$$A = \frac{\overline{Y}_2 + \overline{Y}_4}{2} - \frac{\overline{Y}_1 + \overline{Y}_3}{2}$$

Therefore the variance of each factorial effect in the case of r replications per treatment is given by:

$$\sigma^2_{\text{effect}} = 4\frac{\sigma^2_\varepsilon}{rT} \tag{5.107}$$

A full factorial plan has the property called *orthogonality*, that is: however we choose two columns of the plan, all possible combinations of factor levels appear the same number of times (equally often). This means that the estimates of the various effects are independent of each other.

EXAMPLE

A few years ago (around the 1970s) in the laboratories of the National Research Council (CNR) – Engines Institute of Naples, researchers experienced the effect of three factors on some performances in terms of pollution and fuel consumption of a new car engine. The factors taken into consideration are described in Table 5.23.

Table 5.23   Description of experimented factors and levels.

Factor	Description	Level '−1'	Level '+1'
A	Air filter dirtiness	As new	At 20 000 km
B	Spark plugs wear	As new	At 20 000 km
C	Circuit breaker gap	As new	At 20 000 km

Four responses were measured (Table 5.24). No trace remained of the order of carrying out the tests, but we know that a complete randomisation was made. This is a rather complex multi-response case, as it is doubly multivariate (both factors and responses). Let us for the moment analyse only the fuel consumption response. The analysis that we will do for this response can be repeated step by step for the other responses. We will see later (Section 5.3.5) how multiple responses could be considered together.

The following analyses were done using MINITAB software.

Figure 5.55 presents four graphs. The first two refer to estimates of factorial effects. In the Pareto chart of standardised effects, the effects are taken into an absolute value and divided by their standard deviation. The graph also shows a vertical line. This is a threshold to consider a significant effect. The graph shows that the main effect of factor C is highly significant, while the main effect of factor B is barely significant.

The second graph shows the Gaussian plot of standardised effects. The plot shows that the two main effects of B and C are distant from the line, which represents a Gaussian with zero mean and variance equal to that of the standardised effects. The graph shows that the effect of B is negative, this means that a slight wear of the spark plugs, which is determined in the first 20 000 km of engine life, can improve the fuel consumption.

Table 5.24 Experimental data of pollution and fuel consumption of a new car engine.

Treatment	A	B	C	Emissions of carbon monoxide			Emissions of hydrocarbons			Emissions of nitrogen oxides			Fuel consumption (S)		
1	−1	−1	−1	119.2	122.1	121.1	4.80	4.39	4.96	2.45	2.05	2.63	92.83	93.54	93.89
2	+1	−1	−1	115.2	105.1	132.3	4.41	4.71	5.19	2.44	2.62	2.64	95.60	93.39	92.25
3	−1	+1	−1	138.1	112.8	117.8	4.46	3.78	4.40	2.16	2.20	2.47	94.58	91.45	91.46
4	+1	+1	−1	128.6	116.0	121.9	4.53	3.44	3.37	2.13	1.83	2.07	90.81	92.64	92.47
5	−1	−1	+1	157.1	136.2	140.4	4.12	4.08	3.69	1.69	1.81	1.80	102.73	103.62	102.24
6	+1	−1	+1	135.5	146.7	139.3	4.29	3.85	3.97	2.11	1.76	1.94	100.57	106.25	101.13
7	−1	+1	+1	160.8	133.2	158.3	4.48	4.85	5.08	1.88	2.13	1.67	103.59	99.74	99.26
8	+1	+1	+1	155.5	114.1	127.2	4.41	4.16	4.76	1.83	2.24	2.15	101.50	99.02	100.55

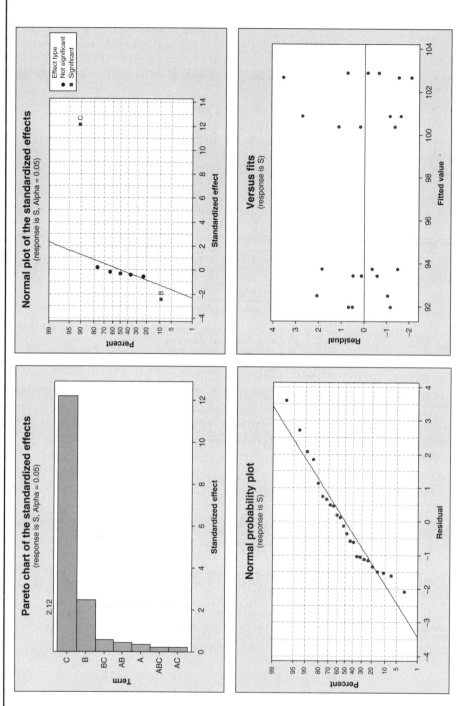

Figure 5.55   Analysis of the data from the full factorial plan.

> The normal plot of residuals shows no anomalies or outlier data, so it supports data analysis based on the assumption of Gaussian errors. The fourth graph shows that the residuals are distributed in two clouds, but this is due to the strong influence of the factor C.

In the case of full factorial designs, as the number of factors grows the number of experimental treatments grows exponentially (Table 5.25). It should also be borne in mind that as the number of factors grows the full factorial plans show a 'redundancy' of effects. For example, a full factorial plan 2^7 with seven factors at two levels requires 128 experimental treatments to estimate 127 effects:

- 7 main effects;

- 21 interactions between two factors;

- 35 interactions between three factors;

- 35 interactions between four factors;

- 21 interactions between five factors;

- 7 interactions between six factors;

- 1 interaction between seven factors.

Table 5.25   Number of experimental treatments required for a full factorial plan.

Factors	Treatments (2^k)	Treatments (3^k)
2	4	9
3	8	27
4	16	81
5	32	243
6	64	729
7	128	2187
8	256	6561

The questions arise: 'are all really necessary treatments foreseen by a full factorial? Is it really necessary to estimate all these effects?'

Experimental experience over the years has shown three major aspects:

- in many situations only a few experimental effects are of appreciable size (*effect sparsity*);

- there is often a *hierarchy* of effects, so the main effects are generally more influential than two-factor interactions, which in turn are generally more influential than three-factor interactions, and so on;

- it is good practice to adopt a sequential approach, so experiments should be done in stages, spending small amounts of budget in the early stages.

5.3.4.4   Fractional factorial plans symmetrical with two levels

A fractional factorial plan is a subset (a fraction) of treatments of a full factorial plan. These plans allow a saving of cost but inevitably result in a loss of information.

In fractioning the plan, it is desirable to retain the orthogonality property of the experimental plan. The fractioning of the plan leads to a phenomenon called *confounding* of effects (also *aliasing*).

Let us examine this issue through the example of the construction of the plan 2^{3-1}. Consider the full factorial 2^3 plan (Table 5.26).

Table 5.26   Fractioning of a full factorial plan 2^3.

Treatment	A	B	C	ABC
1	−1	−1	−1	−1
2	1	−1	−1	1
3	−1	1	−1	1
4	1	1	−1	−1
5	−1	−1	1	1
6	1	−1	1	−1
7	−1	1	1	−1
8	1	1	1	1

Suppose we want to carry out experiments for only half of the treatments foreseen by the full factorial plan, that is four treatments.

If we unwisely choose the first four, we would fall into error because we would not be able to estimate the main effect of factor C.

So we construct the column needed to estimate the interaction ABC and consider only the treatments 2, 3, 5 and 8 having a +1 in that column (see Table 5.26).

We immediately realise that if we choose these treatments columns A, B and C are still orthogonal (Table 5.27), but let us see what happens when – having acquired the experimental data – we want to analyse all the factorial effects.

Table 5.27   Confounding of effects with the fractional plan.

A	B	C	AB	AC	BC	ABC
1	−1	−1	−1	−1	1	1
−1	1	−1	−1	1	−1	1
−1	−1	1	1	−1	−1	1
1	1	1	1	1	1	1

Table 5.27 shows that the column BC is equal to column A. It means that when we estimate the main effect of A, it will be the same as estimating the interaction effect BC. In statistical language we say that there is a *confounding* (in common language we would

say confusion) between the two factorial effects. The same goes for B with AC and C with AB. Obviously, the interaction effect ABC is not estimable, because of how we built the plan.

The fractional factorial plan so constructed is denoted by 2^{3-1}. This is a 'half-fraction' that is equivalent to half of the corresponding full factorial plan 2^3 (note that $2^3/2 = 2^{3-1}$). In general, the half fraction plans are denoted as 2^{k-1}.

The reduction in the number of experimental treatments (in this case a half with respect to the full factorial plan) will ensure cost savings, but will determine confounding of effects which we will have to consider in the analysis and interpretation of results.

In theory, we could rely on the aforementioned principle of hierarchical ordering of effects, but this is not enough in a case like this. Often the interactions between two factors are important and significant. It must be said also that in a case like this the assignment of factors to columns is of little importance. Here, we only know that the main effect of each factor will not be confounded with an interaction that involves itself.

The situation is different if we consider the half fraction of a plan 2^4. The construction is made with the same technique, that is starting from the full factorial (see Table 5.28).

Table 5.28   Construction of the half fraction 2^{4-1}.

Treatment	A	B	C	D	ABCD
1	−1	−1	−1	−1	1
2	1	−1	−1	−1	−1
3	−1	1	−1	−1	−1
4	1	1	−1	−1	1
5	−1	−1	1	1	1
6	1	−1	1	−1	1
7	−1	1	1	−1	1
8	1	1	1	−1	−1
9	−1	−1	−1	1	−1
10	1	−1	−1	1	1
11	−1	1	−1	1	1
12	1	1	−1	1	−1
13	−1	−1	1	1	1
14	1	−1	1	1	−1
15	−1	1	1	1	−1
16	1	1	1	1	1

Selecting only the treatments with ABCD $= +1$ we obtain the following plan (Table 5.29).

Here the main difference compared to the previous case is that the main effects of factors will be confounded with interaction effects between three factors which, according to the hierarchy principle, are generally much less significant than the former. In addition, the plan generates confounding between two factors interactions.

For a symmetrical fractional factorial design at two levels of type 2^{k-p} (in the two illustrative cases seen before $p = 1$) we use the concept of *resolution*.

Table 5.29   Half fraction 2^{4-1} and confounding.

A	B	C	D	AB	AC	AD	BC	BD	CD	ABC	ABD	ACD	BCD	ABCD
−1	−1	−1	−1	1	1	1	1	1	1	−1	−1	−1	−1	1
1	1	−1	−1	1	−1	−1	−1	−1	1	−1	−1	1	1	1
−1	−1	1	1	1	−1	−1	−1	−1	1	1	1	−1	−1	1
1	−1	1	−1	−1	1	−1	−1	1	−1	−1	1	−1	1	1
−1	1	1	−1	−1	−1	1	1	−1	−1	−1	1	1	−1	1
1	−1	−1	1	−1	−1	1	1	−1	−1	1	−1	−1	1	1
−1	1	−1	1	−1	1	−1	−1	1	−1	1	−1	1	−1	1
−1	−1	1	1	1	−1	−1	−1	−1	1	1	1	−1	−1	1
1	1	1	1	1	1	1	1	1	1	1	1	1	1	1

In general, we say that a fractional factorial plan symmetrical with two levels is of resolution R if no effect of order j is confused with effects of order lower than $R - j$ ($j = 1, \ldots, R - 1$).

- In plans of resolution R = III, the main effects (order 1) are not confounded with other main effects, but they are confounded with interactions between two factors (order 2).

- In plans of resolution R = IV, the main effects are not confounded with each other or with interactions of order 2, but they are confounded with the interactions of order 3. In addition, interactions of order 2 are confounded with each other.

- In plans of resolution R = V, interactions of order 2 are not confounded with each other.

The plan 2^{3-1} seen as the first example has a resolution R = III, while the plan 2^{4-1} seen as the second example has a resolution R = IV.

As a further example of fractional factorial plan symmetrical with two levels, we will see below the case of the catapult, another educational toy widely used in the teaching of experimental design.

5.3.4.5   The blocks

As previously mentioned, to ensure the quality of the experimental results, conditions must be kept consistent during the tests. Sometimes, however, it is not easy, or it is impossible, to guarantee this, for example because the tests cannot be made during a single day. In other cases, we want to control the variability of some noise factor, for example the ambient temperature.

In such cases, there is a useful technique to prevent the occurrence of unwanted changes resulting from the inhomogeneity of the test conditions: the allocation of experiments in blocks.

This method explains how to select a certain number of blocks of the experimental treatments. The blocked plan is said randomised if the order of execution of the treatments is randomised within each block.

5.3.4.6   Some important prescriptions for a good experimental design

- Define the work team.

- Give a title to the experimentation and set clear objectives: what do we aim to achieve? Formulate a clear description of the problem.

- Search for previous documented experience. It could be information derived from previous experiments and/or the current level of informal knowledge on the involved phenomena, ideas and advice of experts, and so on.

- Describe the experimental unit: what is the system under study?

- Choose the response variables: these are the output of the experiment, the key variables of interest, the performances of the system.

- Choose the control factors: factors need to be observable, measurable and controllable and are suspected of having influence on the response variables.

- Choose the constant factors: that is factors, measurable and controllable, whose values are defined in advance and fixed for all experimental phases.

- Identify noise factors: that is factors that are not controlled in the real use of the system, but that may have influence on the responses.

- Choose levels of control factors, the constant factors and possible noise factors;

- Choose the experimental plan. It must be related to methods for the analysis of results and to the inferential aspects.

5.3.4.7   Illustrative example: The catapult

The catapult is an educational toy designed to teach how to design, implement and analyse a real factorial experimental plan. A drawing of the catapult is shown in Figure 5.56.

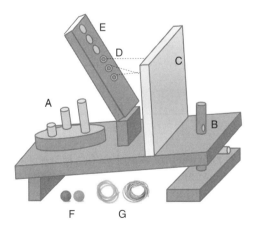

Figure 5.56   Drawing of the catapult used to practice DOE.

A few years ago, together with two students of the University of Sannio, we wanted to adopt the catapult to implement the design of experiments (DOE), trying to follow all best practices.

5.3.4.7.1   Problem definition and objectives   As objectives of the experimentation we set the attempt to better understand the 'system' catapult and as a practical objective we chose to get launches of the ball very powerful and accurate.

5.3.4.7.2   Response variable/s   We chose a single response variable: the distance of the landing point of the ball from a point taken as reference. It is called the range for a real catapult. However to avoid confusion with our use of the word 'range' (see Section 3.1.2.4.1), we will use the term 'distance'.

5.3.4.7.3   Selection of control factors and their levels   The catapult has a variety of factors, and almost every one can have at least three levels. We chose seven factors and for each of them we chose two levels. The factors (A, B, ..., G) are indicated in Figure 5.56 and described in Table 5.30.

Table 5.30   Description of control factors and levels chosen for the catapult experiments.

Factor	Description	Level '−1'	Level '+1'
A	Rung to delimit and support the movable arm before launch	The shortest rung	The tallest rung
B	Inclination of the catapult respect to the horizontal	No inclination: (horizontal)	Maximum possible inclination
C	Position of the rubber band on the fixed support	Lowest level	Highest level
D	Position of the rubber band on the mobile arm	Lowest level	Highest level
E	Compartment for housing the ball on the mobile arm	Lowest compartment	Highest compartment
F	Type of ball	Ping pong ball (light)	Rubber ball (heavy)
G	Type of rubber band	Very flexible	Poorly flexible

5.3.4.7.4   Choice of the experimental plan   A full factorial plan would require $2^7 = 128$ treatments. Obviously, it would be tedious, time consuming and probably of little use to experiment the full factorial.

So we decided to reduce the number of tests and carry out experiments on a maximum of 16 treatments. We could then use a fractional factorial plan 2^{7-3}.

For the definition of the plan, we used a synoptic table where, based on the number of factors and the number of experimental treatments, the *generators* to complete the plan were given, starting from the full factorial for four factors. Such tables are given on several books on design of experiments (see the references). The chosen plan had the highest possible resolution. It is shown in Table 5.31. The first four columns of the plan are a full factorial 2^4. The columns E, F and G are obtained using the generators (E = ABC, F = BCD, G = ACD).

Table 5.31   Adopted experimental plan.

Treatments	A	B	C	D	E	F	G
1	−1	−1	−1	−1	−1	−1	−1
2	1	−1	−1	−1	1	−1	1
3	−1	1	−1	−1	1	1	−1
4	1	1	−1	−1	−1	1	1
5	−1	−1	1	−1	1	1	1
6	1	−1	1	−1	−1	1	−1
7	−1	1	1	−1	−1	−1	1
8	1	1	1	−1	1	−1	−1
9	−1	−1	−1	1	−1	1	1
10	1	−1	−1	1	1	1	−1
11	−1	1	−1	1	1	−1	1
12	1	1	−1	1	−1	−1	−1
13	−1	−1	1	1	1	−1	−1
14	1	−1	1	1	−1	−1	1
15	−1	1	1	1	−1	1	−1
16	1	1	1	1	1	1	1

Generating the columns E, F, G via the formulas above determines a complex confounding pattern reported in Table 5.32.

As can be seen from the scheme, the plan has a resolution R = IV.

The assignment of specific factors to the columns of the plan was made by looking at the confounding pattern, having no previous idea about the interactions between three factors and, with regard to the two-factor interactions, imagining that the interaction BC could be particularly important.

5.3.4.7.5   Running the experiments   It was decided to carry out three tests for each treatment. Overall this results in 48 throws. For practical reasons it was decided not to completely randomise the order of execution of all tests, but more simply just to randomise the order of execution of the treatments. In this case more than talking of *replications*, we can talk about *repetitions* of the experimental treatments. This is a case of restriction on complete randomisation, which must be taken into account in data analysis.

Table 5.33 shows the plan with the actual order for carrying out the experimental treatments and obtaining measurements of the distance (in centimetres).

Table 5.32   Confounding pattern of the adopted
fractional factorial.

$$A = BCE = BFG = CDG = DEF$$
$$B = ACE = AFG = CDF = DEG$$
$$C = ABE = ADG = BDF = EFG$$
$$D = ACG = AEF = BCF = BEG$$
$$E = ABC = ADF = BDG = CFG$$
$$F = ABG = ADE = BCD = CEG$$
$$G = ABF = ACD = BDE = CEF$$
$$AB = CE = FG$$
$$AC = BE = DG$$
$$AD = CG = EF$$
$$AE = BC = DF$$
$$AF = BG = DE$$
$$AG = BF = CD$$
$$BD = CF = EG$$

Table 5.33   Experimental plan for the experiments with the catapult.

Treatments	A	B	C	D	E	F	G	Run order	Y_1	Y_2	Y_3
1	−1	−1	−1	−1	−1	−1	−1	10	65.3	53.9	52.6
2	1	−1	−1	−1	1	−1	1	12	49.3	44.3	44.2
3	−1	1	−1	−1	1	1	−1	9	36.9	41.6	46.0
4	1	1	−1	−1	−1	1	1	3	18.4	18.7	15.4
5	−1	−1	1	−1	1	1	1	6	204.7	213.1	216.2
6	1	−1	1	−1	−1	1	−1	8	59.2	64.8	67.6
7	−1	1	1	−1	−1	−1	1	2	158.6	141.3	146.3
8	1	1	1	−1	1	−1	−1	14	72.3	71.2	68.9
9	−1	−1	−1	1	−1	1	1	5	207.9	211.8	213.6
10	1	−1	−1	1	1	1	−1	7	107.5	119.6	110.4
11	−1	1	−1	1	1	−1	1	1	236.2	235.8	235.3
12	1	1	−1	1	−1	−1	−1	11	77.1	81.0	78.8
13	−1	−1	1	1	1	−1	−1	15	243.4	200.2	217.5
14	1	−1	1	1	−1	−1	1	4	155.0	148.3	155.8
15	−1	1	1	1	−1	1	−1	13	221.2	209.8	217.1
16	1	1	1	1	1	1	1	16	202.3	194.4	200.2

For each run, after setting the level combinations of the seven factors, and after checking that the catapult was in its proper launch point, we proceeded to launch the ball (after having slightly dipped it with white powder in order to mark its landing point).

Some preliminary tests soon showed that at the maximum inclination of the catapult (Factor B) certain combinations of the levels of other factors meant that the ball did not overcome the initial point of measurement, in other words we can say that the experiment 'failed'.

What to do in cases like these? Unfortunately, even after a careful choice of factors and levels, such situations can occur. A conscientious and careful investigator knows that these situations can occur even with some prior knowledge about the phenomenon under study. Preliminary experiments, perhaps in conditions considered most critical, should be run, and if the experiments fail, then it is necessary to change some selected factor levels.

In our case we decided to opt for the intermediate inclination (Factor B) and the intermediate level for the ball compartment (Factor E), which we later renamed as levels '+1'.

After these necessary changes, the experimentation had no further hitches, the initial tests were redone with the new factor levels and following the prescribed run sequence. The data collected are shown in Table 5.33.

5.3.4.7.6 Analysis of experimental results   Both the preparation of the experiments and the analysis of experimental results for this seemingly simple example of a catapult has strikingly interesting and often crucial aspects in respect of the assumptions necessary to support certain methods of analysis.

Two aspects of concern which we have already discussed are the allocation of factors to columns of the plan and the randomisation of the order of testing.

We could add another interesting thought on the obvious fact that among the seven factors selected there are three for which is particularly easy to change the level, that is the height of the rung to delimit and support the mobile arm (Factor A), the compartment housing the ball (Factor E) and the type of ball (Factor F), while for the remaining five factors, the reset operation is more complicated. If we want to take this into account, then we could think to conduct the experiments imposing a further restriction on the randomisation and exploiting the so-called split-plot technique. We omit this aspect here and refer the interested reader to specialised texts on DOE.

Now, concerning the analysis and interpretation of the data we collected, we will do a brief summary and present it to the reader, who is invited to exert maximum critical skills, to possibly repeat the analysis, to think about and develop them, trying to find in the presented data and experimental context even the most hidden items.

Let us give a look at the graphical analysis that we easily obtained with the software MINITAB.

Figure 5.57 presents the Pareto chart and Gaussian probability plot of the estimated effects (main effects and two-factor interactions). In interpreting these graphs we always keep in mind the confounding pattern (Table 5.32).

All the main effects are significant. We tend to attribute this result to the main effects, not to the three-factor interactions confounded with them.

We note that three two-factor interactions are more important than two of the main effects. The most important of these is the interaction AG with a negative effect (the sign of the effect can be deduced from the Gaussian probability plot). From the confounding pattern, we know that AG = BF = CD.

To which of the three interactions do we attribute the significance of this result? Before the experiments we had some prejudice on the interaction BC. However, what about AG, which might also be important? In short, the confounding is a knot that should be untied for a better understanding of the phenomenon. *Resolving the confounding* is another well-known issue within DOE to which we refer the interested reader to specialised texts.

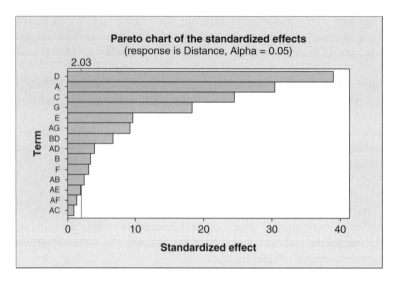

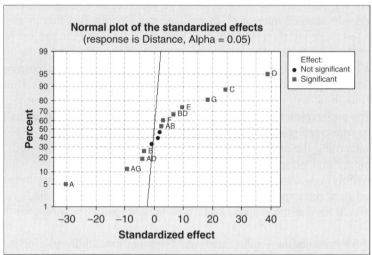

Figure 5.57   Graphs of the estimated factorial effects.

It is noted that almost all the effects that can be estimated are significant, except for three (AC, AF, AE, with their aliases).

Now we examine the diagnostic graphs of the residuals (Figure 5.58). They show that:

1. no trend with the order of execution of experiments is evident;

2. the assumption of Gaussian distribution of experimental errors is adequate;

3. residuals do not grow with the increase of the estimated value of the response, that is the variance of experimental error can be assumed constant.

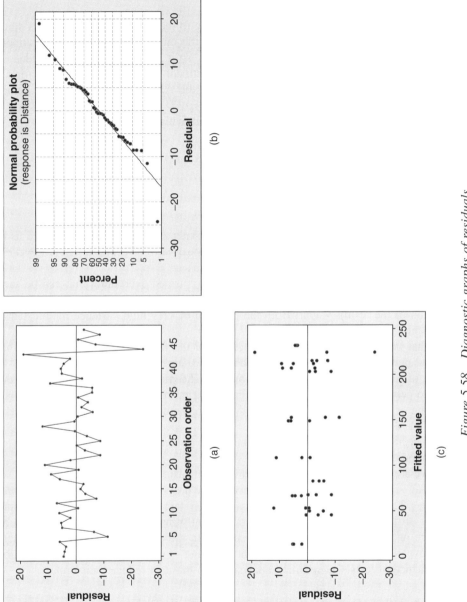

Figure 5.58 Diagnostic graphs of residuals.

All three graphs show us, however, the presence of two outliers or, we could say, unusual results. These are the observations corresponding to standard order 13 (see Table 5.33), where there is a very high average response (positive result in reference to the objectives of the experimentation), but also a high variance between the repetitions. This would deserve a deeper investigation.

The treatment 11 is the preferable one according to the experimental objectives since it has the highest average response value and also the minimum variance, but treatment 13 is the second best treatment in terms of average, so it should not be overlooked.

The fact that some treatments, particularly treatment 13, show too high a variance undermines the use of analytical methods, such as the ANOVA, that require the homogeneity of variance.

If we make a *test of homogeneity* we get the following output:

Bartlett's Test (Normal Distribution)
Test statistic = 33.44; p value = 0.004
Levene's Test (Any Continuous Distribution)
Test statistic = 1.42; p value = 0.196

We see that the Bartlett's test, based on the Gaussian assumption, recommended us to reject the null hypothesis of homogeneity of variances, but the second test, the Levene's test, is not of the same opinion.

However, once we had made the ANOVA, it confirmed everything that can be seen from the charts and tells us that our model based on the fractional plan explains quite well the variation of the results – apart from the problem of very high variance manifested at the treatment 13.

Note that the treatment 13 has the factor F and the factor G both at level '−', that means ping pong ball and very flexible rubber band, and both factors can be subject to the action of factors that will disturb the results. The ping pong ball is very light and sensitive to perturbations of air (just a small draught in the room, for example due to air conditioning or a window opening), while the flexible rubber band is made of a material that could have plastic deformation in the sequential repetition of tests and then lose the elastic effect necessary for a successful experiment (which is returning to the subject of complete randomisation as a good practice!).

Here we are in the presence of noise factors not measurable in the experimental phase, but that may significantly affect performance. They are usually defined as *experimental risk factors*. Their 'action' during the tests may in fact distort the results, increasing the weight of the experimental error and consequently making it difficult to analyse the significance of the effects of 'design parameters'. In the presence of these factors there are some experimental strategies (such as the blocking of experiments, the randomisation of the sequence of tests, the stratified analysis, etc.) that can reduce their impact.

5.3.4.7.7   Conclusions of the case study   After analysing the data, the experiment must lead to practical conclusions, theoretical conjectures and recommendations on a course of action for future experiments or investigations in general.

From the analysis of the data it was observed that the most influential factors on our experimental results were D, A, C and G, although many of the factors and their interactions have played an important role.

Regarding factor A, its effect on the response is very negative, so this means that increasing the height of the rung supporting the mobile arm before launch, from the short one to the tall one, significantly reduces the expected performance.

So, aiming for a possible improvement of the response (i.e. to throw the ball as far as possible from the catapult), strategies should be developed, starting from a consideration of the factors and their levels that have the most significant results in the experiments. Keep in mind not only the average value of the response, but also the variance, which in a case like this is an indication of the *precision*. So, a parameter that can be even more important than the distance capability.

To validate the conclusions of the experiment and to search for further improvements we should run confirmation tests and design additional experiments. All this is left to the interested and scrupulous reader.

5.3.5 Four case studies of robustness thinking

To show how robustness thinking can take place in the real world, in this section we present four case studies based on our direct experience. We will see that the idea of robustness can be applied in very different contexts and with very different facets.

5.3.5.1 Improving our teaching process

Over the last few years we have developed a methodology called *Teaching Experiments and Student Feedback* (TESF). We have shown how a teacher can improve his teaching by using DOE. The response is student feedback collected through an *ad hoc* feedback tool inspired by the SERVQUAL model (see Section 2.3.3) and submitted to students attending a course. The SERVQUAL model was adapted to the service 'university course', leading to the course quality determinants reported in Section 2.3.3.

To carry out teaching experiments, a teacher has preliminarily to: (i) identify the variables that may have an impact on the course quality determinants; (ii) select some of them as control factors; (iii) select the levels for each control factor and (iv) define the experimental plan. The execution of the formulated experimental plan implies the segmentation of the course in sections. Each section is associated with an experimental treatment. There should be the least possible number of sections so as to reduce the experimental burden for the teacher.

In an experimental campaign carried out in 2004, four control factors potentially having influence on the determinants were selected. They are reported in Table 5.34. For each control factor, two levels were identified.

The course was split into sections of equal number of hours. The chosen experimental plan was a fractional factorial at two levels 2_{IV}^{4-1} (resolution IV) consisting of eight experimental treatments. The time sequence of the course sections was predetermined and unalterable. It is reported in Table 5.35. The associations between the experimental treatments and the course sections was randomly selected.

Table 5.34   Description of the adopted control factors for the teaching experiments.

QD	Control factor	Levels
Responsiveness	*A – Modality of practices/laboratory work* The practices/lab work is designed by the teacher to facilitate student learning of the theoretical contents. The teacher can require that the students work in autonomy in small groups (two to three students) giving them some help during their work (level −1). Otherwise the teacher can solve the exercises at the board (level +1)	−1 Autonomy to students +1 Autonomy to teacher
Tangibles	*B – Board type* The board represents the primary material tool for aiding the teacher in the explanation of the lesson. The teacher can use the traditional blackboard, inducing students to take note of what he/she presents (level −1); otherwise, he/she can use already prepared slides presented by the overhead projector (level +1). Copy of such slides are delivered to the students	−1 Traditional blackboard +1 Overhead projector
Assurance	*C – Case studies* They are real examples to which is applied what has been presented in theory. The teacher uses them in order to make clear the contents and to show their practical usefulness.	−1 Absence +1 Presence
Empathy	*D – Teacher – students' interaction* Such interaction can be solicited by the teacher or not, by calling students to the blackboard to solve an exercise, a question made to a student on a presented topic, and so on, in order to keep high the students' attention and reduce the physiological distance between teacher and students and to verify their understanding	−1 Absence +1 Presence

Table 5.35 Course sections and their association with the experimental treatments.

Course section	Presentation order	Standard order (first edition)	Standard order (first edition)
Exploratory data analysis	1	7	6
Calculus of probability	2	2	3
Random variables	3	3	1
Statistical inference	4	5	8
Regression	5	1	2
Longitudinal data analysis	6	6	4
Design and analysis of experiments	7	8	7
Multivariate statistics	8	4	5

Data collected through the feedback tool completed by a selected sample of students at the end of each course section were analysed by statistical techniques. In particular for each of the items of the questionnaire, it was possible to make an ordinal logistic regression analysis (Section 5.1.2). It helped establish the effects of the control factors. Table 5.36 shows one of the outputs of the analysis made by the software MINITAB. This analysis concerns the questionnaire item 1.1 'Teacher's support to students during class'. The table initially presents descriptive information on the analysed data and the indication of the adopted link function (logit). Then it provides information on the effect of single factors (A, B, C, D) and on the model fitting to the collected data. We can observe that only factor B appears to have significant influence, with a positive effect (the shift from the traditional blackboard to the overhead projector has a significantly positive effect on the student satisfaction about the teacher support during class). The goodness-of-fit tests also confirm that the model adequately fits data, regardless of the limited sample.

After a thorough examination of all obtained results we draw opportune conclusions, and decided to redo the experiment in the following course edition (year 2005).

The basic idea here was to improve some aspects of the teaching process, basically improving the consistency on determinants having impact on the student satisfaction. Obviously we were and we still are fully aware that the quality of teaching is not only measured by 'student satisfaction'.

We also developed a robustness indicator named the student satisfaction index, which can also be used when no experiments are made during the course (e.g. for setting a benchmark or monitoring the course performance over the years).

5.3.5.2 Emotionally robust products and services

The emotional sphere of customers is increasingly considered by developers of products and services. A methodology called Kansei Engineering (KE) was born in Japan in the 1970s thanks to Mitsuo Nagamachi. 'Kansei' is a Japanese word that means 'feelings/emotions'.

Table 5.36   Ordinal logistic regression for the experimental results (year 2004, item 1.1).

Link function: logit
Response information

Variable	Value	Count
Score	25	2
	50	18
	75	30
	100	3
	Total	53

53 cases were used
3 cases contained missing values

Logistic regression table

Predictor	Coefficient	SE coefficient	Z	P	Odds Ratio	95% CI Lower	95% CI Upper
Const (1)	−3.3819	0.7252	−4.66	0.000	–	–	–
Const (2)	−0.5007	0.2969	−1.69	0.092	–	–	–
Const (3)	3.0095	0.6168	4.88	0.000	–	–	–
A	−0.0987	0.2761	−0.36	0.721	0.91	0.53	1.56
B	0.5501	0.2840	1.94	0.053	1.73	0.99	3.02
C	−0.1884	0.2770	−0.68	0.496	0.83	0.48	1.43
D	0.2198	0.2781	0.79	0.429	1.25	0.72	2.15

Log-likelihood $= -49.170$
Test that all slopes are zero: $G = 5.021$; $DF = 4$; p value $= 0.285$

Goodness-of-fit tests

Method	Chi-square	DF	P
Pearson	13.379	17	0.710
Deviance	13.792	17	0.682

Today in our competitive marketplace, the identification of strategically important attributes of products and services, even the emotional ones, is a key to success.

Since the early phases of system development, an important issue is to identify the best product profiles (combinations of product attribute levels) in terms of their impact on customer feelings and preferences. *Conjoint analysis* is a methodology developed to that purpose in the 1970s by Paul Green. Basically, conjoint analysis means design of experiments applied to marketing research.

In a conjoint analysis (CA) study we should decide how many attributes to consider. We normally want to consider as many as possible, without increasing respondent fatigue. Therefore experimental designs with few treatments should be used. Once the experimental design is prepared, the profiles are presented to respondents for their judgement. Collected data are then analysed to estimate respondent preferences.

KE experiments are essentially CA experiments where the respondents are called to give their judgement on product profiles in terms of Kansei words.

In relation to the way the selected profiles are presented to the respondent when the product is a physical good, we see three possible strategies.

- *S.1*. The experimenter builds physical prototypes, allowing respondents to interact with them.

- *S.2*. The experimenter builds virtual prototypes (digital mock-ups), allowing respondents to interact with them in a virtual environment.

- *S.3*. The experimenter uses products from those already existing in the market.

Building physical prototypes requires time and resources. Hence, the strategy S.1 is not suitable, especially in very early phases of product development. Conversely, virtual prototypes are able to simulate characteristics and physical behaviours of the product, saving time and resources, and obtaining other advantages, for example early ergonomic evaluations. Sometimes in very early development phases, experimenters may prefer not to spend resources to build any prototype at all, either physical or virtual. Therefore, in these cases products representative of the selected profiles can be chosen from those already existing in the marketplace and presented to respondents for their evaluation (strategy S.3). This solution is the cheapest and simplest to realise, but the respondent feedback will be inevitably affected by experimental risk factors which can heavily bias the analysis of results.

Some years ago we applied the methodology to the design of a new model of cell phone. We investigated the preferences of students. We selected four Kansei words which best summarised the Kansei of a cell phone: *Appealing, Comfortable handling, Stylish, Durable*. We also selected six product attributes potentially affecting those responses (see Table 5.37). For each attribute, two levels were chosen.

Table 5.37   Description of the chosen attributes and levels.

Attribute	Description	Levels	
		0	1
A	Integrated antenna	No	Yes
B	Dimensions	Small	Very small
C	Internal memory	Small	Big
D	USB port	No	Yes
E	Music support	No	Yes
F	VGA camera	No	Yes

A fractional factorial design $2^{(6-3)}_{III}$ was prepared (Table 5.38), consisting of eight profiles only. This is a reminder that a very limited run size is crucial in this type of experimentation.

According to the strategy S.3, eight phone models already in the marketplace were chosen. Each model was consistent with an experimental treatment. The experimental plan and the selected phone models are shown in Table 5.38.

Table 5.38   The experimental design and mobile phone models selected for the evaluation.

Std	Run	Attribute						Concept	Std	Run	Attribute						Concept
		A	B	C	D	E	F				A	B	C	D	E	F	
1	4	0	0	1	1	1	1		5	7	0	0	0	1	0	0	
2	1	1	0	1	0	0	1		6	5	1	0	0	0	1	0	
3	8	0	1	1	0	1	0		7	2	0	1	0	0	0	1	
4	6	1	1	1	1	0	0		8	3	1	1	0	1	1	1	

The eight mobile phone models were shown to the respondent in a well-prepared interview. The respondents were asked to give their evaluation on a five-point Likert scale and the collected data were analysed also with the ordinal logistic regression (Section 5.1.2).

Table 5.39 provides a final summary of the obtained results in terms of the estimated impact of the design parameters on the Kansei words. In other words this is a way for cell phone designers to see how to make a new cell phone model more emotionally robust.

Table 5.39   Estimated impact of attributes on Kansei words.

	Integrated antenna	Dimensions	Internal memory	USB port	Music support	VGA camera
Appealing	**	***	−	−	−	***
Handling comfortable	−	**	−	*	−	*
Stylish	**	**	−	−	**	*
Durable	*	−	**	*	*	−

*Weak impact.
**Moderately strong impact.
***Strong impact.

This methodology was successfully tested also for other products (wine bottles, cameras, motorcycles) and services (tourist services, hospital services).

5.3.5.3   Robust ergonomic virtual design

This is a quick mention on a robustness thinking case study concerning the development of a new motor vehicle for city transportation. We started from a concept design of the vehicle: some sketched ideas and CAD models (see Figure 5.59).

Figure 5.59   Sketched ideas and CAD models for the new city vehicle.

The CAD model could be imported to a virtual environment software program called *Jack*, where it was possible to virtually measure the sitting comfort of the vehicle, thanks to a digital manikin simulating a human being (Figure 5.60).

The main objective of the study was to find a robust design in terms of ergonomics, particularly referring to comfort when sitting. We identified five performances, that is the postural angles involved in sitting. They are reported in Table 5.40 together with their specification limits and target values.

We started by measuring the performances of a benchmark design setting. We adopted three levels of human body height, the main noise factor, intentionally introduced in the experiments.

Figure 5.60   The digital manikin seated on the vehicle concept allows for comfort evaluations.

Table 5.40   Performances selected for the robust ergonomic assessment.

Performance	Description	LSL	Target	USL	Dimensions
Y_1	Upper arm flexion	19	50	75	Degrees
Y_2	Elbow angle	86	128	164	Degrees
Y_3	Trunk-thigh angle	90	101	115	Degrees
Y_4	Knee angle	99	121	136	Degrees
Y_5	Foot-calf angle	80	93	113	Degrees

As the robustness index we formulated an expected total asymmetric loss (ETAL). For the benchmark design setting it was $ETAL_0 = 3.08$.

On the base of experts' judgement, five design parameters were chosen, each one potentially affecting driver comfort. They are described in Table 5.41.

An experimental design was prepared. It was a crossed design with the design parameters in the inner array and the unique noise factor in the outer array. The inner array had the five design parameters with three levels each. It was a fractional factorial symmetric at three levels 3^{5-2} consisting of 27 treatments, that is design settings (Table 5.42).

Figure 5.61 shows the average effects of the design parameters on the ETAL. Based on that analysis we could immediately foresee a better design setting: $A = 1$, $B = 2$, $C = 2$, $D = 2$, $E = 2$.

However, the experimentation went on to a second phase where a response surface design was performed (see Figure 5.62). This allowed for the prediction of the optimal design setting.

In fact, based on that prediction, we checked that we had really found a substantial improvement. In terms of the robustness index ETAL, we had a 77 % improvement respect to the initial benchmark setting.

Note that the initial setting of the design parameters was based on experts' judgement (designers). It was not too bad, but it was only due to the scientific approach we followed that we could attain considerable improvements with a very limited experimental effort.

Table 5.41   Design parameters and levels chosen for the experiments.

Design parameters		Levels				
		0	1	2	Dimensions	
A	Handlebar arc length	500	600	700	Millimetre	
B	Seat arm length	430	445	460	Millimetre	
C	Handlebar angular position	−15	0	15	Degrees	
D	Seat arm angular position	−15	0	15	Degrees	
E	Seat angular inclination	−15	0	10	Degrees	

Table 5.42   Experimental plan of the inner array.

Run	A	B	C	D	E	ETAL
1	0	0	0	0	0	23.34
2	0	0	1	1	2	2.50
3	0	0	2	2	1	2.90
4	0	1	0	1	2	2.77
5	0	1	1	2	1	2.32
6	0	1	2	0	0	19.22
7	0	2	0	2	1	4.00
8	0	2	1	0	0	17.71
9	0	2	2	1	2	2.34
10	1	0	0	1	1	5.33
11	1	0	1	2	0	4.37
12	1	0	2	0	2	10.75
13	1	1	0	2	0	3.64
14	1	1	1	0	2	8.64
15	1	1	2	1	1	4.16
16	1	2	0	0	2	5.19
17	1	2	1	1	1	2.23
18	1	2	2	2	0	3.19
19	2	0	0	2	2	3.12
20	2	0	1	0	1	19.66
21	2	0	2	1	0	13.34
22	2	1	0	0	1	20.54
23	2	1	1	1	0	8.91
24	2	1	2	2	2	2.36
25	2	2	0	1	0	5.71
26	2	2	1	2	2	4.47
27	2	2	2	0	1	9.57

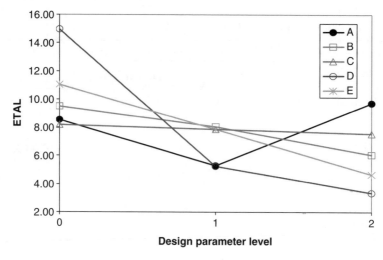

Figure 5.61   Estimated effects on the ETAL.

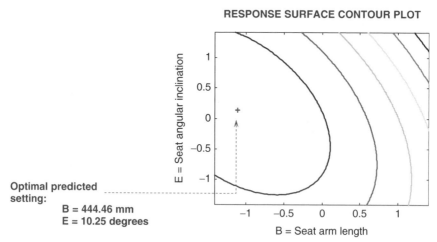

Figure 5.62   Contour plot of the response surface related to the second experimental campaign.

5.3.5.4   Robust on-board diagnostics

Technologically advanced systems are increasingly equipped with self-diagnostic subsystems aimed at monitoring in real time the 'state of health' of the system. If these diagnostic subsystems monitor critical system functions, they are doubly critical.

A diagnostic system generally works with signals coming from some monitored functions of the complex system. Its diagnostic output is generally related to some 'physical output' of the complex system (Figure 5.63). Sometimes, but not necessarily, the

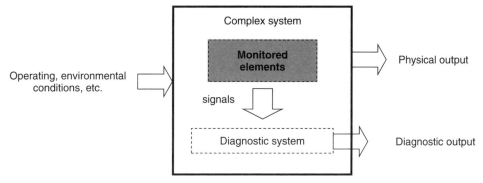

Figure 5.63   Condition monitoring of a complex system thanks to an on-board diagnostic subsystem.

diagnostic output is used as a feedback to adapt the behaviour of the monitored elements (closed loop).

In industry, some diagnostic systems are designed to monitor the condition of manufacturing plant, for example the slow degradation or sudden failures of components. In this case the diagnostic output can give indications for optimal maintenance policies.

The diagnostic output is often expressed in terms of *indexes* related to the state of health of the monitored elements.

By calibration we refer to the diagnostic system parameters setting, usually made during development phases. For example, the setting of a threshold value for a diagnostic index (to be related to the physical output of the complex system) is a calibration problem.

Random variation of the input determines random variation to the physical output and to the diagnostic indexes (see Figure 5.64).

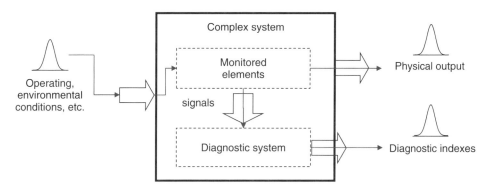

Figure 5.64   Variation: from the input to the output.

By robust calibration we refer to the diagnostic system parameters setting under a robustness perspective. The objective is to make the performance of the diagnostic system as insensitive as possible to the sources of variation.

Since 2001 European regulations prescribe that new motor vehicles must be equipped with on-board diagnostic (OBD) systems aimed at monitoring the state of the health of

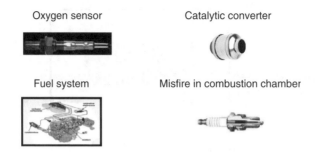

Figure 5.65   Elements monitored by the automotive on-board diagnostics.

engine components or functions affecting polluting emissions. An OBD system should monitor four vehicle elements (Figure 5.65)

During the time that the engine is working, the OBD system calculates diagnostic indexes correlated to the state of health of the monitored elements. The algorithms of the OBD system are characterised by some parameters and index thresholds. The diagnostic index thresholds need to be related to emission regulatory limits. OBD parameters and thresholds will be the same for all examples of the same vehicle model to be produced. They must be set during the engine development phase.

The main objective of our study was to find the combination of calibration parameters and thresholds that would make the OBD system robust, meaning that it should be:

- as sensitive as possible to the degradation/failure of the monitored elements;

- as insensitive as possible to the action of noise factors.

For the oxygen sensor diagnostics we developed a one-component-at-a-time degradation methodology. A scheme of the methodology is provided in Figure 5.66. We wanted to take into account sources of variation due to production processes, engine operating conditions and environmental conditions.

The aim was also to evaluate realistic risks (false alarm and failure-to-detect) related to an adopted calibration (specifically the setting of the indexes thresholds).

We performed the experiments we designed, both in the lab and on-road.

We determined the optimal threshold for a lambda sensor diagnostic index. The methodology was initially applied to an index of the lambda sensor diagnostics measuring one of the possible anomalous behaviours of the sensor. An experimental plan was designed and performed (Table 5.43 and Table 5.44).

The graphs in Figure 5.67 represent some important results. They were the basic tools for engineers to set threshold values for an OBD diagnostic index.

For the catalytic converter diagnostics we developed a multiple-component degradation approach. In fact, although the one-component approach gave good results, a further substantial improvement was imagined by taking into account the concomitant degradation of several monitored functions.

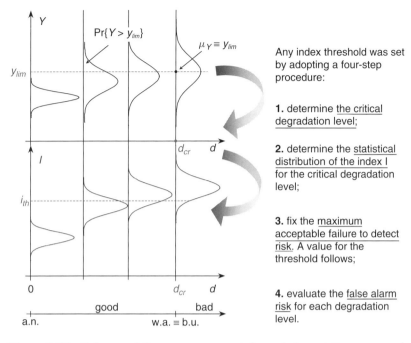

Any index threshold was set by adopting a four-step procedure:

1. determine <u>the critical degradation level</u>;

2. determine the <u>statistical distribution of the index I</u> for the critical degradation level;

3. fix the <u>maximum acceptable failure to detect risk</u>. A value for the threshold follows;

4. evaluate the <u>false alarm risk</u> for each degradation level.

Figure 5.66   Scheme of the one-component degradation at time approach.

Table 5.43   Factors chosen for the experiments.

Factor	Description	Coded levels		
A	Oxygen sensor degradation	0	1	2
B	Oxygen sensor specimen (noise factor)	0	1	–
C	Noise factor	0	1	–
D	Noise factor	0	1	–
E	Vehicle specimen (noise factor)	0	1	–

In this approach both physical and computer experiments were designed and performed, the latter by using simulation algorithms of the OBD system working off-line.

This combination greatly reduced the experimentation time, which would have been unaffordable otherwise.

The basic idea is shown in Figure 5.68. We wanted to search for the calibration setting that best approached such an ideal situation.

5.3.5.4.1 Robustness indicator   The performance of a calibration setting was assessed by means of signal-to-noise ratios for the natural effects and loss functions for the induced effects. An overall performance criterion and measure was formulated.

Table 5.44   The adopted experimental design: OA (2^4, 3^1, 24).

Std	Run	A	B	C	D	E
1	9	0	0	0	0	0
2	6	0	1	0	1	0
3	2	0	0	1	1	0
4	10	0	1	1	0	0
5	11	1	0	0	0	0
6	3	1	1	0	1	0
7	5	1	0	1	1	0
8	4	1	1	1	0	0
9	7	2	0	0	0	0
10	1	2	1	0	1	0
11	8	2	0	1	1	0
12	12	2	1	1	0	0
13	20	0	0	0	1	1
14	13	0	1	0	0	1
15	24	0	0	1	0	1
16	23	0	1	1	1	1
17	21	1	0	0	1	1
18	16	1	1	0	0	1
19	15	1	0	1	0	1
20	19	1	1	1	1	1
21	14	2	0	0	1	1
22	18	2	1	0	0	1
23	22	2	0	1	0	1
24	17	2	1	1	1	1

The experimental approach is depicted in Figure 5.69. The experimental factors were:

- calibration parameters (Θ);
- degradations (Δ);
- noise factors (N).

Peculiarities of the problem were:

- the calibration parameters affect the OBD indices only (not the emissions);
- OBD algorithms working off-line were available.

Therefore, the experiments could be split in two phases.

1. In phase I we carried out physical experiments on vehicle prototypes by using one combination of calibration parameters.
2. In phase II we carried out computer experiments on PCs using the simulation algorithms off-line.

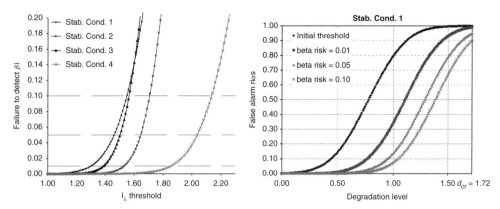

Figure 5.67   (a) Failure to detect risk versus threshold values. (b) False alarm risk versus degradation level.

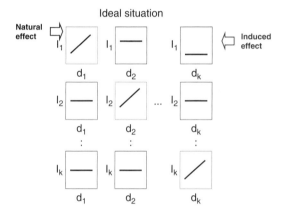

Figure 5.68   Basic idea of a good diagnostic system.

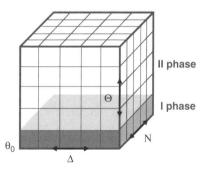

Figure 5.69   The experimental approach for the multi component degradation methodology.

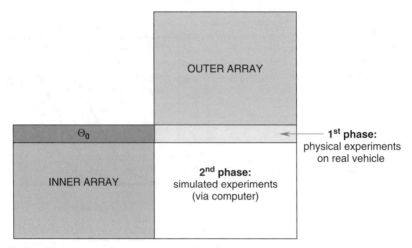

Figure 5.70   Experimental arrangement for the multi component degradation methodology.

Table 5.45   Degradations adopted for the physical experiments.

		Units	Levels
A	Downstream oxygen sensor degradation	ms	3
B	Upstream oxygen sensor degradation (mode1)	mV	3
C	Catalyst degradation (OSC)	mg	3
D	Upstream oxygen sensor degradation (mode 2)	ms	3
E	Upstream oxygen sensor degradation (mode 3)	ms	3

Figure 5.70 is another way to schematise the experimental approach.

For the physical experiments we adopted the degradations and levels shown in Table 5.45.

The selected plan was a fractional factorial 3^{5-2}. It is a plan of resolution III and *minimum aberration*. All main effects and an interaction of specific interest (the interaction CE) were clearly possible to estimate. The plan required 27 working days to be finished.

The adoption of the methodology gave several advantages:

- experimental economy (accurately planned physical and computer experiments);

- highly improved calibration (OBD parameters and thresholds): more sensitive to degradations affecting the monitored functions; less sensitive to noise factors.

- the possibility to set threshold values based on risks;

- forecasting and reduction of false alarms and missed alarm risks;

- development of a dedicated calibration software which is now currently adopted for any vehicle model.

5.4   To assess the measurement system

Variation is inherent to all processes, not excluding the measurement process itself. It is a common experience that if the same characteristic is measured several times it is unavoidable to find different results.

In business and industry, but also in science and more in general wherever we need to measure, variation of the measurement system needs to be understood, evaluated and controlled.

In fact among the possible effects of a poor quality measurement system we may find that:

- the actual variation of the system of our interest remains unknown;

- good products can be wrongly rejected and bad products wrongly accepted;

- process capability (see Section 5.2.1) appears better or worse than actually it is;

- decision making is more difficult;

- problems with suppliers and customers arise.

The measurement system is the technical apparatus used for making measurements and all factors associated with the measuring situation. A chosen measurement system should be suitable for the intended function.

5.4.1   Some definitions about measurement systems

The ability of the measurement system to detect small differences in the characteristic of interest to be measured is usually denoted as *discrimination*. As a general rule, the measurement system should be able to discriminate at least to 1/10th the specification interval (see Section 5.2.1).

When studying measurement uncertainty it is advisable to make a list of the possible causes of measurement error. Such a list may be created by a group of experts on the measurement system, operators, technicians, and so on. An Ishikawa diagram is a useful tool for this scope (Figure 6.71).

The *true value* (also *actual value*) of the characteristic to be measured is unknown and is unknowable in most cases. The *reference value* is used as a surrogate of the true value, but it needs an operational definition.

The common problems encountered with measurement systems are:

- the poor resolution;

- low *accuracy* (also *bias*);

- low *precision*;

- lack of reproducibility and repeatability;

- instability and drift.

The concepts of accuracy and precision are well explained graphically in Figure 5.72.

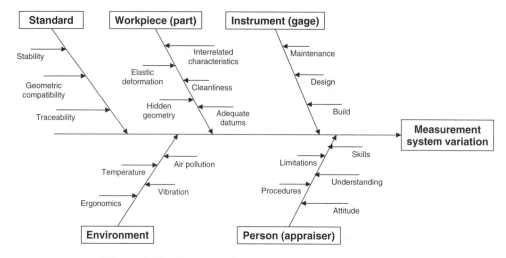

Figure 5.71   Sources of measurement system variability.

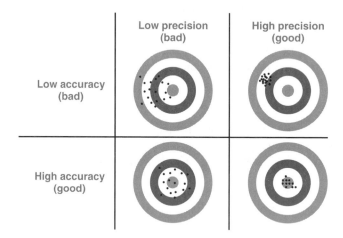

Figure 5.72   Graphical representation of the concepts of accuracy and precision.

Measurements can be *direct* or *indirect*. In direct measurement the characteristic of interest is directly measured. For example, the diameter of a shaft measured by a simple caliper is an example of direct measurement (Figure 5.73).

Conversely, for measuring the height of a tree, for example, we can measure the length of its shadow and the length of the shadow of a reference object of known length (see Figure 5.74). This is an example of indirect measurement.

5.4.2   Measurement system analysis

The main aim of measurement system analysis (MSA) is to split variation into 'components' due to different sources. A study aiming to identify and calculate components of variation is called a 'gauge repeatability and reproducibility' (R&R) study.

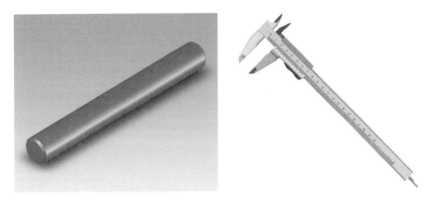

Figure 5.73   A simple caliper (b) for the direct measurement of the diameter of a shaft (a).

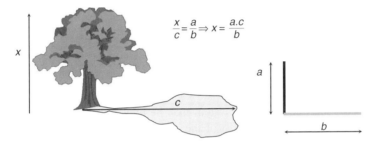

$$\frac{x}{c} = \frac{a}{b} \Rightarrow x = \frac{a.c}{b}$$

Figure 5.74   An example of indirect measurement.

The *lack of repeatability* is the variation arising when one operator is repeatedly measuring the same object with the same gauge.

The *lack of reproducibility* is the variation arising when different operators are measuring the same object with the same gauge.

Possible causes of poor repeatability are:

- *within-part*: shape, position, surface finish;

- *within-instrument*: repair, wear, poor quality, poor maintenance;

- *within-method*: variation in setup, technique, holding;

- *within appraiser*: technique, position, lack of experience, feel, fatigue;

- *within environment*: short-term fluctuation in temperature, humidity, vibration, lighting.

Possible causes of poor reproducibility are:

- between methods;

- between appraisers;

- between environments.

The main goals of gauge R&R analysis are:

- to identify and quantify the dominant causes of measurement variation;
- to compare the variation of the measurement system with the variation of the measured object.

An index used in gauge R&R analysis is:

$$\Upsilon = \frac{\sigma_{R\&R}^2}{\sigma_{R\&R}^2 + \sigma_{product}^2} \tag{5.108}$$

where:

$\sigma_{R\&R}^2$ is the variance due to lack of reproducibility and repeatability. It is normally assumed that: $\sigma_{R\&R}^2 = \sigma_{repeatability}^2 + \sigma_{reproducibility}^2$;

$\sigma_{product}^2$ is the variance of the characteristic of interest (e.g. the diameter of shafts produced in a certain factory).

As a rule of thumb we can say that:

$\Upsilon < 0.1$: the measurement system can be considered satisfactory;

$0.1 \leq \Upsilon < 0.3$: the measurement system may be questioned, depending on the importance of the application, the cost of the gauge and the cost of its maintenance;

$\Upsilon \geq 0.3$: the measurement system is not reliable at all, and we should make any effort to improve it.

If lack of repeatability is large compared to the lack of reproducibility, we can deduce that:

- the gauge needs maintenance;
- the gauge can be redesigned;
- there is excessive within-part variation.

Conversely, if lack of reproducibility is large compared to the lack of repeatability, we can deduce that:

- the operators need to be better trained;
- instructions on the use of the gauge may be not clear;
- there is a need to have some fixture to help the operators.

5.4.3   Lack of stability and drift of measurement system

Environmental conditions, for example room temperature fluctuations, can cause a random bias in the measurement system. This is called 'lack of stability'.

Drift is different and is due to the possible wear of the measurement instruments, which causes a systematic degradation and a consequent bias over time.

The following gives ideas on how to detect lack of stability and drift:

1. take some items to be used as 'standards';

2. measure the standards with a 'perfect' measurement system to obtain reference values;

3. measure the standards by using the measurement system under evaluation, and repeat the measurements a number of times;

4. estimate the bias by using: $\hat{b} = \overline{y} - y_{\text{ref}}$;

5. repeat steps (3) and (4) at possibly equally spaced time lags;

6. plot the bias as a function of time – let b_i be the bias at time $i = 1, \ldots, T$;

7. if no drift is detected in (6), the lack of stability may be estimated as:

$$s = \sqrt{\frac{\sum_{i=1}^{T} (b_i - \overline{b})^2}{T - 1}}$$

If s is too large, then you need to search for the root causes and put the measurement system under statistical control.

If a drift is detected, it is necessary to 'calibrate' the measurement instrument. In the worst case the measurement system may need to be scrapped.

5.4.4   Preparation of a gauge R&R study

- Plan the study carefully.

- Select parts and appraisers (choose appraisers who normally use the gauge, or at least are well trained on using it).

- Ensure that the gauge has a good resolution which is sufficient to discriminate between measurements.

- Is it a destructive test? If yes, then think about the implications.

Before running the gauge R&R study, it is a good practice to label all parts and to randomise the order of execution of measurement sessions.

5.4.5   Gauge R&R illustrative example

For an illustrative example of an MSA and gauge R&R study we will use the paper helicopter.

The aim of this illustrative example is to apply MSA to a well-known device, in a way that can also be performed in a classroom.

The paper helicopter is shown in Figure 5.32.

In a brainstorming session we will:

1. make some trial experiments to get acquaintance with the system under study;

2. define the performance(s) of the system, which are the characteristics to be measured;

3. identify possible sources of variation;

4. think about the design parameters, for example:

   a. paper density,

   b. blade length,

   c. setting angle,

   d. body bending type,

   e. bottom bending,

   f. width.

We will keep the following factors constant, for example:

- height of the helicopter: 29.7 mm (A4 format);

- the material used: commercial paper;

- diaphragm height (30 mm).

Three helicopter specimens of the same identical design were built. The release height was decided: the launcher was a tall student who was standing on a chair, in that way he could easily touch the ceiling of the room (Figure 5.75). Also the modality of release was decided: the launcher had to keep the helicopter still with the blades on the ceiling and release it as soon as one of the appraisers says 'go'.

Three appraisers were selected, each one with his own chronometer.

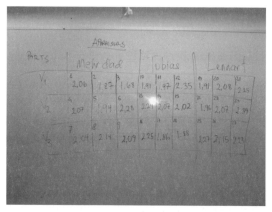

Figure 5.75 Execution of the experiments in the classroom. Helicopter flight times reported on the whiteboard.

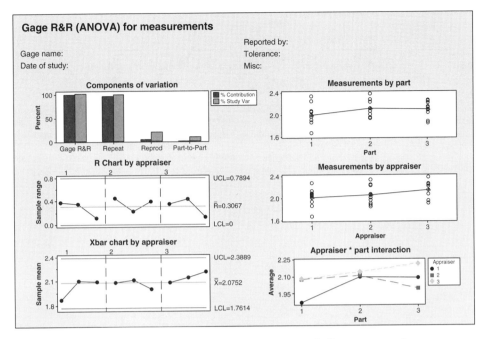

Figure 5.76   Gauge R&R results for the paper helicopter experiments.

Twenty seven launches were made, and the order was randomised. The results of the data analysis are reported in Figure 5.76.

References

Box, G.E.P., Hunter, H. and Hunter, G. (1978) *Statistics for Experimenters*, John Wiley & Sons, Inc., New York.

Draper, N.R. and Smith, H. (1998) *Applied Regression Analysis*, 3rd edn, Wiley-VCH Verlag GmbH.

Green, P.E. and Rao, V.R. (1971) Conjoint measurement for quantifying judgmental data. *Journal of Marketing Research*, **8**, 355–363.

Hosmer, D.W. and Lemeshow, S. (2000) *Applied Logistic Regression*, 2nd edn, John Wiley & Sons, Inc., New York.

Kotz, S. and Lovelace, C.R. (1998) *Process Capability Indices in Theory and Practice*, Arnold, London.

Montgomery, D.C. (2001) *Introduction to Statistical Quality Control*, John Wiley & Sons, Inc., New York.

Park, S.H. (1996) *Robust Design and Analysis for Quality Engineering*, Chapman & Hall, London.

Phadke, S.M. (1989) *Quality Engineering Using Robust Design*, Prentice-Hall International Edition.

Taguchi, G. (1986) Introduction to Quality Engineering – Designing Quality into Products and Processes, Asian Productivity Organization, Tokyo.

Wu, C.F.J. and Hamada, M.S. (2009) *Experiments*, 2nd edn, Wiley-VCH Verlag GmbH.

6

Six Sigma methodology in action: Selected Black Belt projects in Swedish organisations

In this final chapter of the book we present a summary of a selected sample of Six Sigma Black Belt projects carried out in the last three editions of the Six Sigma Black Belt course offered by the Chalmers University of Technology in Gothenburg, Sweden.

In fall 2007 Chalmers launched an international master's programme on Quality and Operations Management (QOM). The programme includes a number of courses, among which is a Six Sigma Black Belt course. To complement the theoretical knowledge in the course, private and public organisations are invited to send in participants.

The Chalmers Black Belt course is based on traditional Black Belt course material. It is necessary that participants face real industrial problems during the course. Thus, the industrial participants work in groups together with two or three students. This arrangement is a win-win solution – the organisations get access to a number of problem-solving tools, get some of their employees trained as Black Belts and get important problems solved; from the educational point of view the students get access to real world problems and get valuable input to the programme from the industrial participants.

As with many other Black Belt courses, the course is planned according to the DMAIC framework (see Chapter 1). Illustrations are taken from product and service industry environments and it is advantageous if the participating organisations provide examples and cases in order to increase participant engagement. Graphical methods are emphasised and a critical attitude is taken in order to avoid some easily avoidable pitfalls. Facilitation

Statistical and Managerial Techniques for Six Sigma Methodology: Theory and Application, First Edition.
Stefano Barone and Eva Lo Franco.
© 2012 John Wiley & Sons, Ltd. Published 2012 by John Wiley & Sons, Ltd.

and support for the selection of projects and feedback in the DMAIC process is provided within the frame of the course.

It is important that the industrial participants have some basic mathematical and statistical knowledge. Also, they should be prepared to devote time not only to study the course material between course sessions, but also to apply the ideas to problems within their organisation. The industrial participants make suggestions of possible Black Belt projects early before the course start. These projects should have the potential to reduce costs and increase customer satisfaction considerably.

The participating organisations are required to provide their selected Black Belt candidates with the time and resources needed to successfully work with the Six Sigma projects, where considerable cost savings and increasing customer satisfaction could be gained, if the project is successful. Also, the participating companies will provide the group of QOM students possibilities to work together with the industrial participants within the company.

A formal examination is held at the end of the course and a Black Belt certificate is issued once a candidate has passed the examination.

The case studies collected, edited and reported in the chapter by the authors are a selection of unpublished reports written by the Black Belts listed below.

List of the selected projects

1. 'Resource planning improvement'. Project carried out in 2009 at SAAB MICROWAVE SYSTEMS by Johan Hammersberg, Johan Holgård, Hamed Hakimian and Alireza Habibzadeh.

2. 'Improving capacity planning of available beds: a case study of the Medicine/ Geriatric Wards'. Project carried out in 2009 at SAHLGRENSKA & ÖSTRA HOSPITALS by Racelis Acosta, Marcus Danielsson, Anita Lazar and Claes Thor.

3. 'Controlling variation in play in mast production process'. Project carried out in 2010 at ATLET by Anders Lindgren, Rickard Eriksson and Ellen Hagberg.

4. 'Optimising the recognition and treatment of unexpectedly worsening in-patients'. Project carried out in 2010 at SKARABORG HOSPITAL by Jana Hramenkova, Maria Dickmark, Arash Hamidi and Amirsepehr Noorbakhsh.

5. 'Optimal scheduling for higher efficiency and minimal losses in Warehouse'. Project carried out in 2011 at STRUCTO HYDRAULICS AB, by Therese Doverholt, Anna Errore and Sepideh Farzadnia.

6. 'Reducing welding defect rate for a critical component of an aircraft engine'. Project carried out in 2011 at VOLVO AERO CORPORATION by Sören Knuts, Henrik Ericsson and Arnela Tunovic.

7. 'Attacking a problem of low capability in final machining for an aircraft engine component'. Project carried out in 2011 at VOLVO AERO CORPORATION by Johan Lööf, Giovanni Lo Iacono and Christoffer Löfström.

6.1   Resource planning improvement at SAAB Microwave Systems

Project purpose: to minimise sudden variations in the order stock for the storage department at SAAB Microwave Systems (SMW)

Organisation: SAAB Microwave Systems, Göteborg, Sweden

Duration of the project: five months (2009)

Black Belt candidates team:
Johan Hammersberg, SAAB Microwave Systems, Gothenburg, Sweden
Johan Holgård, Hamed Hakimian and Alireza Habibzadeh – masters students.

Phases carried out and implemented tools: Define (project charter, five whys, Gantt chart, SIPOC, process map) – Measure (interviews, observation, histograms) – Analyse (ANOVA test, cause-and-effects diagrams, Pareto charts).

6.1.1   Presentation of SAAB Microwave Systems

SAAB Microwave Systems is a world-leading company in supplying radar and sensor systems both for military and civil security. These systems are used to generate information superiority in order to generate the capability of predicting opponents' intentions in the air, on land and at sea.

The manufacturing methods of SAAB Microwave Systems are more like project-oriented companies. In other words, when a customer puts in an order for a product, a project is defined for that order until the end product is delivered to the customer. Since each customer has high importance in this manufacturing environment and it is critical not to lose even one customer, production planning is done based on an unlimited amount of resources.

6.1.2   Project background

The overall planning philosophy of the production at SMW (SAAB Microwave Systems), which is all planning is made based on the assumption that the resources are unlimited, leads to the effect that several production projects and orders can be initiated and issued at the same time. This, however, does create some problems in planning resources in storage department and consequently for the production lines to deliver the product on time. These delays occur when the storage department becomes incapable in providing requested order stocks from the production line due to high and sudden changes in pre-forecasted figures for order stocks. In such cases the storage department does not have enough time to react to changes and to hire the needed resources for performing orders.

For these reasons, the company needs to find methods and suggestions to reduce the effects of this philosophy on the storage department.

6.1.3 Define phase

The frame of the project is outlined in a project charter and in a Gantt chart (Figures 6.1 and 6.2). The project owner Göran Berggren experienced problems with variation in the workload for the storage department. It is not the variation itself that caused the problems but the short notice in which it appeared. Today, he plans his resources on the information he gets from a computer system called C:M. The information consists of the number of orders that are planned for each upcoming week until the end of the year. The main

Business Unit	SAAB Microwave System	Product/Service Family	
Name	Göran Berggren	Telephone Number/e-mail	
Sponsor & Process Owner		Site or location	
Project Start Date	2009 - 02 - 06	Project Target Completion Data	
Element	**Description**	**Charter**	
1. Process Impacted	Where opportunity exists	In the forecasting process	
2. Benefit to Internal/External Customers:	Define the customer, most critical requirements and note internal customers.	To be able to trust the forecast with one week notice	
3. Business Case: Benefit to the business:	Define the improvement in business performance that is anticipated and when.	Better planning of resources	
4. Purpose/Objective /Problem Statement	State project's purpose, overall objective/problem	An improved forecast	
5. Project Scope	Define the part of the process that will be investigated.	The forecasting process and input to forecast	
6. Team Members	List Core Team and technical experts	Team plus Göran Berggren and Robert Larsson	
7. Goals	Define the Baseline, your Goal for the project and the Best Case target for improvement.	Actual Value (Baseline)	Your Goal by Project End Date (This is the net Improvement)	Best Case Goal
		2	2009 - 05 - 20	1

Figure 6.1   Project charter.

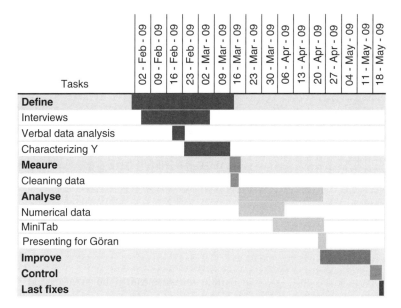

Figure 6.2   Gantt chart.

problem with this information is that the orders are not set; they can be moved in time by any planning department involved until one day before green light and execution. This is the reason for the changes in very short notice, and therefore Göran cannot trust the order/stock information given to him each week. The planning philosophy that SAAB uses makes it possible for order planners to plan unlimited number of orders on the same week. There is nothing in the system stopping an order planner from moving all orders to the same week one day before execution. This is because the planning philosophy assumes unlimited resources, so that SAAB never has to say 'no' to an order. Göran has no tool other than the order/stock information from the C:M system to help him predict the future workload. Figure 6.3 is an example of how it can look when this problem occurs. The peak on the picture went up by 2000 orders during one night and left Göran with less than two days to acquire the required resources.

Definition of the problem was done through interviews with Göran, a technician Robert, one of the order planners and the head of project planning. 'Five whys' were used to break down this problem into the core problem. Some of the questions were:

• why can't Göran trust the order/stock information?

• why does it change with short notice?

• why do peaks occur?

• why isn't there more communication between departments?

The core problem for the storage department and Göran was identified as lack of communication between departments, and that Göran cannot plan his resources with the

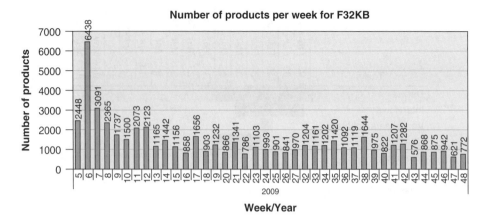

Figure 6.3   Order stock for 2009, there is a large peak in week 6.

information from C:M only. He can handle the variation if he gets around one week's notice but at the moment the variation is a problem since it may occur with short notice.

As mentioned above, Göran can handle peaks if he receives information about them at least one week before they occur. The reason why there are big peaks still had to be investigated. During interviews with personnel from other departments a list of causes for peaks was collected. The three main ways peaks occur are:

- several big orders are started at the same time;
- order planners that are away from work start a whole list of orders when they are back at work or before they go away for a vacation or a period of leave;
- if a key article has been delayed, many orders get started upon its arrival.

With this information, the planning philosophy was identified as the main reason for peaks, which is outside the scope of this project. Because of the planning philosophy used, a part of Göran's problem is symptoms of problems caused in other departments. However, there are some warning signs for when a peak may occur. One of these warning signs is that orders move in time between weeks. A reason for this is that the orders are waiting for an article to arrive in storage. To better understand the scope of the project and to identify what can be done to help Göran, a SIPOC (suppliers, inputs, process, outputs and customers) diagram was used (Figure 6.4).

This SIPOC diagram, together with the process map (Figure 6.5) that was done together with storage department personnel and managers, showed the scope of the project. Because the main problem is caused outside this scope, the project involves creating a process that will absorb the variation occurring in the amount of orders for the storage department. The variation itself cannot be solved within the scope.

The initial step is triggered by the customer. According to the customer order a project is defined and a project manager is appointed to control the project. He sets the start and end dates of the project. The project characteristics go into the IFS (Industrial and Financial Systems) system and required parts and resources are estimated. The product

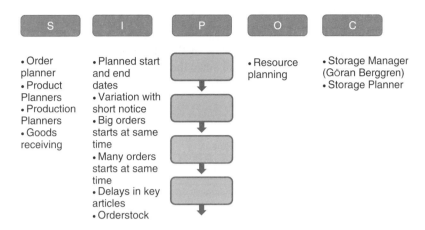

Figure 6.4   SIPOC.

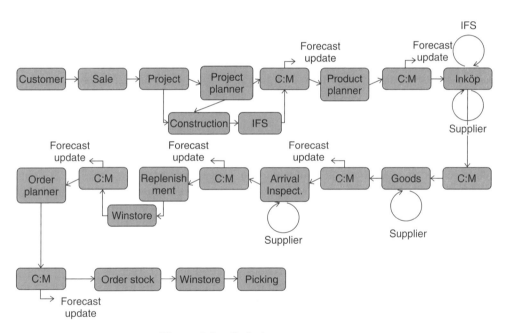

Figure 6.5   Ordering process map.

planners define the production start and end dates for the ordered products. They also arrange the required resources and the dates that these resources should be available. Data for parts required goes to the suppliers and dates they have to be picked from the storage department are also reported to the order planners. Then, when the order planners give the green light for the order the ordered articles are picked from the storage to go to the production line. The complexity of this process shows that a lot of the problems the storage department is experiencing are originated from other departments. This led to the conclusion that the project had to be concentrated within the storage department and should not, therefore, take into consideration the problems in other departments, except if specific root sources for large variations can be identified.

Figure 6.6 clearly demonstrates that most of the variations are originating from the input variables and thus cannot be changed since they are out of the scope of the project. The task here is to make the process used by storage department robust or insensitive to this variation.

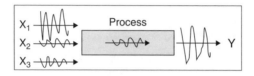

Figure 6.6   Process affected by variation.

Because of the project type, historical data has to be analysed at a very early phase. Historical data can be collected from the two computer systems used by SAAB Microwave Systems; the Winstore and the C:M systems. The C:M system is the production system and the Winstore system is the storage system. The data in the Winstore system consists of information about orders, such as type, date and frequency, and stores data for five years back in time. The size of the file is 1 million rows and 1 GB. The C:M system database only stores a small fraction of that amount of data going back three to six months.

To build this process, patterns have to be recognised of what was occurring when these peaks appeared. The limitations in the Winstore data is that it only has orders that have been picked, so there is no real information about order peaks in the C:M systems that were not handled by the storage department. However, the historical data from the Winstore system is still useful in creating an understanding of the situation. It could be used in building the process for the storage department, especially building a robust process for Göran to help him plan his resources. The data received from the C:M and Winstore system is analysed using MINITAB©. We determine this data to be trustworthy for our project. The reason why we trust it is because all the historical data has been retrieved from the IT systems by the same experienced technician. When verifying the data, reasons for why it may not be trustworthy were discussed, such as different people retrieving data and different methods used in retrieving data. This was discussed with the technician and, after a meeting about how the systems work, we determined the data to be trustworthy.

The three-step method used to design the process is:

1. map the process;

2. make a cause and effect diagram;

3. make a Pareto analysis.

This method is designed to identify and prioritise the factors influencing the process that should be built. In step 1, all factors influencing the process have to be identified. That means identifying factors which are influencing all sub-process steps; see the process map. Step 2, a cause and effect diagram is used to identify root causes of the problem. Gathered data from interviews, process map, Winstore, brainstorming and other sources is used to draw the cause and effect diagram. Finally in step 3, a Pareto chart is used to determine the most important factors and causes of the problem that have a higher influence on system performance. This method will sort out three to five important factors to work with.

There are two types of cost reductions that can be gained from this project. One is the direct cost for the labour in the storage department and the other cost is the cost of delayed deliveries from the storage department. The second one is very difficult to estimate as the storage department supports the whole company and there is a large variation in consequences for the projects in case of a late delivery.

6.1.4 Measure phase

Because of lack of information about historical inputs to the process, that is the historical order stock for the storage department, measurements are mainly done on historical outputs from the Winstore system and orders of stock ahead, registered in the C:M database. In other words, the Winstore database tells us how many orders they have picked each day, each month and each year, and the C:M database tells us how many orders they will have to pick for the coming weeks until the end of the year.

Information about picked orders will be used as feedback for the manager so that they will be able to better trust the order stock forecast.

The historical data in the Winstore database had to be sorted and non-useful data filtered out. Original data consisted of orders from 2001 until today. The storage department used today at SAAB Microwave Systems was opened at the end of 2003. Also, the data consisted of many different types of orders, for example orders for spare parts. Different order types were identified with codes. After filtering out unwanted data from 2004 until today and only using picked orders, a histogram was drawn.

This histogram (Figure 6.7) shows picked orders, with every line representing a 10-day period. A clear annual pattern can be seen. It is important to remember that this is historical data of what has been picked; it does not say anything about possible order peaks showing up with short notice, this is handled by the C:M system with only fractional memory.

During the Measurement phase we also gathered data from the order stock in the C:M database. The database has no memory; it only looks at the order stock until the end of the year. These reports were collected to see if any sudden peaks appeared and to see if there are any large movements in the order stock. The first order stock was collected in week 5 and it is presented in Figure 6.8. From week 10 the order stock reports were collected every second day.

6.1.5 Analyse phase

When previous order stock reports are compared to what actually was picked for the corresponding weeks it became obvious that the actual picking does not follow the peaks and lows of the report (see Figure 6.9). Big peaks are picked over a couple of weeks instead and the workload is kept quite constant. For a period of time after the peak, the actual pickings stored in the Winstore system are a bit higher than what is forecasted in the C:M system.

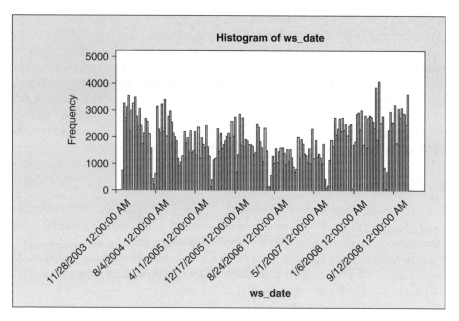

Figure 6.7   Number of picked rows in 10-day periods.

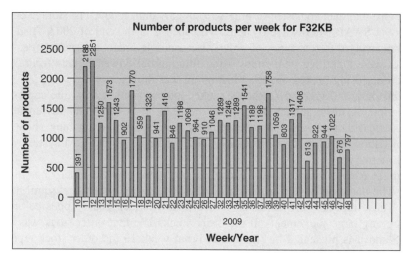

Figure 6.8   Order stock report for week 10.

With this discovery in mind the group decided to investigate if an annual cycle could be found in the data. By using ANOVA test on the five years of data available, an annual cycle that shows the average number of picked rows per day for each month over one year was created (Figure 6.10). This model will be useful even if trends go up or down, since it shows the months relative to each other.

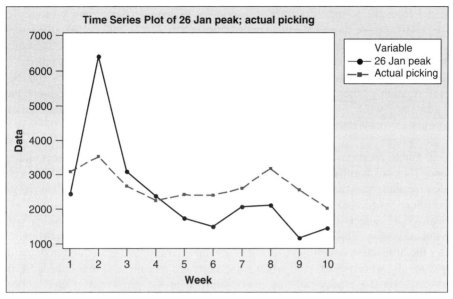

Figure 6.9   The forecasted ordered number of rows to be picked and the picked number of rows per week.

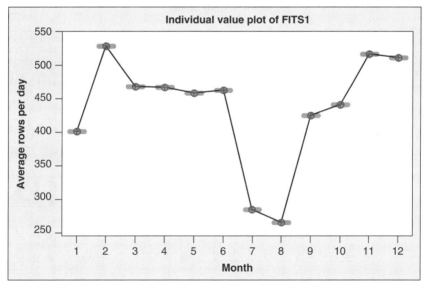

Figure 6.10   The average number of picked rows per day and month during the last five years.

To create better understanding of the problem and to be able to discuss the issue with personnel from the storage department, a number of Pareto charts were created. To be able to discuss problems and to kill so called 'sacred cows' (i.e. beliefs taken for granted), data has to be shown in a way that is easy to understand. One example of a 'sacred cow' can be the belief that one department always changes their orders at the last minute. This can be taken for granted even though no one has ever seen any proof for this belief. The information is there but it is not used to explain the reality, but a simple Pareto chart can show what is really going on. Data presented using the 7 QC tools is often an eye-opener for management, and it often challenges their initial beliefs and the way they view the system.

Four Pareto charts were created showing product ID, user ID, article number and order planner. They all highlighted areas for improvement. Because the Pareto charts can be updated regularly they can be used to help the management investigate the alarms given by the future order stock control chart.

Figure 6.11 depicts the Pareto chart for order IDs. Those order IDs starting with C are requisite orders. This graph can be important not only for the storage department, but also for the production department. C orders are important since the production department charges for the whole order even when some parts are missing and that is because these orders can stop the production line if not fulfilled at the proper time. One important observation from the Winstore data is that the large orders are usually not picked in the storage department in the same week. Instead they are spread out over long time periods, even as long as one year.

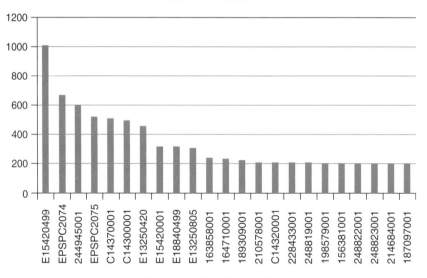

Figure 6.11   Order ID.

The user ID identifies the persons responsible for picking the incoming orders from the storage section for the production line. Figure 6.12 shows the small number of people who have picked large number of orders from January 2008 to April 2009.

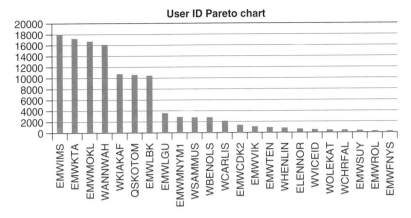

Figure 6.12   User ID.

Figure 6.13 shows the Pareto chart for different parts. By analysing Pareto charts of order IDs and part numbers it is possible to predict larger amounts of orders when the company has received the same types of orders. Here part SCA103030 had the highest frequency of picking. More information coming from this data is that the storage department should almost always have this part on hand since it seems it is a common part in many products and its lack may cause production to stop.

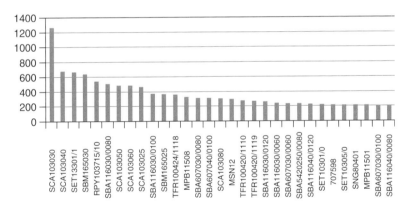

Figure 6.13   Part number.

Figure 6.14 simply shows which order planner released the largest number of orders during the previous half year. The list of order planners contains more than 400 signatures. However, in reality this list shows the major source of the large variation in the input data to the order stock process. JHZ approved the most orders for the storage department during this period.

By analysing different data and results we could gain a general view of the problem, the starting points and the important areas which should be referred to in case of big variation. Among these areas three to five of them were selected as the main causes of

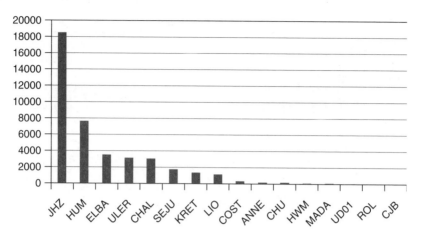

Figure 6.14   Order planner.

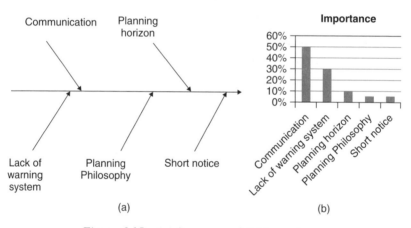

(a)   (b)

Figure 6.15   (a) Sources and (b) importance.

the problem. These causes are shown in Figure 6.15a. We also classified them according to their importance, just to make it clearer for the reader (Figure 6.15b). Ranking was done following interviews with managers and order planners, and by relating their contribution to the scope of the project.

6.1.6   Improve phase (ideas and intentions)

One large difficulty that has been exposed by the analysis is the difficulty of automatically finding and predicting sudden peaks in the order data in the C:M system, that is there is no pattern in the data that can predict the changes that occur at short notice. Since this is difficult to do automatically we suggested that an early warning system was made. By contacting the 10 most frequent order planners in Figure 6.14 and asking them to highlight the major forthcoming orders in advance in an e-mail to the storage department manager Göran, he can plan for extra resources in advance. This will give him a week's notice in the best case scenario. The early warning messages should be highlighted in the subject note in the e-mail, in order for Göran to spot it easily.

For example, investigating the order peak (Figure 6.3), we found out that this order appeared due to the fact that one order planner started a large number of orders before he went on vacation. This high order number was later picked during the week after the order (Figure 6.9). Thus with an early warning system and the annual cycle Göran would know that this is the case and he could plan his resources accordingly instead of reacting to the peak as being urgent.

Göran, the manager at the storage department, cannot trust the order stock given to him by the C:M system and cannot trust it, by itself, as a tool to plan the capacity. However, by comparing the order stock on a daily basis, movement among orders can be detected. As described earlier, orders that get moved from one week to another can be an early sign of problems with variation. In periods of much movement in the order stock more communication than usual is needed. Because this method is not about controlling the incoming orders, but rather identifying situations when the order stock is not stable, a time series graph is used (Figure 6.16). This is just one way of visualising a movement of orders, and there are other ways of doing this. SAAB Microwave Systems should implement a warning method that best fits their IT system. The time series chart will show the proportion of absolute difference between two order stocks. For example, if 150 order rows are moved from week (a) to week (b) the number plotted will be 300 divided by the total amount of orders for 20 weeks ahead. The model will not show where the movement has occurred, just that it has occurred. It is then up to personnel at the storage department to investigate the movements.

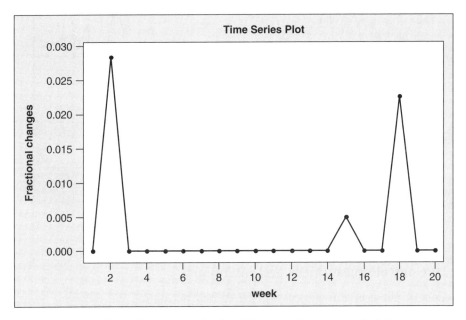

Figure 6.16   Time series plot for difference in amount of pickings.

For this chart a 20-week time period will be used, and orders ahead of this time period will not be in the calculations. The first week will also not be in the calculation, since orders being picked will reduce the orders for that week and may cause a false alarm. However, orders entering the time period of 20 weeks will show up in the control chart once a week. Every Monday, the graph will show a big peak because all orders have been moved one week up. This is why Mondays are not plotted in the time series.

All these calculations and the time series must be done by the IT system, as the order stock histogram is done in real time. Göran should be able to view this chart whenever he needs to. A very important aspect here is that this chart will not reduce the variation of incoming orders, but instead will work as a tool to identify situations where more communication between departments is needed.

If resources are planned according to the annual cycle (Figure 6.10), the storage manager will be better prepared for peaks and more able to work off a backlog of orders. The annual cycle can also be used as a tool to get production planners to place orders when there is available capacity. Here we also suggest a chart that shows the average number of order picked per day in the Winstore database system (Figure 6.17). This chart will help the manager to see that his department's capacity matches the needs for the week. He will also more easily compare it with the annual average per day and month for previous years.

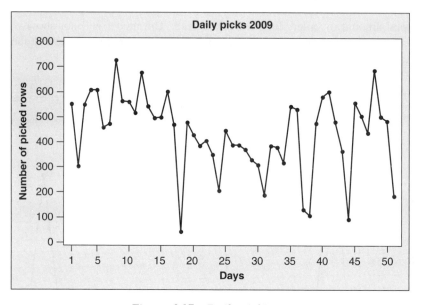

Figure 6.17   Daily pickings.

6.1.7   Control phase (ideas and intentions)

The Control phase of this project will be carried out during the following year after that the suggested improvements have been implemented. During this year Göran should try to execute the control plan in order to improve his forecasting and planning of the department activities. In the control plan the following actions that Göran needs to take when he gets an alarm from the control charts are suggested.

- On a day to day basis, extract information from the C:M database and calculate the order stock and the order stock control chart. The control chart will highlight large changes in the order stock. If these are present, locate the week for the large changes and try to find out the assignable cause to these changes by first contacting the 10 most-frequent order planners.

- Implement the early warning system by contacting the 10 most frequent order planners; ask them to use the early warning message.

- Continuously check the e-mail inbox for early warning messages. If an early warning message is received, take action to handle the increased demand or check that the order can be spread out in time, that is by using the normal capacity.

- Long-term planning should be based on the annual cycle.

- Check if the capacity needed exceeds the annual average number of rows for a long time period by comparing the control chart for the daily capacity with the annual cycle.

- In order to simplify future analysis of the capacity of the storage department, we suggest that the weekly order stock forecast generated by the C:M system is stored each week. This will simplify a similar future analysis to see if the storage department order capacity matches the needs in the order stock.

6.2   Improving capacity planning of available beds: A case study for the medical wards at Sahlgrenska and Östra Hospitals

Project purpose: to optimise the utilisation of existing beds in the medical ward, leading to reduction of variations and waiting times.

Organisation: Sahlgrenska and Östra Hospitals, Göteborg, Sweden.

Duration of the project: five months (2009).

Black Belt candidates team:
Eva Brändström, Quality Manager, Sahlgrenska University Hospital.
Racelis Acosta, Marcus Danielsson, Anita Lazar, Claes Thor – masters students.

Phases carried out and implemented tools: Define (project charter, Gantt chart, P-diagram, SIPOC, process map, CTQs, tollgate checklist) – Measure (interviews, observation) – Analyse (histograms, time series plots, box plots and interval plots, Kawakita Jiro (KJ) analysis, Ishikawa diagrams, ANOM, Pareto charts, process capability analyses, correlation).

6.2.1   Presentation of Sahlgrenska and Östra Hospitals

Sahlgrenska University (SU) Hospital was founded in 1997 when three hospitals merged; Mölndal Hospital, Sahlgrenska Hospital and Östra Hospital.

SU provides emergency and basic care for the Göteborg region, and its 700 000 inhabitants, and highly specialised care for West Sweden, with 1.7 million inhabitants. SU is also the country's centre for certain types of specialised care, especially in paediatrics (paediatric heart surgery, incubator care for premature babies, as well as treatment of paediatric endocrinology). SU is also well known for its successful transplant activity, treatment of cardiovascular diseases, immunology (research into rejection mechanisms) as well as research into vaccines.

6.2.2   Project background

According to both quality managers at Sahlgrenska Hospital and Östra Hospital, the increased demands of healthcare, due to an increased average life length, will naturally increase the flow of patients. This, in combination with a tighter budget, has proven to be a challenge for the hospitals.

Healthcare, just like industry, deals with variations in processes. Bed management and the planning of bed utilisation is one major factor that varies tremendously, and unfortunately leads to longer waiting times and queues. The bed capacity is set according to government legislation and cannot be altered easily. However, it is possible to reduce the so called 'overloading' and 'relocation' of patients by planning the capacity and managing the flow in an optimised way.

Forecasting the demand of beds is limited as well, since it is almost impossible to forecast the amount of sick people in the future. Hospitals certainly utilise previous trends for a rough count, but nevertheless it is hard to predict a specific number.

Another issue that is hard to predict is the length of time the patient stays in hospital. There is a tendency for patients to stay in the hospital for a longer time than planned, because the post-care units in the community are not able to admit them. It is estimated that one in three of all patients are occupying a bed due to unnecessary waits.

There is also a tendency that people go directly to the hospital instead of visiting the primary care units first. The government target is that 80 % of total patients should be handled by the primary care units and 20 % by the hospital, but this is not the case today.

Another factor that has left the quality managers wondering about the increasing waiting times is the poor communication between wards. There are computerised systems working for the entire hospital; however, since hospitals deal with human lives, the first priority is not to update the computer/IT systems but to care for the patient. Leaving the systems incomplete and not up-to-date affects the flow of accurate information.

6.2.3   Define phase

The project definition was stated in a project charter and in a Gantt chart for each hospital. Figure 6.18 shows the project charter for Östra hospital.

The CTQ characteristics were identified by the quality managers of the wards during a special meeting (see Figures 6.19 and 6.20). Then, process maps were used in order

Project Title: Capacity Planning of Available Beds at the Department of Medicine/Geriatrics at Östra sjukhuset			
Business Unit	Sahlgrenska University Hospital / Östra sjukhuset	Product/Service Family	Department of medicine and geriatrics
Name		Telephone Number/ e-mail	
Sponsor & Process Owner	Sponsor: Gun-Lis Olofsson, Owner; Putte Abrahamsson	Site or location	Gothenburg
Project Start Date	2009-01-23	Project Target Completion Date	2009-05-20
Element	**Description**	**Chapter**	
1. Process Impacted	Where opportunity exists	Department of medicine and geriatrics	
2. Benefit to internal/External Customers:	Define the customer, most critical requirements and note internal customers.	Department of medicine and geriatrics, patients	
3. Business Case: Benefit to the business:	Define the improvement in business performance that is anticipated and when.	Improved capacity planning for beds	
4. Purpose/objective/problem Statement	State project's purpose, overall objective/problem	Identify a model for improvement of capacity planning at Östra sjukhuset. Identify costs of poor quality (CoPQs) due to overusage of beds and adverse events as a cause of this.	
5. Puroject Scope	Define the part of the process that will be investigated.	To find a way how to utilize the number of beds available	
6. Team Members	List Core Team and technical experts.	Core term: Racelis Acosta, Marcus Danielsson, Anita Lazar, Claes Thor; Experts: Gun-Lis Olofsson, Putte Abrahamsson, Tobias Karlsson	
7. Goals	Define the Baseline, your Goal for the project and the Best Case target for improvement.	**Actual Value (Baseline)**	**Goal by project end date (net improvement)**	**Best Case Goal**
		Hard to tell	N/A	Optimal utilization of beds capacity

Figure 6.18   Project charter – Östra Hospital.

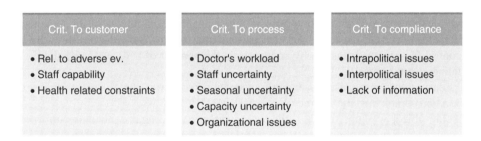

Crit. To customer	Crit. To process	Crit. To compliance
• Rel. to adverse ev. • Staff capability • Health related constraints	• Doctor's workload • Staff uncertainty • Seasonal uncertainty • Capacity uncertainty • Organizational issues	• Intrapolitical issues • Interpolitical issues • Lack of information

Figure 6.19   Critical to quality characteristics for Östra Hospital.

Critical to customer	Critical to process	Critical to compliance
• Patient's safety • Immediate care • Doctor's competence	• Bed availability • Capacity • Staff's competence • Data update • Patient first	• Governmental regulations • Hospital procedures • Environmental requirements • Health care rules and standards

Figure 6.20   Critical to quality characteristics for Sahlgrenska Hospital.

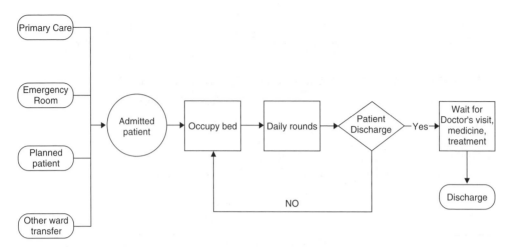

Figure 6.21   Access process map – Östra Hospital.

to describe the access to health care (Figure 6.21). The current way of working of the medical ward was analysed by utilising the SIPOC diagram (see Figure 6.22), following the actual flow and interviewing staff involved throughout the process.

With the objective of identifying causes affecting the output of the process, which is having available beds, the P-diagram was utilised for breaking down the noise factors (NFs) (the number of acute patients that arrive to the hospital needing care at the medical ward; the level of illness – are the patients acutely ill? – and the time that they can wait before receiving care without compromising their healthiness; the age of the patients; the municipality's current capacity to accept patients that are discharged from the ward) and the control factors (the transport by ambulances; the number of beds per ward; the process guidelines) (see Figure 6.23).

To ensure that certain aspects and areas were covered during the process, tollgates were used. After completing the Define phase, the checklist was assessed (see Figure 6.24). If a specific object was not fulfilled, the work was reviewed again and more information was gathered to enable the work to be completed in a satisfactory manner.

6.2.4   Measure phase

After discussions between the parties involved in the project, a decision was made that 2008 available data would be used as a baseline for the analysis. The data needed was collected from the computerised system used at the hospitals. In order to understand the situation at the medical ward and to obtain an accurate image of the different relationships, the following parameters were acquired from the system. View Figure 6.25 for chosen parameters. A decision was made to focus on the hospital stay, that is the length of stay, for each patient. Effort was put on finding different relations between the hospital stay and parameters such as reason for stay, responsible ward, and so on.

Supplier input		Process output customer		
Emergency Room, Primary Care, Other Wards	Patients	Admittance	Occupied Bed	Medicines/ Geriatrics Ward
Doctors, Nurses, Laboratories	Patients (blood or similar)	Bed Usage, Tests, Labs	Test Results, Recovery Plan	Patient
Doctors, Nurses, Pharmacies	Pharmaceuticals	Bed Usage, Treatment	Treated Patients	Patient
Doctors, Nurses, Laboratories	Patients (blood or similar)	Bed Usage, Follow Up Tests	Test Results, Healthier Patients	Patient
Doctors	Results	Discharge	Available bed	Medicine/Geriatrics Wards

Figure 6.22   SIPOC diagram – Östra Hospital.

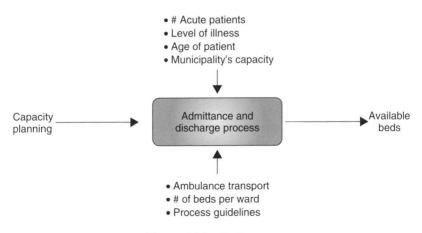

Figure 6.23   P diagram.

Main activities	YES	NO
Definition of the problem		
Definition of business case		
Prepare the Project Charter		
Define the Critical to Quality Characteristics		
Deliverables	**YES**	**NO**
Project Charter		
SIPOC		
CTQs		
P-Chart		
Project goals	**YES**	**NO**
Is the project an improvement priority and is supported by the Hospital Management?		
Has the project's potential impact and cost savings been presented and confirmed with the financial department?		
Has a Problem Statement been done?		
Have the goals been defined, and are this targets feasible?		
Have the leaders assigned the team and roles, a preliminary plan and schedule?		
Has the project charter been reviewed by both sponsors from Chalmers and Hospital to confirm his/her support?		
Has the team identified customer requirements (CTQ) of the process being improved?		

Figure 6.24   Define phase final tollgate review.

Responsible ward	Admittance date	Admittance code	Discharge date	Discharge code
60102	2008-01-01	15	2008-01-08	11
...	...	...	...	...

Figure 6.25   Merged data set layout.

6.2.5   Analyse phase

In order to understand the retrieved data, an exploratory analysis was conducted. Different graphs, such as histograms, time series plots, box plots and interval plots, were plotted to study data concerning the length of stay, the patient's age, the ward affiliation, the admittance date and their relations, overload and relocation. The majority of the patients are admitted for one day, and there is a high percentage staying for less than a day. Furthermore, it can be stated that there is a high demand on average for patients that require a longer stay during winter months, particularly November and March. It can be stated that the length of stay variable does not follow the Gaussian distribution.

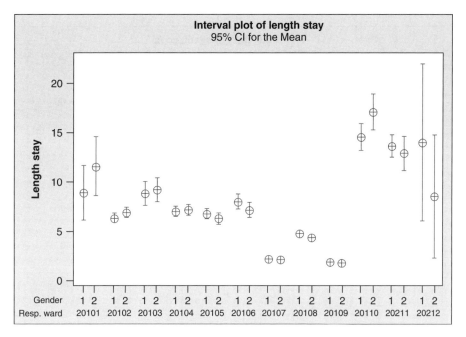

Figure 6.26   Interval plot of length of stay per gender and ward at Östra Hospital.

There is a balanced mix of male and female patients for both hospitals. The plot of Figure 6.26 shows that the Östra Hospital managed approximately the same number of men and women, except for wards 20 101, 20 210 and 20 212. We note that the medical ward is handling a great percentage of patients born in 1920. It can also be stated that these wards are the ones that have the highest variability in the length of stay and are also the ones managing patients that are staying for longer.

Some comparisons were made between the two hospitals. For example, the Sahlgren-ska Hospital's patients stay for a shorter time than at Östra. However, there seems to be less variation in the length of stay at Östra. In the end of the year, Östra's patients seemed to stay for a shorter time.

Probability plots and histograms excluded a normal distribution both for the admittance date and for the discharge date variable. A box-plot showed that admitted patients with the code 22 have the most variation, but it cannot be stated that the data is negatively or positively skewed. A Pareto chart (Figure 6.27) highlights that 79 % of all admitted patients come from the emergency room (code 39 means from ER). 74 % of the discharged patients are sent home after the hospital stay.

Concerning overload and relocation: Östra has a high variation in overload, from 0 to more than 30 on some days. The mean is 9.205, with a high standard deviation. On the contrary, there is almost no relocation (mean of 0.65 and low standard deviation); Sahlgrenska has a lower rate of overloading, it is kept below 10 almost always, its mean is 1.363 and the standard deviation is low. Also relocation at Sahlgrenska is low (a little bit higher, the mean is 1.5, and the standard deviation is a little bit higher compared with Östra).

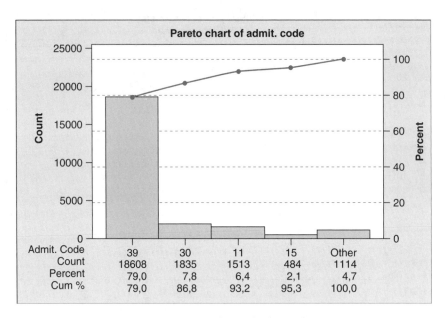

Figure 6.27   Pareto chart of admitted patients.

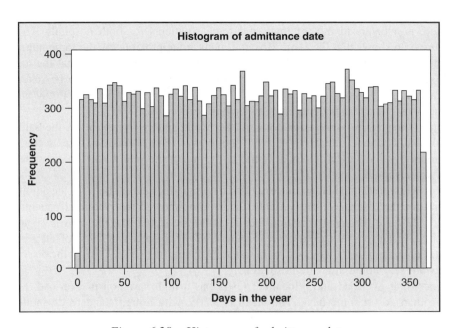

Figure 6.28   Histogram of admittance date.

Afterwards, deeper statistical analyses were carried out in order to understand the behaviour of the above variables. Tests, like T test and normality test, seem to state that length of stay follows a gamma distribution. ANOM (ANalysis Of Means) methodology was conducted for comparing the mean of each ward to the overall process mean to detect statistically significant differences: the main conclusion is that there is a high dispersion of most ward-specific means far from the overall mean. In total, 8 out of 12 wards have means far from the overall mean. Concerning the admittance date and discharge data, starting by the data histograms (Figure 6.28 shows the histogram of admittance date), the chi-square goodness of fit test was applied in order to prove that the data followed a uniform distribution.

The test confirmed the hypothesis.

Correlation analyses were checked between the measurements that were collected for 2008. Generally, no significant correlations were found.

A binomial capability analysis was conducted after having counted the defective units and classified them in two groups, defective or not defective, go/no go. There is a defect when the patient is relocated and also when there is an overload. The results are:

- *For the Östra case*:
  - total defects during 2008 = 3606
  - total patients during 2008 = 12 303
  - total percentage defective = 29.3 %
  - approximately 2.1 sigma
- *For the Sahlgrenska case*:
  - total defects during 2008 = 1058
  - total patients during 2008 = 11 251
  - total percentage defective = 9.03 %
  - approximately 2.9 sigma.

To gain a deeper understanding of the problem and of what factors affected the variation of the bed utilisation, it was decided to execute a KJ analysis including both hospitals, Sahlgrenska and Östra. Four participants were present to contribute to the analysis: one quality manager and one head nurse at the ER from Östra, and one quality manager and one head nurse from the medical ward at Sahlgrenska. The different methods gave in the end these factors to be the ones that affect the bed utilisation and determine its variations (the main key process characteristics, KPC) the most:

- round system (i.e. waiting for doctors) (7p);
- availability of non-institutional care (6p);
- utilisation control (5p);

- doctor competence and emergency room (5p);

- coordinated care planning (1p).

All that resulted in this short answer: is the bed capacity affected negatively by today's round system and the availability of non-institutional care? The affinity diagram in Figure 6.29 was constructed following the results gained from the analysis. Starting from the identified CTQs (Figures 6.19 and 6.20) and the KJ Shiba session results, the sub-KPCs and 'sub–sub KPCs' (two levels) were identified as listed below.

- *Level of care*
  - the public's general knowledge of care;
  - direct control of ambulances;
  - availability of non-institutional care.

- *Capacity planning*
  - utilisation control;
  - seasonal variation.

- *Admittance*
  - admittance routines;
  - doctor competence in ER;
  - double booked doctors.

- *Discharge*
  - doctors' rounds;
  - lack of available beds.

- *Cooperation with municipality*
  - coordinated care planning.

The noise factors (NFs) affecting the levels of sub-KPCs and ultimately the KPC can be found in the cause-effect diagram in Figure 6.30.

It was agreed that the 'most important' sub KPCs were (in descending order of importance):

1. the doctors' round system (7p);

2. availability of non-institutional care (6p);

3. utilisation control (5p);

4. doctor competence and emergency room (5p);

5. coordinated care planning (1p).

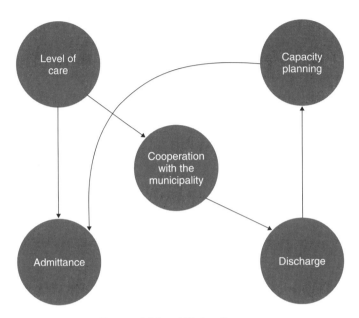

Figure 6.29   Affinity diagram.

The 'importance' is based on the fact that variation is very high with these factors, thus causing significant impact on the quality. Due to lack of quantitative measures, a qualitative point system was used, where each member placed 1, 2 and 3 points on three sub-KPCs in order of significance.

6.2.6   Improve phase (ideas and intentions)

By utilising the results from the KJ tool it was concluded that the biggest problem in the medical wards affecting the availability of beds in the daily work was the doctors' rounds. By having standardised times for the rounds, staff handling the admittance of patients have real time updates of what beds are available and how many there are going to be in a certain time. With the planning about patients improved, there is also a better possibility to plan staff between the different departments in the medical wards. Along with the standardisation of the rounds, it is very important that the meeting for discharge of patients is also standardised, in combination with the rounds, due to the fact that a lot of patients wait to be discharged just because of the meeting.

Some identified wards presented less variation in the length of stay of patients. Therefore, the other wards should take time to go through the processes in the above wards that present less variation, and ask the question: 'What is different between the processes?' 'Out of these differences, what applies to each specific case?'

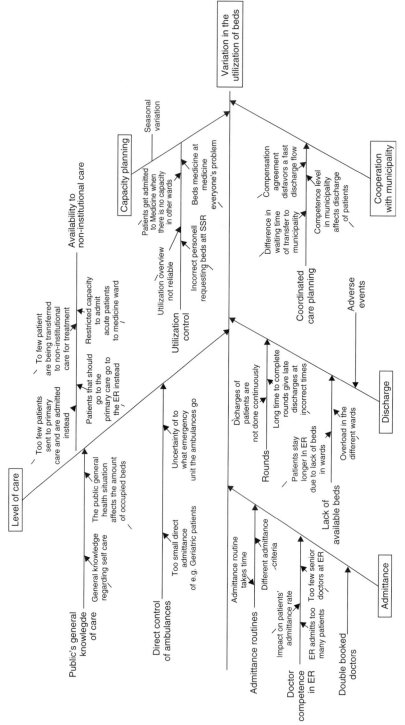

Figure 6.30 Hishikawa diagram.

The proof of uniformity in the admittance date data and gamma distribution followed by the length of stay data is very important for the quality managers to acknowledge. With this discovery, probabilities or changes in the pattern could be somewhat forecasted.

One of the tasks besides analysing the data was to provide a model that could improve the situation at hand and serve as an aid to better planning of beds. It can be useful to consider, as a way of working, the guidelines published by the Ontario Hospital Association in 2004: 'Improving Access to Emergency Care: Addressing System Issues'.

- An appropriate admission process should be able to reduce or eliminate delays for the initiation of the care service. Being able to ensure that all patients have a fair access to the wards while minimising the length of stay is the key aim.

- Besides the complexity of the situation due to the nature and context of the problem at hand, in terms of processes there is space for improvement. In that sense an efficient information system is fundamental; there is a lot of documentation and administrative work needing to be done, but it is important to avoid overlaps of information caused by double documentation.

- It is important to understand that excessive workload due to overcrowding decreases staff satisfaction, and therefore has a direct impact in the overall efficiency. Staffing levels should also be studied.

- The fact that there is an increase in the number of elderly patients coming in, and also that they are the ones staying longer, as can be confirmed in the age versus length of stay plots, has a direct impact on overcrowding. The consideration of delivering information about health-related services via telecommunications technologies is important for this situation, in combination with the improvement of the communication and relationships between the community primary care units and the hospital.

- Discharge planning should be an integrated activity throughout all departments in the hospital. Estimations should be donebeforehand of a patient's length of stay; as shown before, this estimate should use a uniform distribution.

- There should be a team assigned to the coordination and management of beds. Capable and knowledgeable people can assign properly a patient to where he/she fits the best, ensuring the correct treatment will be delivered. This team should include key stakeholders from different wards.

- The admission procedure should be immediately designed where the policies and priorities for available beds are established.

- Home care teams should be considered and act in combination with the wards; this will enhance the planning of patients.

- There should be established a distinct procedure for these short patients stay (zero, one or two days in the wards). For example, short stay units.

- Because of the fact that the highest percentage of the patients who are discharged from the ward go home, moving the doctors' rounds for discharge of this group of patients to early in the day should have a great impact on the bed availability.

- If patients who are scheduled to be discharged leave the hospital before the daily demand peak and, with regards to annual planning, they leave before the peaks due to seasonality, then the availability of beds will increase.

- Ensure that the discharge of patients is done before other patients are admitted during that day.

- The flow should be clearly understood by all involved, including the patients.

6.2.7   Control phase (ideas and intentions)

Possible cost savings should be estimated, as well as institutionalising and documenting the results. Also, a control plan should be constructed, with the purpose of monitoring and making sure the improvement targets have been achieved and are sustained.

An important part of the Control phase is to document the new ways of working and institutionalise them, so everyone is involved and follows the new ways. A way of addressing this point is by having a good communication plan. This communication plan should state the results and must be shared throughout the organisation. In the end, it is very important to not forget a very important part of this methodology, which is to celebrate the efforts that have been made and the results that have been obtained.

6.3   Controlling variation in play in mast production process at ATLET

Project purpose: to reduce variation in the play between mast sections by investigating the welding and assembly process.

Organisation: ATLET AB

Duration of the project: five months (2010)

Black Belt candidates team:
Anders Lindgren – Quality Director, ATLET AB, Göteborg, Sweden.
Rickard Eriksson and Ellen Hagberg – masters students.

Phases carried out and implemented tools: Define (project charter, Gantt chart, 'Y' and 'y' definitions, SIPOC, process maps, product flow-down tree) – Measure (cause-and-effect-diagram, interviews, measurement system analysis: testing, protocol and gauge R&R) – Analyse (process capability analyses, control charts).

6.3.1   Presentation of Atlet AB

Atlet AB is one of Sweden's leading producers of electrical warehouse trucks for indoor usage. Since 2007, it has been owned by Nissan. Atlet's offer to the customer is a complete logistic offer including trucks and service. For Atlet AB the mast system is a core component and it is produced in-house.

6.3.2   Project background

Although the Atlet trucks are of high quality, there is room for improvement in both the production process and in the end result reaching the customers. The studied truck model is shown in Figure 6.31. The lift system on the trucks is named 'Mast'. Today, the mast's assembly process is to a very high extent dependent on the skills of the operators and sequences of the process are time consuming.

The mast system consists of two or three mast sections (inner, middle and outer), depending on the lift height (see Figure 6.32). On the biggest model, the Reach Truck,

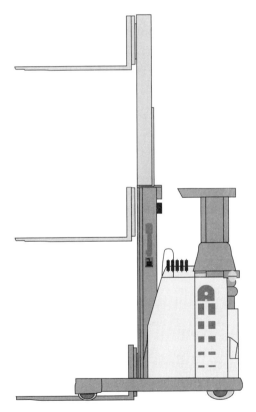

Figure 6.31   Atlet truck type U.

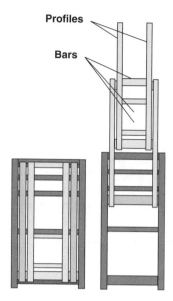

Figure 6.32   An assembled mast package in lowered/raised position.

the lift capacity is 1000 kg with up to 12 m lift height. When assembling the profiles together, the play between the mast sections is very important. The play between the mast sections must be 0.7–1.0 mm when the mast is shimmed correctly. Too much play causes too much movement between the mast sections when lifting, and they rattle while the truck is driving. Too little play causes wear in the mast profiles. After the welding robot, all profiles are measured according to drawing. Depending on variations, the shape of a welded mast section can have an A-shape, a V-shape, an O-shape or an X-shape. When assembling the mast sections together it is difficult to get the correct free play between the mast sections over the whole length of the mast. This is depending on the shape of the different mast sections. The project task is to control the variances of width of the mast section from the welding robot and to make the shimming more effective. Also the measurement method needs to be analysed.

Today, shimming the mast takes between 2 and 3 hours according to operators. According to the operators, the time is very much dependent on the shape of the mast sections as they come from the welding processes. Our estimation of best case potential is to save approximately 1 hour per each assembled mast package. A financial consequence would be a cost saving for 625 tkr (thousand Swedish krona) in cost of labour in one year assuming 2500 trucks passing through production. Moreover, the quality of the masts would increase, which would increase the end customer satisfaction. The full economic effect of this is difficult to estimate.

6.3.3   Define phase

In the Define phase, the project charter (Figure 6.33) and the Gantt chart were developed (Figure 6.34). Moreover, y's for improvement were identified.

Company/ Organization	ATLET AB	Unit/ Department	Quality department
Executive	Keiji Ikeda/ Martin Björkroth	Senior Deployment Champion	
Deployment Champion	Anders Ytterberg/ Fredrik Vestlund	Project Champion	
Master Black Belt		Finance Champion	
IT Champion		HR Champion	
Industrial participant (Black Belt candidate)	Anders Lindgren	Telephone/e-mail	
Sponsor & Process Owner	Anders Lindgren	Site or location	ATLET AB, Mölnlycke
Project Start Date	18/2-2010	Project completion Date	26/5-2010
Expected impact level		Expected financial impact (savings/ revenues)	625 tkr
Element	**Description**	**Charter**	
1. Project description	A short description of the project		
2. Impacted process	The specific process/es involved and where opportunity exists	The specific process the project address starts with welding of bars onto the mast to the assembly of the mast. An opportunity exists to increase quality of mast sections and to save time in assembly due to less rework.	

Figure 6.33   Project charter.

1. Benefit to customers	Define internal and external customers (most critical) and their requirements	Internal customer = Employee working in assembly of mast section. Requirements = Mast sections with correct tolerances over the whole length of the mast		
2. Benefit to the business	Describe the expected improvement in business performance	Expected improvements are savings in assembly time due to less rework and improved quality of mast sections. 1h saved in assembly line for all trucks = 625tkr		
3. Project delimitations	What will be excluded from the project?	The project will only address one type of mast system: U** 200 DTFV		
4. Required support	Support in terms of resources (human and financial) required for the project	Required support for the project is interview time with concerned employees working with the process.		
5. Team members (students BB candidates)	List names of the master students who will take part in the project	Rickard Eriksson Ellen Hagberg		
6. Team members (others)	List technical experts and other people who will be part of the team	Marcus Drevik Bengt Davidsson Anders Ytterberg		
7. Specific goals	Define the Baseline, your realistic goals for the project and the best case target for improvement.	**Actual value (Baseline)**	**Realistic goal by project end date (net Improvement)**	**Best case goal**
		2-3h to assemble each mast	Save 1h per each assembled mast package	The assembler can choose type of shims from information given about the specific mast sections
DEFINE phase completion date	April 12th, 2010	MEASURE phase completion date	April 20th, 2010	
ANALYZE phase completion date	May 17th, 2010	IMPROVE phase completion date	May 24th, 2010	
CONTROL phase completion date	June 2nd, 2010	PROJECT results presentation date	May 25th, 2010	

Figure 6.33 (continued).

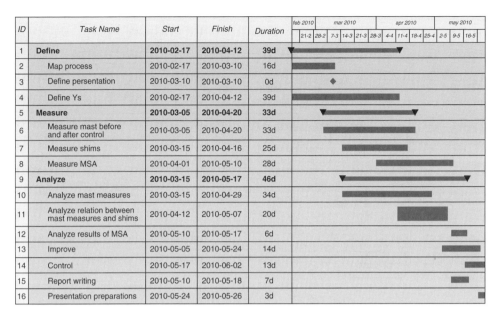

ID	Task Name	Start	Finish	Duration
1	**Define**	**2010-02-17**	**2010-04-12**	**39d**
2	Map process	2010-02-17	2010-03-10	16d
3	Define persentation	2010-03-10	2010-03-10	0d
4	Define Ys	2010-02-17	2010-04-12	39d
5	**Measure**	**2010-03-05**	**2010-04-20**	**33d**
6	Measure mast before and after control	2010-03-05	2010-04-20	33d
7	Measure shims	2010-03-15	2010-04-16	25d
8	Measure MSA	2010-04-01	2010-05-10	28d
9	**Analyze**	**2010-03-15**	**2010-05-17**	**46d**
10	Analyze mast measures	2010-03-15	2010-04-29	34d
11	Analyze relation between mast measures and shims	2010-04-12	2010-05-07	20d
12	Analyze results of MSA	2010-05-10	2010-05-17	6d
13	Improve	2010-05-05	2010-05-24	14d
14	Control	2010-05-17	2010-06-02	13d
15	Report writing	2010-05-10	2010-05-18	7d
16	Presentation preparations	2010-05-24	2010-05-26	3d

Figure 6.34   Gantt chart.

The production process at Atlet is customer order driven in a one-piece operation flow. The mast package is one of the main parts of an indoor truck (for a product flow-down tree, see Figure 6.35).

In the mast production, the two key processes have been studied, the welding process and the mast assembly process. A SIPOC process map describes the process in general steps (see Figure 6.36).

A more detailed view of the welding process and the assembly process is illustrated by the two process maps in Figures 6.37 and 6.38, respectively. Moreover, the two processes are described in more detail in the Section 6.3.4.

When looking at the mast production process at Atlet, several improvement areas can be identified; the variance of the width and the parallelism of the masts from the welding process, the time spent in the assembly process and the shimming technique in the assembly process. However, from the customer perspective, the most important Y to be improved is that the masts have the right shim and play. If not, the customer might experience noise, wear and decreased lowering speed on the lighter trucks. Looking at the internal process, time can be saved both by reducing the number of adjusted masts in the welding department and by reducing the time needed for shimming in the mast assembly process. To sum up, two y's are identified:

- the geometric variations of the mast sections after adjustment;

- the time of the mast assembly.

To determine the geometric variations, the data from the welding inspection station are studied during the Analyse phase.

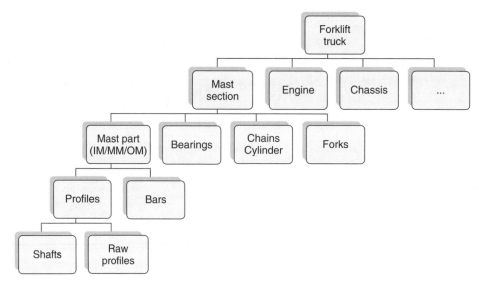

Figure 6.35   Product flow-down tree.

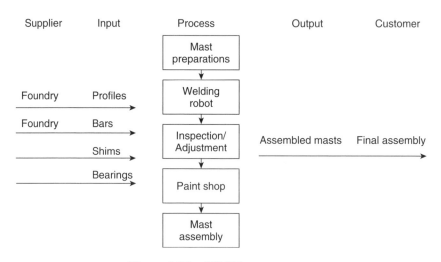

Figure 6.36   SIPOC process map.

To determine the time it takes to assembly the mast packages is difficult, but there exist earlier time studies and these, together with interviews conducted, tell us that it usually takes 2–3 hours to assemble a complete mast package.

6.3.4   Measure phase

With the y's from the Define phase together with the information that was recorded during the initial work, a cause-and-effect-diagram was created (see Figure 6.39).

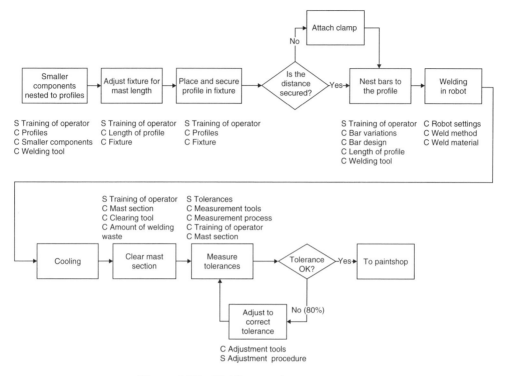

Figure 6.37   Welding station process map.

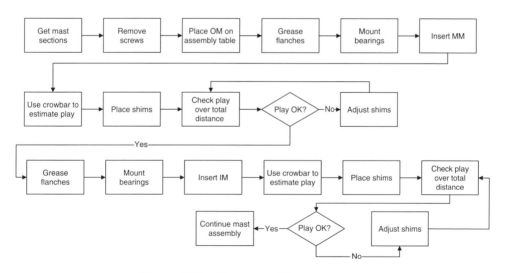

Figure 6.38   Mast assembly process map.

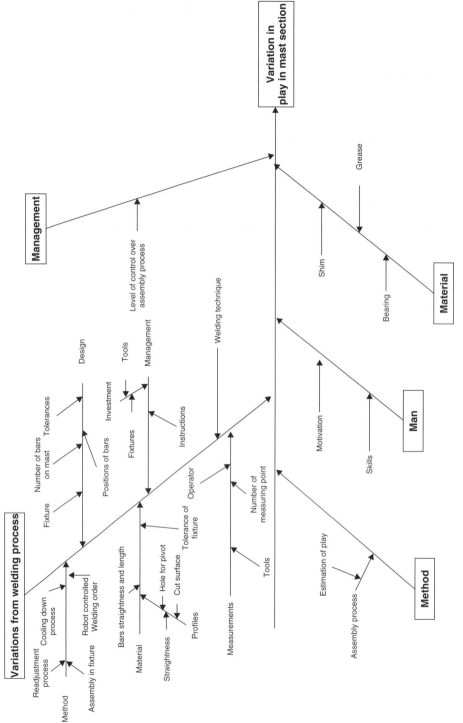

Figure 6.39   Cause-and-effect diagram.

When searching for x's to improve the y's, the focus can easily be put on the welding process to improve it in order to provide the assembly station with more accurate masts. However, to fully understand the whole process and its demands, the project group decided to investigate the requirements that the mast assembly station has on the mast welding station. The reason for this is to prevent us optimising processes at the welding station that then have no effect on the output in form of quality of the assembled mast and the time spent in the assembly process. In order to identify the most important factor in the internal customer relationship between the welding process and the assembly process, interviews with the operators and the responsible production engineer was conducted and drawings were studied. The most important factor that was identified was that the mast should be parallel. It takes more time to choose the right shim, and the play over the whole length of the mast is less satisfying if the mast is not parallel. Therefore, data was collected in two ways:

1. the character of the mast before and after any adjustment;

2. how many shims are used in the assembly of the mast.

Below, the processes around the measurement points and the measuring system analysis (shown in Figures 6.37 and 6.38) are described in detail.

When the mast section has left the welding robot, the distance between the profiles is measured at three specified points: 500 mm from the top of the mast section, at the centre of the mast section and 500 mm from the bottom of the mast section. The measurements are noted in a specific protocol for the truck that is stored in a binder at the welding work station. This is done after the mast sections have cooled. The measurement points together with the specifications can be seen in Figure 6.40.

If the mast section is outside specification it is adjusted. This is done at the welding station and can be done cool (small adjustments) or the mast section can be re-heated

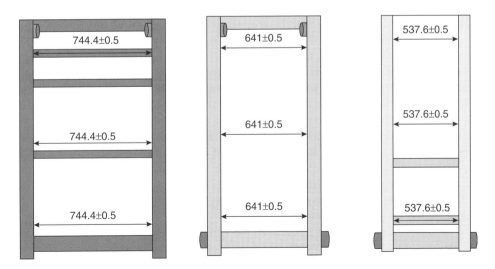

Figure 6.40   The three mast sections with points of measurements and specification limits.

(major adjustments). The outer masts, which have more bars than the inner and the middle masts, are more difficult to adjust. When the mast section is cool the new measurements are noted in the truck's protocol.

When the Six Sigma project was initiated the measurement protocols were not filled in completely even if the measurements were done correctly. A more accurate recording of the measurements was started, which is why the number of data points in the project are somewhat limited. The distributions of data from the welding station are included in the Section 6.3.5.

The tool used for the measurements at the welding work station is a digital micrometer (see Figure 6.41). The tool is calibrated each morning by using a reference. At each end of the micrometer there is a bearing that is assumed to either support a right angle when measuring and/or make it possible to move the micrometer while measuring. The two bearings are marked with arrows in the picture.

Figure 6.41   Digital micrometer used at the welding working station.

To examine the reliability of the measuring system at the welding station a measurement system analysis was conducted. This is essential in order to understand what variations can be derived from the measurement tool and operator, and which variations originate from the process itself. Designs with 27 tests (3 × 3 × 3) were developed. The test included three parts, with three operators randomly repeating three measures. The tests were conducted at the same place the measurements are usually made and in a randomised order. To make sure that the operators were not influenced by each other they did not witness their colleagues' tests. At the mast assembly station the three different mast sections are fitted together to create a complete mast package that is later mounted on the truck in the final assembly. The three different mast sections can move relative to each other due to bearings, which are positioned at the bottom of the inner mast, at the top and at the bottom of the middle mast and at the top of the outer mast. When the mast sections are lifted or lowered it is important that the play between the sections is 0.7–1.0 mm. Because of variations in play, shims are used inside the bearings. These are provided in sizes of 1 mm and 0.5 mm and can be combined in different ways to make the play between the masts as optimal as possible. However, the total amount of shims used in each of the four points has to be between 2 mm and 6 mm. If the operator thinks that the play is not sufficient even after shimming, the mast section is returned to the welding station for further adjustments.

To be able to seek relationships between the variations in shimming and other sources of variation in the process a protocol was developed to record how the shimming of the mast varied with order, operator and truck type. Altogether shimming data from 16 trucks was noted.

6.3.5   Analyse phase

The purpose of the Analyse phase is to learn about the y's and what x's influence them, and to set improvement targets. Because the project group decided to focus on the internal relationship between the welding and the assembly process, the analysis will look at the measures of the mast and whether there is a relationship with the use of shim.

The results from the measurement system analysis are shown in Figure 6.42. The results show that the measurement system variation is 19.82 % in total, which means that the measurement contributes to a large extent to the variation of the measures.

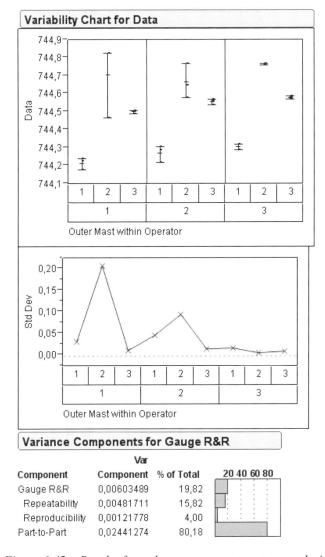

Figure 6.42   Results from the measurement system analysis.

Of the total gauge R&R, repeatability stands for 15.82 % of the variation. This means that the variation arises when the same operator repeatedly measures the same object with the same measurement tool. In Atlet's case this means that it is the digital micrometer's inability to repeat a measure that contributes to most of the measurement variation. Reproducibility, which describes the variations that can be derived from variations among the operators, is small in comparison. In general, a total gauge R&R result between 10 and 30 % could be acceptable depending on the nature of the process and the cost of investment. However, if Atlet wants to improve the welding process, a measurement system that can detect small variations and changes is important. Therefore the measurement system should be improved. Even if the measurement system analysis showed that the measuring gauge contributes to the variation, data from the inspection station will be analysed.

Distributions of measures taken after the welding robot were analysed by comparing data collected before and after any adjustments. The measures represent the shape of the masts that the assembly station receives. Looking at the Cp/Cpk (all below 1, see Figure 6.43), a conclusion is that the inspection station does not provide products that are satisfying according to the specification limits. The Cp/Cpk values are improved due to the adjustment but the still poor Cp/Cpk values means in reality that the operators let masts pass the inspection station although the measures are outside specification limits. The reason for this could be that the operators know from experience that the assembly station can handle that the middle mast and the outer masts being above specification as long as the masts are parallel. Moreover, comparing the results before and after the adjustment one can notice that the masts are often widened but rarely narrowed. A reason for this could be that it is more difficult to decrease the width of the mast compared to expanding it.

Due to the fact that the distribution from the welding station was considered relatively stable, control charts were constructed to describe the data. Because of the small sample size $\overline{X}$ charts and r charts were used. The $\overline{X}$ charts describe how the average measures of the masts vary and the r charts describe how the range of the measures varies (Figure 6.44).

The $\overline{X}$ value in this case describes the averages of the measured widths of each mast section. Three measures are noted by the operators and the $\overline{X}$ value shows the average of the three. There are some masts that fall outside the calculated control limits but after the adjustments most of these points are eliminated. A recommendation is to investigate those points, especially those that are narrower than they should. Because of the constraints of the fixtures used at the welding station, a more logical outcome would be that some masts are bigger than the others.

The r value describes the range of the three measures taken on each mast section and can therefore be a good indication about the parallelism of each section. The design specification is ±0.5 mm which means that a range of 1 mm is acceptable, apart from the position of the distribution. The averages of the ranges all seem acceptable but the upper control limit is greater than the design specification for both the middle and the outer mast. This is even though the masts have been adjusted. This indicates that work on making the mast more parallel should be done, at least according to the design specifications.

The analysis of the measurements from the welding working station shows that the mast sections are relatively stable but not always according to specification limits. To

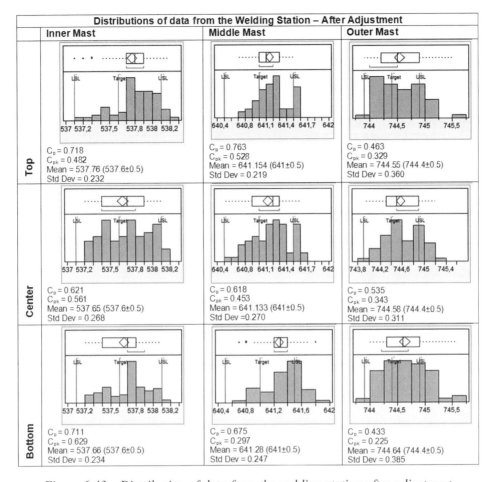

Figure 6.43   Distribution of data from the welding station after adjustment.

further examine how the assembly is influenced by variations of the mast geometry, data was collected at the mast assembly station. This data can later be used in combination with data from other sources of variation to be able to find correlations. Below is an analysis of how the shimming of the masts is correlated with the measurements of the mast leaving the welding working station.

The analysis of the shimming was focused on four different points, which are shown in Figure 6.45 – these were called:

1. mast lowered – outer shim;

2. raised mast – outer shim;

3. mast lowered – middle shim down;

4. raised mast – middle shim down.

Outer Mast – Before Adjustment

Outer Mast – After Adjustment

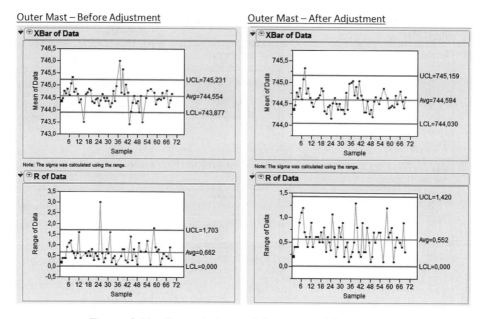

Figure 6.44   Control charts of the mast welding process.

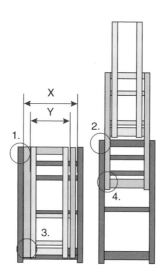

Figure 6.45   Points where shimming is analysed.

According to design specification, the number of shims used in a specific point should be dependent on how much space exists between the outer and the middle mast. This space is dependent, among other things, on the width of both the outer and the middle mast and therefore an applicable measure that could represent both the outer and middle mast needed to be used. The project team decided to use the difference between the corresponding measurements from the outer and the middle mast, X and Y in Figure 6.45. This measure was plotted against the number of shims used in the specific point.

The result of the analysis shows that there are no significant linear relationships between the measured variations from the welding station and the amounts of shims used (no or very small correlation exists). An important conclusion is that the differences between measures from the welding station, which from the beginning was believed to be a major factor in the mast assembly, seem to be very small when comparing them with the differences in shimming. This means that the variations in geometry that can be derived from the welding process are not, at least not alone, the major factor when looking at the variation in time for the shimming operation.

With the limited number of observations from the shimming operation it was not possible to find any significant linear relation between existing measurements on mast geometry and the amount of shims used. But the analysis of the shimming in the previous section showed that there could be some continuity in the amounts of shims used. To examine this, diagrams showing the shimming of each truck were constructed, shown in Figure 6.46. To easier understand which bearing the charts refer to, please see Figure 6.40.

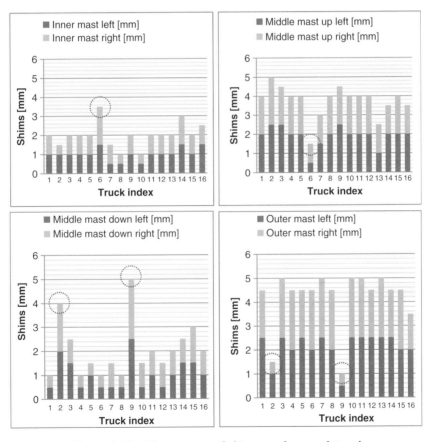

Figure 6.46   The amount of shims used on each truck.

The charts show that some amounts of shims are more common than other. Of the 16 different mast packages that were studied, 56 % had 2 mm of shims on the inner mast. On the outer mast 50 % had 4.5 mm and another 31 % had 5 mm of shims. On the upper bearings on the middle mast, 50 % of the trucks had 4 mm of shims. When considering that the shimming can vary between 1 and 6 mm with 0.5 mm steps it seems like an obvious result. The reason for this continuity in the shimming of the different mast sections needed to be examined further.

When comparing the charts there could be some interesting relations between the different shimming points. In the two bottom charts, which describe the shimming points between the middle and the outer mast, it is possible to see that a small amount of shim on the outer mast coincides with a big amount of shim on the middle mast (truck numbers 2 and 9). This means that the total amount of shim between the middle and the outer mast could be relatively constant. This is in line with the two upper charts where the case is the same for truck number 6.

6.3.6   Improve phase (ideas and intentions)

In the Analyse phase there were many reflections on how the process, from incoming to components to assembled mast system, are all connected. Below is a summary of the most important ones.

- *Significant variations in the analysis of the measurement system*. The measurement system analysis conducted at the welding station showed that 19.82 % of the variations in the welding process are due to variations in the measurement system. Of these, 15.82 % is because of the lack of repeatability, which means that efforts should be made to try to improve the measurement tool used, before looking at variations in technique between the operators.

- *Stable welding process with poor capability*. The results from the measures out from the welding robot show that the process is stable but has low capability, Cp's and Cpk's (all smaller than 1). In the project 72 trucks have been studied and a more extensive study is recommended to do a more exact evaluation of the distributions.

- *Adjustments of mast with poor measures only in some cases*. Analysing the capability after inspection and any possible adjustment shows that the inspection process/adjustment process does not improve the capability of the process significantly. Only too tight masts are widened, but only few of the opposite case. The reasons for this could be for practical reasons or based on an idea of such demand from the assembly process.

- *Large shimming variations compared to welding variations*. The variations in shimming are very big compared to the measured variations from the welding station. The views that the assembly process is made more difficult because of the variations from the welding process appear incorrect.

- *The amount of shim used is not related to the widths of the mast sections*. The analysis made to investigate if the amount of shims used is linked to the width of the masts did not show any such relationship.

- *Continuity in the amounts of shims used*. The recording of the assembly process showed that there is continuity in the amount of shims used by the operators. This is even though the amount of shims can vary between 1 mm and 6 mm to compensate for the variations from the ingoing components and the previous processes.

These findings will lay the base for two fundamental improvement proposals.

- *Improve the measurement system*. The measurement system analysis clearly showed that the digital micrometer used at the welding station does not provide sufficiently accurate measurements and more modern measurements systems should be evaluated. The purpose is twofold; a new measurement system will probably give more exact and robust measurements than the micrometer used today and an electrified measurement system will allow for more online process monitoring, information that can be useful in other parts of the process. A first step could be to invest in a more modern version of the micrometer but in the long term bigger investments should be examined. If the welding robots vision system can be utilised then the cost could probably be reduced considerably. If the present measurement tool is kept, Atlet will have high internal cost because of a large amount of unnecessary rework in the mast assembly as well as trying to improve and control a process without being able to measure the impact.

- *Examine possibilities of using standard shims*. The analysis of the shimming operation shows that it is somewhat standardised even though it is not supposed to. This result should be discussed with the design division to get an understanding of how the amounts of shims should vary according to the drawings. But when looking at the service issues there are no significant amounts of service issues related to these kinds of mast problems and because of that these standard amounts of shims seem to work fairly well in the field. Experiments can be conducted to see if a standard amount of shim works. If standard shims are introduced a more extensive monitoring of the process should be carried out, and when masts that need other amounts of shims are identified, these should be examined more thoroughly to find the reason for this variation. If successful, this improvement could mean a considerable reduction of the time it takes to assemble the mast package and it will also help Atlet when they are changing their layout which will require a more standardised assembly. The improvement potential increases further if implemented solutions are expanded also to other truck types.

6.3.7   Control phase (ideas and intentions)

The data that is gathered in the process today are stored in a binder at the welding working station. This data is not followed up unless there is a problem with a specific truck. Atlet should instead introduce routines that involve monitoring this data more regularly to be able to recognise problems in the process at an earlier phase then today. This could at an early stage mean that the data is stored in an Excel template with attached control charts but could also, in the long term, mean a more computerised process where the data is monitored online. In this case the process itself can signal if there are signs of assignable causes. In the case where Excel templates are used the production leader or the operators themselves can fill in their measurements in the computer immediately and can themselves monitor if there are trends in the data. This solution can probably be adopted without any significant resources but will add a lot of understanding of the current state of the welding process and how it can be improved.

If Atlet conducts experiments and then also implements standard shims at the assembly station, it is of high importance that the effects of that are studied. The effects should be studied in-house, further down in the process, but it is also important to monitor unexpected effects that can occur first in the use environment of the product.

6.4   Optimising the recognition and treatment of unexpectedly worsening in-patients at Kärnsjiukhuset, Skaraborg Hospital

Project purpose: to optimise the process of taking care of in-patients with unexpectedly worsening conditions in order to reduce the mortality rate of the patients at regular wards of Kärnsjukhuset Hospital in Skövde.

Organisation: SKARABORG HOSPITAL

Duration of the project: five months (2010)

Black Belt candidates team:
Jana Hramenkova – MD, Cardiology department, Kärnsjukhuset, Skaraborg Hospital, Skövde, Sweden.
Amirsepehr Noorbakhsh, Arash Hamidi and Maria Dickmark – masters students.

Phases carried put on and implemented tools: Define (project charter, Gantt chart, 'Y' and 'y' definitions, P-diagram, SIPOC, process map) – Measure (interviews, observation, KJ method, Ishikawa diagrams, voting, Pareto charts, VMEA).

6.4.1   Presentation of Skaraborg Hospital

Skaraborg Hospital consists of four different hospitals in Skövde, Lidköping, Falköping and Mariestad, which all are cities in one of Sweden's regions – Västra Götaland. The biggest of four hospitals is located in Skövde and is called Kärnsjukhuset. It has about 3200 employees and there are 454 beds. About 260 000 people are living in the area which Kärnsjukhuset is responsible for. There are nearly 30 different medical specialities active in the hospital and most of challenges are solved without sending patients to the region's hospital Sahlgrenska Hospital in Gothenburg.

6.4.2   Project background

In 2009, a research was made at Kärnsjukhuset by Dr. Jana Hramenkova to investigate the treatment of in-patients with sudden cardiac arrests during the year of 2008.

There were 35 patients in total who suffered an unexpected cardiac arrest while staying in the hospital for some other reasons, and 57 % of these patients had abnormal vital parameters 24 hours before the alarm. There are strict definitions in medical science for classifying vital parameters as normal or abnormal/pathological. The research showed that in 40 % of cases with identified pathological vital parameters no contact was initiated with the responsible physician. Mortality rate in this group was higher than in the group where nurses contacted the responsible doctor for treatment.

The research consisted of too few patients to show a statistical significance, but with reliance on other research, combined with the result of this study, the hospital considered the process of taking care of worsening patients to be a serious problem and decided to start a Six Sigma project in order to improve that process.

6.4.3   Define phase

The frame of the project is outlined in a project charter and planned in a Gantt chart. Also the customer is identified and the impacted process visualised through different tools such as process map, SIPOC and P diagram. The charter defines the scope of the project, the impacted process and delimitations (Figure 6.47). Also the project group is defined here, consisting of two nurses, one doctor and two black belts apart from the project team of the course. In the project charter the different phases of the project are outlined with completion dates. Based on the dates presented in the project charter, a project plan was designed, in which the different phases of the project were planned. In parallel with the progress of the project, it was revised to be as accurate as possible. The final one, which represents the way the project proceeded is shown in the Gantt chart (Figure 6.48).

Usually the purpose of the Six Sigma project includes the so-called 'big Y', that is the high level measure that the project aims to improve. Big Y is often non-measurable, and therefore a number of small y's that indicate the CTQ characteristics should be outlined as links to the big Y. The big Y was defined as: 'taking care of in-patients with unexpectedly worsening condition at Kärnsjukhuset'. This indicates that the goal of this project is to improve the process of how unexpectedly worsening patients are recognised and taken care of. Based on the literature and the previous research made by Dr. Hramenkova, two aspects play important roles in this process. Those are: checking the parameters and calling the doctor. Parameters such as blood pressure, pulse and temperature are effective indicators about the patients' condition. By continuously taking these parameters the nurse has a good chance of recognising when a patient gets worse. The other important factor for the process of taking care of unexpectedly worsening patients is the doctors' involvement in the process. The two small y's are therefore defined as: 'frequency of parameter check-ups' and 'rate of calls to the doctor in those cases that the patient gets worse'. These small y's were approved by the hospital team and then set the base for the rest of the project.

A P diagram is an analysis of the elements of a process (see Figure 6.49). The process is represented by a box where a signal factor from the outside starts the process. On the opposite side, a response is created. Along with the process, a number of factors are influencing. Some of them are controllable, others are noise factors.

Project title:	Optimized recognition and treatment of unexpectedly worsening in-patients at Kärnsjukhust		
Company/Organization	Skaraborgs	Unit/ Department	Kärnsjukhust
Executive	Birgitta Molin	Senior Department	Stefan Håkansson
Deployment Champion	S.Lifvergren	Project Champion	
Master Black Belt	A. Chakhunashvili	Finance Champion	
IT Champion		HR Champion	
Industrial participant (Black Belt candidates)	Jana Hramenkova	Telephone/e-mail	
Sponsor & Process	Skarborgs	Site or location	Kärnsjukhust, Skövde
Project Start Date	2010/02/17	Project completion	2010/05/25

Element	Description	Charter	
1. Project description	A short description of the project	Our project concerns all regular wards at Kärnsjukhust. It is based on previously collected data which showed correlation between delayed or absent reaction in case of new symptoms and/ or other alarming signs in an inpatients condition and higher mortality in cardiac arrest. The purpose of the project is to reduce the mortality rate of the patients at regular wards of Kärnsjukhust in Sköved by optimizing the process of taking care of in-patients with unexpectedly worsening condition. This to avoid unwanted variation between different clinics and hopefully to reduce mortality.	
2. Impacted process	The specific process/es involved and where	The process of taking care of in-patients on regular wards whose condition is unexpectedly worsening	
3. Benefit to customers	Define internal and external customers (most critical) and their requirements	In healthcare systems identification of customers is challenging, because the whole system incorporates a large number of customers, any of which are related to one or more steps in the whole process. But considering all the customers, it could be mentioned that patients are the prime customers in the health care systems. The benefit for them is to get optimal treatment in time to prevent more severe condition compared to their current situation and to have higher probability to survive. The internal customers are the employees if the hospital, and in focus of our project are the nurses. The benefit for them is a process that makes it natural and easy for them to react in the correct way when a patient's condition is unexpectedly getting worse.	
4. Benefit to the business	Describe the expected improvement in business performance	Lower hospital mortality	
		Better working environment for the employees	
		Hopefully lower amount of days per patient in the hospital	
5. Project delimitations	What will be excluded from the project?	Outpatients	
		Accompanying people	
		Intensive care unit patients	
		Emergency unit patients	
		Childrens (<18years)	
6. Required support	Support in terms of resources (human and financial required for the project	Management support	
		Human resources (hospital team members)	
		Travel expenses for the industrial participant and the students	
7. Team members (students BB candidates)	List names of the master students who will take part in the project	Amirsepehr Noorbakhsh	
		Arash Hamidi	
		Maria Dickmark	
8. Team members (others)	List technical experts and other people who will be part of the term	Lena Landenmark	
		Weimar Ahlgren	
		Daniel Rodriguez-Santos	

	Define the Baseline, your realistic goals for the project and the best case target for improvement	Actual value (Baseline)	Realistic goal by project end date (net improvement)	Best case goal
9. Specific goals			Optimized process of taking care of in-patients on regular wards whose conditioning is worsening	Reduced hospital mortality

DEFINE phase	26/03/2010	MEASURE phase	07/03/2010	
ANALYZE phase	21/05/2010	IMPROVE phase	25/05/2010	
CONTROL phase	25/05/2010	Project presentation	May 25th, 2010	

Figure 6.47   Project charter.

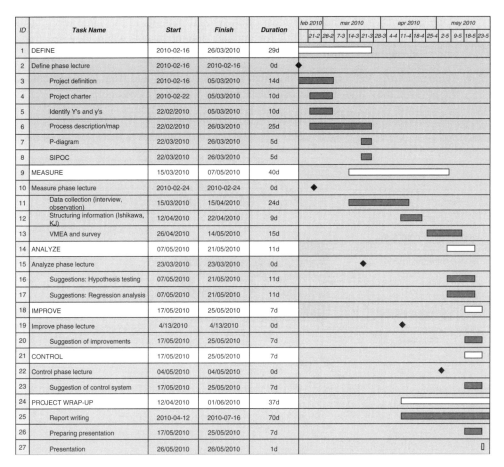

ID	Task Name	Start	Finish	Duration	feb 2010	mar 2010	apr 2010	may 2010
1	DEFINE	2010-02-16	26/03/2010	29d				
2	Define phase lecture	2010-02-16	2010-02-16	0d				
3	Project definition	2010-02-16	05/03/2010	14d				
4	Project charter	2010-02-22	05/03/2010	10d				
5	Identify Y's and y's	22/02/2010	05/03/2010	10d				
6	Process description/map	22/02/2010	26/03/2010	25d				
7	P-diagram	22/03/2010	26/03/2010	5d				
8	SIPOC	22/03/2010	26/03/2010	5d				
9	MEASURE	15/03/2010	07/05/2010	40d				
10	Measure phase lecture	2010-02-24	2010-02-24	0d				
11	Data collection (interview, observation)	15/03/2010	15/04/2010	24d				
12	Structuring information (Ishikawa, KJ)	12/04/2010	22/04/2010	9d				
13	VMEA and survey	26/04/2010	14/05/2010	15d				
14	ANALYZE	07/05/2010	21/05/2010	11d				
15	Analyze phase lecture	23/03/2010	23/03/2010	0d				
16	Suggestions: Hypothesis testing	07/05/2010	21/05/2010	11d				
17	Suggestions: Regression analysis	07/05/2010	21/05/2010	11d				
18	IMPROVE	17/05/2010	25/05/2010	7d				
19	Improve phase lecture	4/13/2010	4/13/2010	0d				
20	Suggestion of improvements	17/05/2010	25/05/2010	7d				
21	CONTROL	17/05/2010	25/05/2010	7d				
22	Control phase lecture	04/05/2010	04/05/2010	0d				
23	Suggestion of control system	17/05/2010	25/05/2010	7d				
24	PROJECT WRAP-UP	12/04/2010	01/06/2010	37d				
25	Report writing	2010-04-12	2010-07-16	70d				
26	Preparing presentation	17/05/2010	25/05/2010	7d				
27	Presentation	26/05/2010	26/05/2010	1d				

Figure 6.48   Gantt chart.

The signal factors that influence the process are of five different kinds.

- *Patients' call or complaint*. This signal requires for the patient to be conscious enough to be able to call, which is not always the case. A button is located close to the patient's bed for the ease of calling the nurse in case of pain.

- *Notice during check-ups*. The nurses do some regular check-ups of the patients where they use all of their senses in order to evaluate the patient's condition. Some examples are to talk to the patient, look at the skin colour, listen to their breath, and so on. Some of these check-ups are at fixed times, others are done more sporadically.

- *Notice by chance*. This signal refers to the times when the nurse notice the patient's condition by chance while passing the corridor or talking care of another patient in the same room.

- *Other patients' call*. There are most often several patients in the same room at the wards. This gives the possibility for another patient to call when they see that a patient in the neighbouring bed is not feeling well.

- *Relatives' call*. Visitors in the hospital, such as relatives and friends, could contact the nurse if they find that the patient is getting worse or needs help.

Figure 6.49   P diagram.

During the process there are several factors that influence what kind of treatment the patient will get. Some of these are the noise factors that are uncontrollable. These could be what disease the patient has and the demographics of the patient. Also some less apparent factors influence the process, for example the nurse's mood or relatives' influence. When relatives are present, the situation can get more stressful for the nurse because of their needs, thoughts or concerns about the patient. Some of the controllable factors are availability of equipment or instructions for the nurse.

SIPOC (see Figure 6.50) was used in early steps of process improvement to identify the relevant elements of the process. It is most easily started with the output of the process and the customers, to later deal with what inputs are needed and who the suppliers are.

- *Outputs*: The main output of the process is a patient who is getting treatment or is feeling better. Either patients get the treatment needed from the nurse, or they get contact with doctor. But they could also be sent to the emergency unit, which would not necessary mean that they are feeling better, but means that they get the right treatment for further recovery. Other outputs are experience of the nurses and doctors. After treating one patient they get more skills and practice of how to treat a patient in a certain condition, which can be used in later situations. Finally, the last output is a healthier society.

- *Customers*: From the reasoning of the outputs, the customers are identified as the patients who get treatment, the employees who get experience and the whole society.

- *Process*: The process is described according to the process map (see Figure 6.51).

- *Inputs*: It was chosen to use the relevant signals from the P diagram of the process as inputs in the SIPOC. They are the intangible factors that start the process, such as a patient's feeling, abnormal vital parameters, nurses, senses and relatives' feelings. On the other hand, many other aspects could also be mentioned as inputs. Some examples are the education of the employees, the equipment for taking parameters, the patient's bed in the hospital, and so on. The ones mentioned in the SIPOC were chosen because they were considered the most relevant ones in this case where the process is looked at from a qualitative and psychological point of view.

- *Suppliers*: From the description of the inputs, the suppliers are easily identified as the patient, the nurse, the relatives and other patients in the ward.

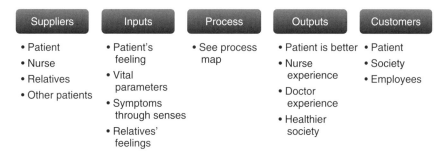

Figure 6.50   SIPOC.

A closer outline of the different steps of the process: 'recognition and treatment of in-patient's conditions getting worse' can be found in Figure 6.51. The four signals from the P diagram are now the starting points of the process that all lead to the nurse's recognition of the worsening condition of the patient. If it is a cardiac arrest, the nurse will immediately press the emergency alarm. If not, the process can take two turns. If the parameters have been checked recently, the nurse will evaluate whether it is necessary to call the doctor or not. If the parameters are not checked recently, the nurse decides if it is necessary to check them and depending on the decision the parameters will be checked or not checked. The next activity implies another decision by the nurse: whether to call the doctor or not. Sometimes the nurse does not call the doctor, and consequently chooses to handle the situation alone. Other times the nurse does call the doctor. To summarise, there are three different endings of the process. The nurse uses the emergency alarm, the nurse handles the situation alone or he/she calls for the doctor. This is where the impacted process ends. But as shown in the picture, it was chosen to investigate one

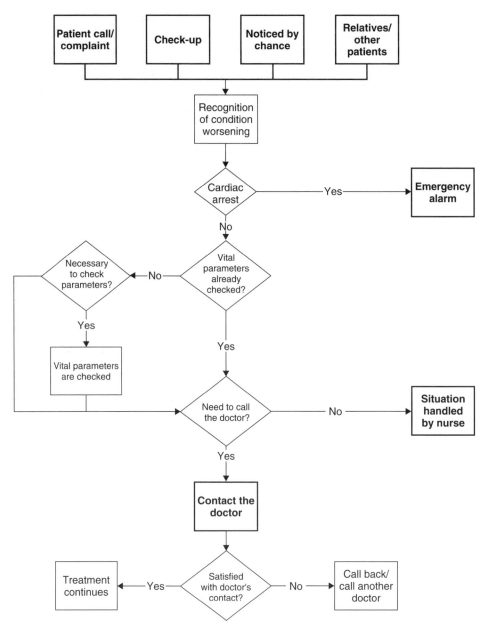

Figure 6.51   Process map of 'recognition and treatment of in-patient's conditions getting worse'.

step further after calling the doctor. This is because 'need to call the doctor' is one of the small y's of the project where it can be interesting to find reasons for its variation. In order to investigate it further it was chosen to keep it in the process map. The result shows that depending on the outcome of the call for the doctor the nurse will continue with the treatment or call the doctor again. It also happens that the nurse could choose to call another doctor when unsatisfied with the first doctor's response. Both of the small y's are the main conditional situations in this process. These are the critical situations in the process where the variation in nurses' reaction should be reduced and/or predictable based on the situation.

The purpose of the Measure phase is to gather baseline information about the process, from which it is possible to understand what is happening in the hospital. The first activity of this phase is to find a number of x's for each of the y's. Therefore, key measures will be identified and the data collection planning and execution method will be decided. For the collection of data it was decided to conduct as many interviews and observations as possible in different wards of the hospital. Eventually, due to the limitations and tight working schedule of the employees at the hospital, interviewees were chosen based on convenience sampling by selecting on the basis of their availability and willingness to participate in the research. Therefore, the interviews and observations were conducted in four wards of the hospital considered as representative of almost all the regular wards of the hospital.

These four wards are:

- *MAVA* – general medical department, for patients with all kind of medical problems. Patients usually stay in this department only for a couple of days and then either get discharged or moved to specialised departments when longer treatment is needed.

- *KAVA* – general surgical department, which is organised in the same way as MAVA but for the patients with surgical diseases.

- *Cardiology* – specialised department for patients with heart problems, such as myocardial infarction, heart failure, heart rhythm disturbances, and so on.

- *71–72 (Surgery)* – specialised department for patients who need advanced surgical treatment, mostly abdominal surgeries, obesity surgery and surgery for different types of cancer diseases.

6.4.4 Measure phase

The Measure phase started by conducting four semi-structured interviews with nurses in the four different wards of the hospital. The aim of these interviews was to understand the working procedures with focus on the two main issues (the small y's): *checking the parameters* and *calling the doctor* in the situations where a patient's condition gets worse. The process owner considered nurses as the most influential group on the small y's of the project. This was the reason for interviewing only nurses. During the brainstorming sessions before the first interview, the general plan for the interviews was decided, and later a comprehensive interview guide was developed. The interview guide incorporated all the questions and notes useful about the process of taking care of the in-patients at the regular wards. All conversations during the interviews were recorded by a laptop and an analogue sound recorder was used as a supporting system. Moreover, during the interviews, not only the answers for the pre-designed questions were observed, but also nurses reactions to the questions, in order to be able to better analyse the interviews. Before each interview, the project and the main purpose of it were introduced to the nurses. Also interviewees were asked for permission to record the interviews. Right after each interview, the answers were discussed in order to have a first interpretation about the whole process and procedures. Necessary notes were written down as a log of our activities, in order to enable later tracking of our discussions. Later on, all interviews were transcribed in order to make the further analysis possible.

After a thorough study of the interview transcripts, it was concluded that information resulting from interviews could be categorised into two main groups: facts and opinions.

For further investigation in the process of taking care of the in-patients at regular wards of the hospital, an observation was arranged. This observation was conducted at MAVA, one of the wards where one of the interviews was previously conducted. The observer did not participate in any working processes, and she was just following a nurse and taking notes during the shift. The nurse who was going to be observed was informed and the project was introduced for her.

After conducting four interviews and one observation, the next step was to analyse them. A modified conversation analysis (CA) method was used for analysing the interviews. CA is an analysis of talk as it occurs in interaction in naturally occurring situations. What is meant by 'modified' CA is that this method was used for the semi-structured interviews. Moreover, the observation was discussed in brainstorming sessions within the team members. Key concepts and ideas were collected and coded based on the interviews and observation.

Thereafter, principles of the KJ analysis were used in order to categorise those key concepts and put the most relevant ones within one category. By using the KJ analysis, the categories, which made the understanding of the whole process much easier (see Figure 6.52), could be visualised. Further on, the outputs of the KJ analysis were used in the next step, which was building Ishikawa diagrams.

Based on everything collected so far and the categories generated from the KJ method, two Ishikawa diagrams were made; one for parameter checking (see Figure 6.53) and one for calling the doctor (see Figure 6.54). These two diagrams are concerning the two research questions (y's) of the project. By making these Ishikawa diagrams, possible x's of the process are identified. The next step is to determine which of these factors are the most significant ones, and then focus on them.

In order to find out the most significant factors which will be the identified x's of the process, two methods were utilised: voting and variation mode and effect analysis (VMEA).

To find out the identified x's, the two Ishikawa diagrams were distributed between nine nurses from different wards of the hospital. Each nurse had five votes on each diagram, and they could vote for any of the x's they found most significant in the process. After collecting and summing up the votes, a Pareto chart was made to see which x's has got higher values (see Figures 6.55 and 6.56).

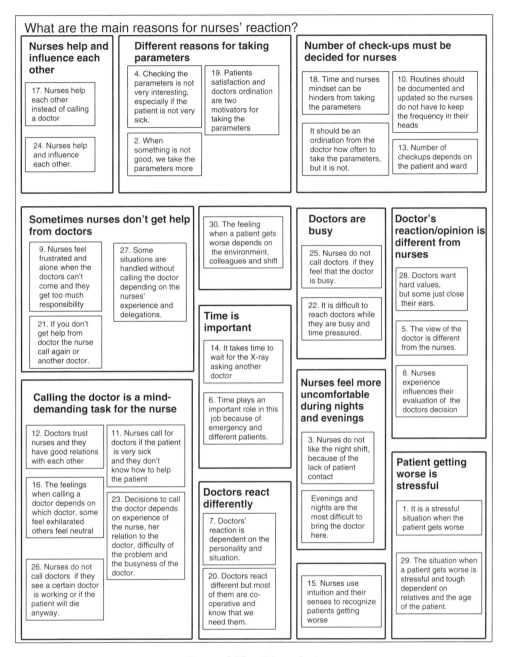

Figure 6.52   KJ analysis.

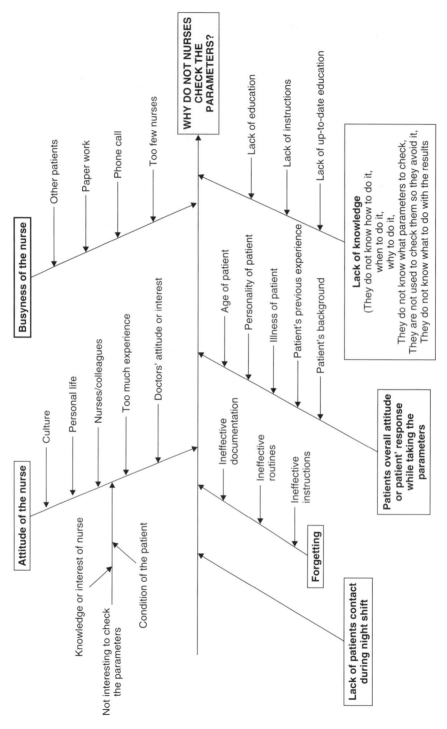

Figure 6.53   Ishikawa diagram for 'checking the parameters'.

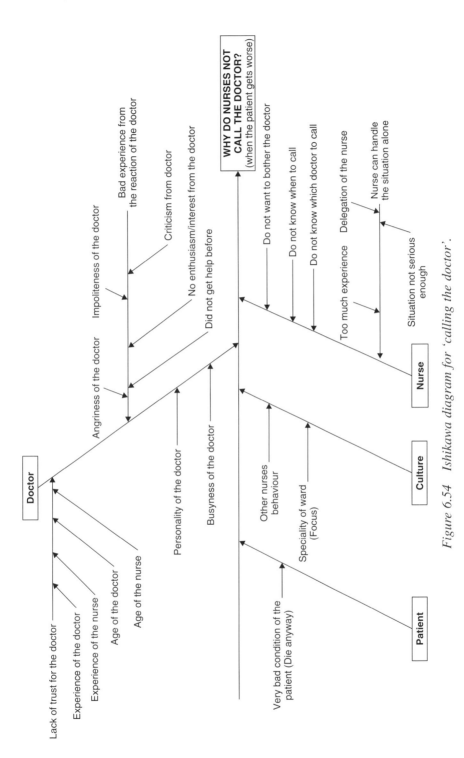

Figure 6.54   Ishikawa diagram for 'calling the doctor'.

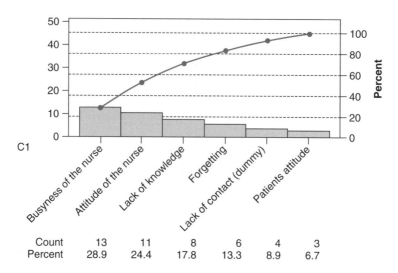

Count	13	11	8	6	4	3
Percent	28.9	24.4	17.8	13.3	8.9	6.7

Figure 6.55   Results of voting for not 'checking the parameters'.

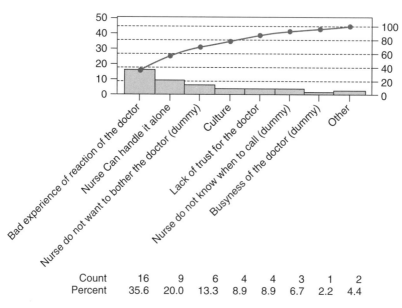

Count	16	9	6	4	4	3	1	2
Percent	35.6	20.0	13.3	8.9	8.9	6.7	2.2	4.4

Figure 6.56   Results of voting for not 'calling the doctor'.

The Pareto chart identifies that for the first Ishikawa diagram ('checking the parameters'): *busyness of the nurse*, *attitude of the nurse* and *lack of knowledge*, are the first three important factors. For the second Ishikawa diagram ('calling the doctor'): *bad experience of reaction of the doctor*, *nurse does not want to bother the doctor* and *nurse can handle the situation alone*, are the three most significant ones.

In order to check the validity of the results from voting it was decided to use VMEA. If the results of VMEA are the same as the results from voting, the identified x's can be judged more confident.

Two VMEA tables were designed, one related to each of the Ishikawa diagrams. Three nurses and one doctor (team members from the hospital) were asked to fill in the table with values according to the scales provided in the paper titled: 'Variation mode and effect analysis: a practical tool for quality improvement' by Johansson, Chakhunashvili, Barone and Bergman (published in 2006 on *Quality & Reliability Engineering International Journal*). The simple average was calculated first, and in order to make it less sensitive to outliers the geometric mean of the four values assigned by the team members was obtained. Eventually, as no significant difference could be identified between the simple and geometric mean of the values, the rest of the calculations for variation risk priority number (VRPN) were done with the geometric mean. VRPN are the final outcomes of the VMEA method which identifies the significance of each of the sub-Key product characteristics and noise factors.

Results from the VMEA (see Tables 6.1 and 6.2) show that *busyness of the nurse*, *attitude of the nurse* and *lack of knowledge*, are the most important factors for not 'checking the parameters'. Furthermore, *lack of trust for the doctor*, *nurse can handle it alone* and *culture*, are the three most significant factors for not 'calling a doctor', followed by *bad experience of reaction of the doctor*.

Table 6.1   Results of VMEA for not checking the parameters.

Sub-KPC	VMEA	
	VRPS	Percentage (%)
Attitude of the nurse	348 707.85	49
Busyness of the nurse	123 521.26	18
Lack of contact (dummy)	13 953.52	2
Forgetting	90 360.33	13
Patient's attitude	1907.79	0
Lack of knowledge	126 274.16	18
Total	704 724.90	100

Comparing the results from voting and VMEA depicts that for the first Ishikawa diagram for not 'checking the parameters' the results are exactly the same and they highly support each other. For the second Ishikawa diagram for not 'calling the doctor' two of the factors are exactly the same: *nurse can handle it alone* and *bad experience of reaction of the doctor*.

After finding out the identified x's from the Ishikawa diagrams, the next step is to investigate each factor further and explore its root causes. The results from VMEA are used to find out what are the most important causes for the factors found. The summary of the root causes is shown in Table 6.3.

Table 6.2   Results of VMEA for not calling a doctor.

Sub-KPC	VMEA	
	VRPS	Percentage (%)
Lack of trust for the doctor	35 254.53	19
Bad experience of reaction of the doctor	33 419.59	18
Personality of the doctor (dummy)	14 722.78	8
Busyness of the doctor (dummy)	7465.81	4
Vary bad condition of the patient (dummy)	8033.61	4
Culture	38 283.68	21
Nurse does not want to bother the doctor (dummy)	7683.26	4
Nurse does not know when to call (dummy)	6105.14	3
Nurse does not know which doctor to call (dummy)	168.55	0
Nurse can handle it alone	34 549.51	19
Total	185 686.45	100

Table 6.3   Root causes for sub-KPCs based on VMEA.

Sub-KPC	Root cause
Not checking the parameters	
Attitude of the nurse	Culture of the ward
Lack of knowledge	Lack of instructions
Not calling the doctor	
Culture of the ward	Speciality of the ward (focus)
Bad experience of the nurse	Criticism of the doctor
Lack of trust for the doctor	Experience of the nurse
Not wanting to bother the doctor	–
Handling the situation by the nurse	Delegation of the nurse

6.4.5   Analyse phase (ideas and intentions)

In this phase of the Six Sigma project the identified indicators from the Measure phase should be quantified in order to determine how to achieve the process improvement goals. By the end of this phase of the Six Sigma DMAIC framework, it is possible to disclose which of the x's influence y, and as a result, an improvement solution can be easily designed. Tools and techniques that can be used in this phase are hypothesis testing and regression analysis. By using these two tools, it is possible to assess the interactions between the identified indicators and their correlations with the y's, which will help us to recognise the most effective factor to change and improve in the next phase.

6.4.6   Improve phase (ideas and intentions)

In the Improve phase, x's and their interaction with y's are taken into account and possible solutions to improve the y's and ultimately improve the big Y are designed. It is important to avoid implementing solutions before evaluating the possible disadvantages and re-work they may cause.

One idea in this regard is to graphically show the vital parameters of the patient. This can be done by several methods. For instance, the nurse can be provided with pre-designed paper charts on which he/she can plot the vital parameters taken. Plotting the vital parameters on one single chart facilitates the recognition of trends and patterns in changes in the patient's vital parameters for the nurse, it can serve as a single data recording sheet which prevents data loss in different places, and finally it can be an effective way of communication between nurses and doctors.

The presence of the human factors, such as the nurses' attitude towards taking the parameters including the influence of the self-confidence of the nurse in his/her decision to involve the doctor, the culture of the ward, and so on, necessitates some kind of communication and campaigning programmes aiming at organisational change in terms of the culture and routines. Some examples of a possible campaigning programme can be the hospital director's talks, columns in the hospital newsletter, associating the vital parameter taking attitude with symbols and labels, distributing flyers and posters, communicating the idea in group meetings, and so on. A new process should be designed and implemented which ensures that manuals and instructions are regularly reviewed and updated properly. Moreover, some educational activities such as an on-the-job training programme can be useful in order to both refresh the nurses' mind and update their knowledge.

In order to proceed further, one intention is to consider and promote the robust thinking for redesigning and evaluating processes. Robust thinking consists of three main principles: awareness of variation, insensitivity to the noise factors, application of various methods and in all stages of the process design.

6.4.7   Control phase (ideas and intentions)

The Control phase consists of four main tollgates, including monitoring the y's, documentations, visualisation and communication, in order to make sure that the results have been spread throughout the organisation and they are kept properly.

In the Improve section, one idea was to visualise the vital parameters. The visualising methods can be used in the Control phase too. Specifying limits on these visualised charts, that is the use of the control charts with the limitation lines, can assist nurses both to identify the possible patterns and trends in the patient's vital parameters, and to more easily recognise if the vital parameters are altering in a warning zone which needs more attention. A plotted line over time also makes it easier for the nurse to see if the parameters have been checked recently, and if it will be obvious who did not check them, since they know who worked the last shift. It could in this way also be an incentive to actually take them, which is one of our y's.

Regular check-ups by using historical data can be of help to make sure if everything has been done as they have been supposed to. This can include regular investigations of y's of the project, for example every six months.

During the brainstorming sessions within the group, the possibility of utilising a mistake-proofing approach has been discussed as an effective solution for preventing the mistakes and causes of unwanted variation from occurring again. Any mechanism that helps an equipment operator avoid mistakes by preventing, and/or correcting human errors as they occur, is called 'Poka-yoke', from *yokeru* (avoiding) and *poka* (inadvertent errors). One possible Poka-yoke project can be the hospital software. Currently, nurses can enter patient's check-up reports and other related information freely as text into the software. This can be improved by modifying the software in a way that does not allow the user to enter more information unless he/she has entered some compulsory required items, which in our case can be the patient's vital parameters.

6.5 Optimal scheduling for higher efficiency and minimal losses in warehouse at Structo Hydraulics AB

Project purpose: to reduce the amount of waste in the cutting process at the warehouse department of Structo Hydraulics AB.

Organisation: Structo Hydraulics AB

Duration of the project: five months (2011)

Black Belt candidates team:
Therese Doverholt, Structo Hydraulics AB.
Anna Errore and Sepideh Farzadnia – masters students.

Phases carried put and implemented tools: Define (project charter, Gantt chart, 'Y' and 'y' definitions, SIPOC, process map) – Measure (process map, measurement system analysis: exploratory analysis) – Analyse (affinity diagram, Ishikawa diagrams, cause-and-effect matrix, Pareto charts, histograms, probability plots, scatter plots, pie charts) – Improve (improvement solutions rating, prioritising tool, cost-benefits analysis table, software, Excel matrix) – Control (control plan, control chart).

6.5.1   Presentation of Structo Hydraulics AB

Structo Hydraulics AB manufactures tubes for the hydraulic industry. The main customers are manufacturers of hydraulic cylinders and construction machines. The product range include cold drawn seamless tubes, cold drawn welded tubes, skived and roller burnished tubes and components. It also covers a wide range of dimensions up to 264 mm in outer diameter. The tubes are sold in both random lengths and cut lengths. Random is when the customer orders a total quantity in metres, and each single tube can be delivered in lengths between 4 and 12 m (according to European standards). Most of the biggest customers, however, order in cut lengths. The warehouse is the department for cutting the fixed lengths according to requirements from customer.

6.5.2 Project background

The project has been performed in the warehouse is about reducing the amount of waste in the cutting process. The objective of the project is to increase yield by 1 %, reducing costs by 600 000 SEK. By looking at the current situation in the company we see that there is a considerable amount of waste caused by the cutting process. There has been made many efforts in trying to reduce the yield losses in the warehouse. However, none of the projects and/or activities has been completely successful. Figure 6.57 is a pie chart which illustrates the different causes of scrap in the company.

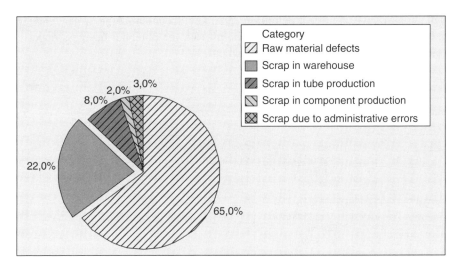

Figure 6.57   Scrap distribution at Structo Hydraulics AB.

The total scrap in the company can be grouped into five different types.

- Raw material defects – scrap due to supplier non-complying with specification.

- Scrap in warehouse (in the cutting process).

- Scrap in tube production – defects caused in the production of cold drawn tubes.

- Scrap in component production – defects caused in the production of components.

- Scrap due to administrative errors/root causes.

Raw material defects could be deeper divided into several reasons and there is a dedicated database for this data. The main supplier is also the owner of Structo Hydraulics AB, and there are several ongoing projects at the supplier in order to solve the problem.

For this reason the focus of this Six Sigma project is the second largest piece of the pie chart, the warehouse. This scrap is pure waste; the material is complying with quality requirements but not, however, the length requirement of the customers.

The waste is the largest issue of scrap (internally produced scrap) and it is actually 2.6 % of the annual turnover, approximately 5.3 MSEK. In the cutting process at warehouse the yield is approximately 90.5 % (fourth quarter of 2010).

The efficiency and capacity of the process is currently not entirely considered and this leads to a non-optimal planning process. These are large issues, which sometimes result in a non-complete quantity to deliver to the customer and the failure to meet the required delivery dates.

6.5.3   Define phase

The Define phase gives a project definition and a key document is the project charter (see Figure 6.58), including the main scope of the project, the potential business opportunity, the project team, the plan and the goal statement.

Company (organization)	Structo Hydraulics AB	Unit/Department	Warehouse
Executive	Kishore Bharambe, Managing Director	Senior Deployment Champion	N/A
Development Champion	Göran Larsson, Quality Manager	Project Champion	N/A
Master Black Belt	N/A	Finance Champion	Kerstin Rienas
IT Champion	N/A	HR Champion	N/A
Industrial participant (Black Belt candidate)	Therese Doverholt, Quality Engineer	Telephone/e-mail	Email: Telephone:
Sponsor & process owner	Jerry johansson, Production Manager	Site or location	Storfors, Sweden
Project Start Date	2011-01-19	Project completion Date	2011-05-20
Expected impact level	Increase yield by 1% (from 90,96% to 92%)	Expected financial impact (savings/revenues)	502 000 SEK

Figure 6.58   Project charter.

Element	Description	Charter
1. Project summary	A short description of the project	Structo Hydraulics AB manufactures tubes for the hydraulic industry. The main customers are manufacturers of hydraulic cylinders and construction machines. Tubes are sold in both random lengths (4-12 meter) and cut lengths. Warehouse is the department for cutting the fix lengths according to requirement from customer. In this process Structo have yield of approximately 90,5 % (4th quater of 2010). The yield losses are related to material defects, non-optimal input length for the process and other reasons. In this project the input length will be prioritized. The efficiency and capacity is not properly clarified, which leads to a non-optimal planning process. These are large issues, which sometimes result in a non-complete quantity to deliver to customer and failure of meeting required delivery times.
2. Impacted process	The specific processes involved in the project	CUTTING PLANNING PROCESS (INCLUDING UTILISATION OF MATERIAL AND PLANNING HORIZON AND DISPOSITION OF REMAINING MATERIAL).
3. Benefit to customers	Define internal and external customers (most critical) and their requirements	External customer: Delivery with correct quality and at right time. Internal customer: Operators. Getting the optimal planning to cut tubes in most effective way.
4. Benefit to the business	Describe the expected improvement in business performance	More long term cost effective in the warehouse process and improved performance to customer. Mainly cost reduction and better utilization of cutting capacity.
5. Project delimitations	What will be excluded from the project	Purchase process not included. The technical aspects of cutting and handling process are not included. IMPACTS FROM MATERIAL DEFECTS FROM SUPPLIER SHALL NOT BE INCLUDED.
6. Required support	Support in terms of resources (human and financial) required for the project	Statistics and data from manufacturing orders (yield), customer requirements in terms of size and volumes. Commitment by the management. Involved people (in the company) shall be available when needed.
7. Team members (including students BB candidates)	Names of the master students who will take part in the project	Sepideh Farzadnia Anna Errore Therese Doverholt

Figure 6.58   (continued).

8. Other people involved	List technical experts and other peolpe who will be part of the team	Supervisor at Chalmers: Marcus Assarlind MD (executive): Kishore Bharambe Deployment Champion: Göran Larsson Process owner: Jerry Johansson Operator(s): x

9. Specific goals	Define the baselines, your realistic goals for the project and the best case targets fro improvement.	Actual value (baseline)	Realistic goal by project end date	Best case goal
		Losses 2010: Yield 90,96%	Increase Yield by 1% on annual basis. Savings: 502 000 SEK	Increase Yield by 1% per year over a 3 year period means savings of appr. 1,5 MSEK.

DEFINE phase completion date	2011-02-17	MEASURE phase completion date	2011-03-21
ANALYZE phase completion date	2011-04-15	IMPROVE phase completion date	2011-05-13
CONTROL phase completion date	2011-05-19	PROJECT result presentation date	May 25th, 2011

Figure 6.58   (continued).

The project is planned to be completed in three four in five phases (see Figure 6.59). However, due to lack of time the implementation and Control phases are not planned to be fully done during this period, but there is a plan for the Control phase that should be followed after the implementation.

ID	Task Name	Duration	Start	Finish
1	Six sigma project at Structo Hydraulics	70 days?	Fri 11-02-11	Thu 11-05-19
2	define phase	5 days?	Fri 11-02-11	Thu 11-05-19
3	develop project team	1 day?	Fri 11-02-11	Fri 11-02-11
4	develop project charter	4 days	Mon 11-02-14	Thu 11-02-17
5	problem definition	4 days	Mon 11-02-14	Thu 11-02-17
6	understanding the process	4 days	Mon 11-02-14	Thu 11-02-17
7	define phase completed	0 days	Thu 11-02-17	Thu 11-02-17
8	measure	22 days	Fri 11-02-18	Mon 11-03-21
9	root causes analysis	22 days	Fri 11-02-18	Mon 11-03-21
10	developing measurement plan	5 days	Fri 11-02-18	Thu 11-02-24
11	data collection	22 days	Fri 11-02-18	Mon 11-03-21
12	measurement phase completed	0 days	Mon 11-03-21	Mon 11-03-21
13	analyse	19 days	Tue 11-03-22	Fri 11-04-15
14	data analyse	14 days	Tue 11-03-22	Fri 11-04-08
15	defining improvement tragets	5 days	Mon 11-04-11	Fri 11-04-15
16	analyse phase completed	0 days	Fri 11-04-15	Fri 11-04-15
17	improve	20 days	Mon 11-04-18	Fri 11-05-13
18	defining alternative solutions	10 days	Mon 11-04-18	Fri 11-04-29
19	cost/benefit analysis for solutions	10 days	Mon 11-04-18	Fri 11-04-29
20	implementing best solution	10 days	Mon 11-05-02	Fri 11-05-13
21	improve phase completed	0 days	Fri 11-05-13	Fri 11-05-13
22	control	4 days	Mon 11-05-16	Thu 11-05-19
23	planning for future improvement	4 days	Mon 11-05-16	Thu 11-05-19
24	finish	0 days	Thu 11-05-19	Thu 11-05-19

Figure 6.59   Gantt chart.

The Y to be improved is the yield in the cutting process, defined as follow:

$$\text{yield} = \frac{\text{output(kg)}}{\text{input(kg)}} \ (\%)$$

The yield of 2010 in the cutting process was 90.96 %, which means a material cost of 5.3 MSEK (top line in Figure 6.60). Due to technical requirements of the cutting machines the yearly average yield cannot exceed 96–97 %, due to the fact that the machines needs 30–80 mm to fix the tube when cutting the last piece of each long tube. Figure 6.60 also shows some calculations of possible savings for different levels of improvements in terms of yield. The goal set by Structo Hydraulics AB for this project is an increase by 1 % to 92 % (the high-lighted line in Figure 6.60), which means savings of 608 150 SEK (based on the quantity of 2010). However, in the long term, the goal is to improve by a further 2 % by developing successful improvement solutions from this project.

	Output Kgs	Input Kgs	Yield %	Loss kgs	Savings SEK
Actual	2 743 632	3 016 353	90,96%	272 721	0
	2 744 881	3 016 353	91%	271 472	24 184
	2 759 963	3 016 353	91,5%	256 390	316 167
Target	2 775 045	3 016 353	92%	241 308	608 150
	2 805 208	3 016 353	93%	211 145	1 192 115
Long Term Target	2 835 372	3 016 353	94%	180 981	1 776 081
	2 865 535	3 016 353	95%	150 818	2 360 047
	2 895 699	3 016 353	96%	120 654	2 944 013

Figure 6.60   Current performance (actual) and expected impact level (target).

The planning of cutting is normally based on a two-week period (see Figure 6.61). The orders for cutting are printed and sorted by dates, customers and product (input product). The planning is not being done with help of any structured system or software. In the past the operator has had a list with commonly sold lengths per product in order to check whether the remaining piece is suitable to use for a smaller cut length. However, this control of suitability is not done now in any structured way.

The SIPOC diagram in Figure 6.62 shows how the process is actually working. This SIPOC was divided into two parts: the administrative planning as a key step and the actual production as the second key step. This was done in order to capture both the flow of information and the actual flow of material. The two key steps are most often running parallel to each other, and that is why the SIPOC is more visual in these two parts. This also shows that there are actually two customers in the process; the internal customer (operator) and the external customer (end customer).

6.5.4   Measure phase

A detailed process map of the cutting process (Figure 6.63), from the customer order to the evaluation of the waste remained after cutting, is made to better understand the process inputs in each step and then use these inputs in a cause-and-effect matrix (see Figure 4.66).

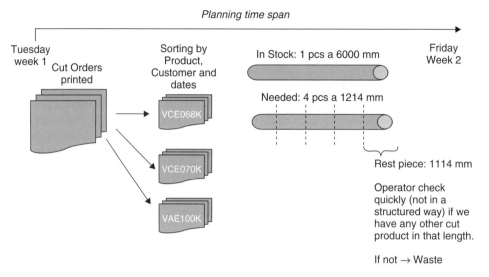

Figure 6.61   Process for planning of cutting.

Suppliers	Input	Process	Output	Customers
Marketing	Customer orders (Fix cut products) Requirements: Part number Length Delivery date Quantity	Delivery date within Tues Week 1 to Fri Week 2 Planning of cutting *When and what to cut*	Schedule Requirements: Yield=output/ input%	Operator-warehouse
1)Production 2)Tube manufactures	Long tubes to cut from Requirements: Size matching the required part number etc	Start of cutting	Fix cut products Requirements: customer specifications	Final customer of Structo

Figure 6.62   SIPOC – cutting process.

In order to better understand and analyse the process, the measurement plan developed takes into consideration the whole year 2010. The data collection consists of an Excel spreadsheet extracted from the company's database, and contains in total 3290 orders of the year 2010.

Input	class	Step	Output	Notes
customer specifications: e.g product type, length, pieces, due date	N	get customer orders	customer orders' specifications	
tolerances	S			
customer orders' specifications	N	planning for cutting and send the orders to the cutting warehouse	order at the cutting warehouse	
order management policy	C		manufacturing/purchase decisions	
customer orders to be processed	C	prioritize the orders	priority order/ cutting schedule	
operators' knowledge/experience	C			
priority order	C	check available of product type in the stock	orders available to be executed	
stock data available on the information system	C			
orders available to be executed	C	get the tube from the stock	long tubes to be cut	
cutting machine available	S	cut the tubes in fixed lenghts	tube cut in fixed lenghts according to the specifications	
cutting machine set up	C		remaining material	
long tubes to be cut	C			
tube cut in fixed lenghts	S	evaluate the waste	order date	
remainig material	C		waste date	
order data	N	register data	data registered	
waste data	N			
waste	C	register data	waste	
waste data	N		tube's remaining part to be considered again for future orders	
customer orders	N			
operator's knowledge/experience	C			

Figure 6.63 Cutting process map.

The most important pieces of information we need for the analysis are the following.

- The part numbers: the output part number is a specific code which identifies one product with specific physical and process characteristics (inner and outer diameter, wall thickness, surface characteristics, etc.); the input part number identifies the long tubes used to cut the final products.

- The customer: identified by his name and code, in the following analysis only the customer codes will be used for confidentiality reasons.

- The quantity of pieces required: when the sales unit is not pieces but metres, the quantity will be calculated dividing the invoice quantity by the length (data column added by the group).

- The length specified by the customer.

- The input and output kilos and metres: used to calculate the yield and the amount of waste.

- The date of order and due date (required delivery date).

From a first look at the data set it was evident that there are some orders with a yield over 100%. In order to understand what was the problem when they were registered, the first analysis was an exploratory analysis of the customer orders.

6.5.5   Analyse phase

An 'Affinity' session was held with four operators and the production leader of the warehouse. The session started with the simple question: 'What is causing waste?' It was found as a result of this session (see Figure 6.64) that some of the reasons behind cutting problems had not been mentioned by the managers, such as unreadable handwriting and communication problems. These findings are later considered in the Improve phase.

In order to consider all the possible causes for the waste due to the cutting process, a brainstorming session was arranged to bring out all the thoughts about the causes of the problem. Some inputs from the affinity diagram were also taken into account. The causes of the problem are identified in seven groups: management, marketing, planning, machinery, stock, operators and material, by a fish bone diagram (see Figure 6.65).

A cause-and-effect matrix (Figure 6.66) was used to prioritise potential causes according to what is critical for customers. The matrix showed that the first four process steps related to the most significant inputs are: getting customer orders, planning for cutting, prioritising the orders and checking the availability of the tube in stock; the critical inputs are: customer order's specifications, tolerances, customer orders selected to be processed after prioritising orders and stock data available in the information system. According to the analysis these inputs are the most important and have the most influence on customer satisfaction. These have been shown through the rating that was allocated to each parameter. It also shows that amount of waste and meeting product specifications are the most important outputs of the process for the customer, which seems quite reasonable.

One of the other objective of the Analyse phase was to study the customer orders throughout the year 2010 in order to understand which products are the best sellers. The

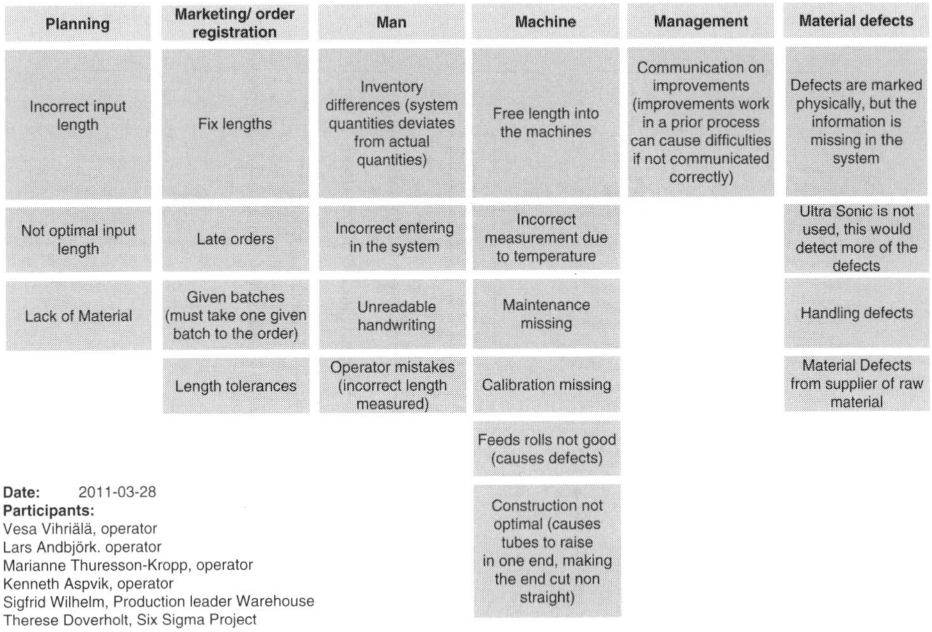

Planning	Marketing/ order registration	Man	Machine	Management	Material defects
Incorrect input length	Fix lengths	Inventory differences (system quantities deviates from actual quantities)	Free length into the machines	Communication on improvements (improvements work in a prior process can cause difficulties if not communicated correctly)	Defects are marked physically, but the information is missing in the system
Not optimal input length	Late orders	Incorrect entering in the system	Incorrect measurement due to temperature		Ultra Sonic is not used, this would detect more of the defects
Lack of Material	Given batches (must take one given batch to the order)	Unreadable handwriting	Maintenance missing		Handling defects
	Length tolerances	Operator mistakes (incorrect length measured)	Calibration missing		Material Defects from supplier of raw material
			Feeds rolls not good (causes defects)		
			Construction not optimal (causes tubes to raise in one end, making the end cut non straight)		

Date: 2011-03-28
Participants:
Vesa Vihriälä, operator
Lars Andbjörk. operator
Marianne Thuresson-Kropp, operator
Kenneth Aspvik, operator
Sigfrid Wilhelm, Production leader Warehouse
Therese Doverholt, Six Sigma Project

Figure 6.64   Affinity diagram.

products sold in the year 2010 are identified in our data collection by a code (output part number) and a product description. There are 216 product codes, each identifying a specific type of tube with specific characteristics and specific dimensions (inner and outer diameter and wall thickness). Each customer specifies in his order the required length of the cut products. Five different Pareto analyses have been made with MINITAB. The first one shows which products are the best sellers in terms of number of orders; then the products are analysed in terms of total amount of pieces (the range in these terms is from 1 piece to 1386 pieces). Then we come into details of the variation in the lengths required by the customers. The third Pareto analysis studies the products in terms of how many different lengths have been found in the orders of the last year. The reasoning behind this analysis is that there are products which are always, or almost always, required in the same length or a few different variants. Other products, with a huge range of length and number, are essentially also the best sellers. Thus, an optimisation study for these products can perhaps be more complex but reasonably more effective for the whole business of Structo Hydraulics. Finally, the last Pareto chart (see Figure 6.67) shows the products which are the best sellers in a specific length and in terms of total amount of pieces required.

On the other side also, the customers were analysed with a Pareto analysis, studying the number of their orders to figure out how much one customer is relevant for the company (other possible analysis of the customers have been done considering the amount of kilos ordered or the quantity of cut products); for confidentiality reason only the customer ID numbers are used. There are two particularly important customers in terms of orders. Moreover, 80 % of the revenues for the company are related to the first eight major customers (which are less than 20 % of the total amount of customers of Structo Hydraulics).

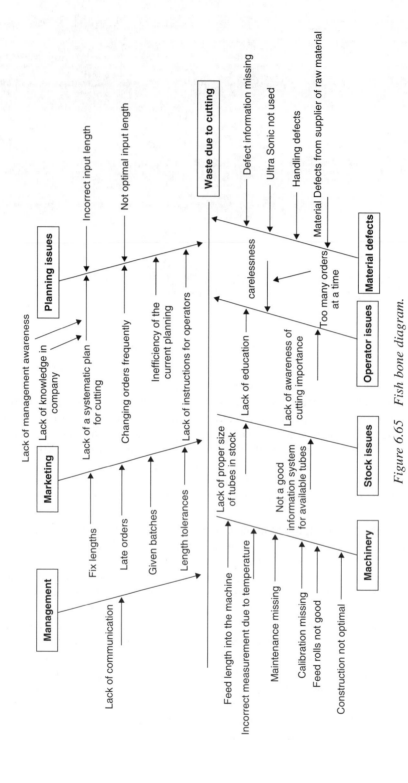

Figure 6.65   Fish bone diagram.

			1	2	3	
		Rating of importance to customer	8	8	10	
		Key Process outputs	Meet product specifications	Meet the delivery data	Waste	
	Process step	Process input				Total
2	get customer orders	tolerances	9	9	9	**234**
3	planning for cutting and send the order to the cutting warehouse	customer orders' specifications	9	9	9	**234**
5	prioritize the orders	customer orders to be processed	9	9	9	**234**
8	check availability of product type in the stock	stock data available on the information system	9	9	9	**234**
9	get the tubes from the stock	order available to be executed	3	9	9	**186**
12	cut the tubes in fixed lengths	long tubes to be cut	9	3	9	**186**
18	evaluate the waste (store it for future orders or throw it away)	waste data	9	3	9	**186**
19	evaluate the waste (store it for future orders or throw it away)	customer orders	9	3	9	**186**
11	cut the tubes in fixed lengths	cutting machine set up	9	1	9	**170**
4	planning for cutting and send the order to the cutting warehouse	order management policy	0	9	9	**162**
6	prioritize the orders	operators' knowledge/experience	3	3	9	**138**
15	register data	order data	3	3	9	**138**
17	evaluate the waste (store it for future orders or throw it away)	waste	3	3	9	**138**
20	evaluate the waste (store it for future orders or throw it away)	operators' knowledge/experience	3	3	9	**138**
14	evaluate the waste	remaining material	0	3	9	**114**
1	get customer orders	customer specifications	9	3	3	**126**
7	check availability of product type in the stock	priority order	3	9	3	**126**
16	register data	waste data	1	1	9	**106**
13	evaluate the waste	tube cut in fixed lengths according to be specifications	0	0	9	**90**
10	cut the tubes in fixed lengths	cutting machine available	1	3	1	**42**
Total			**808**	**760**	**1600**	

Figure 6.66   Cause-and-effect matrix.

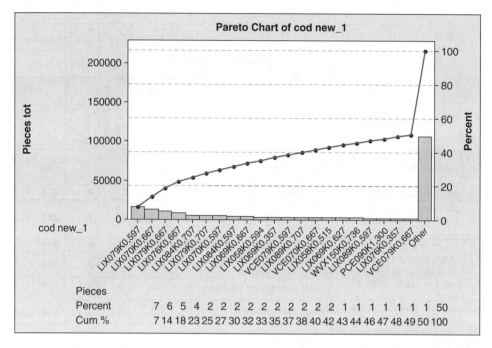

Figure 6.67   Product of a specific length – total number of pieces.

After the previous analysis we could now focus our attention on the best-selling products. Looking at the first product in terms of number of orders and total number of pieces required, the product LIX079K, this one was required last year in 17 different lengths and 3 of them are among the first 20 couples of product length analysed. The length 0.597 m is the most popular. The purchase of this product and this length last year has been made only by the customer 1600. The histogram of the yield and the normality test are shown in Figure 6.68.

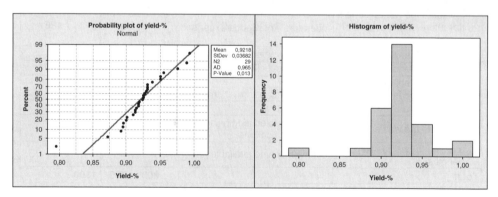

Figure 6.68   Histogram of yield and normality test for LIX079K-0.597 m.

Further analysis of these products were performed in order to study the distributions of all products' orders. The data collection consists of 3290 observations. An observation, in this case, is a particular order. The yield (the big Y) for the whole time period, gives following statistics:

- mean, $\mu = 0.91543$;

- median $= 0.936429$;

- standard deviation, $\sigma = 0.0985$;

- variance $= 0.00971$.

The data collected do not follow any distribution model according to individual distribution identifications that were made to test the distribution of the data. However, considering the histogram of the yield (Figure 6.69a) it can be seen that the data is negatively skewed as a truncated distribution (note: some data are on the right of the value 1, these are the data that showed us the problem with the information system and the registration of the data). Also, when studying the histogram of the different lengths sold (Figure 6.69b), the data can be seen as positively skewed with a bimodal distribution model.

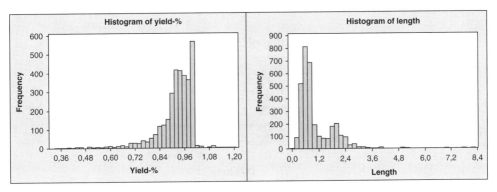

Figure 6.69   Histograms of yield (a) and length (b).

6.5.6   Improve phase (planning)

From the results of the previous phases it seems clear the process can be improved from several points of view: in the cutting schedule planning, the information flow and the orders management.

The solutions need to be classified, some of them are less complex to implement and can be applied to the process with certain simple changes, others are radical and imply substantial improvement solutions. The ratings were given after reflections and discussion within the group and with planning people at Structo Hydraulics AB. In Table 6.4 some solutions are rated using the Likert scale (from 1 to 5). Figure 6.70 shows the solutions in a graph, showing difficulty versus effectiveness. This shows how to prioritise the different solutions. The analysis of tolerances represents an improvement that should be done later

Table 6.4   Improvement solutions rating.

Improvement	Rating of difficulty to implement	Rating of effectiveness	Comment
Add cutting machines to the preventive maintenance programme	2	2	'Just do it' improvement
Calibration programme for the cutting machines	2	2	'Just do it' improvement
Analysis of tolerances	4	2	Starting a new activity/project
Information system	4	4	Starting a new activity/project
Short-term planning (software number 1)	2	4	Short-term planning in warehouse
Long-term planning (Excel spreadsheet)	3	3	Long-term planning of optimal input length

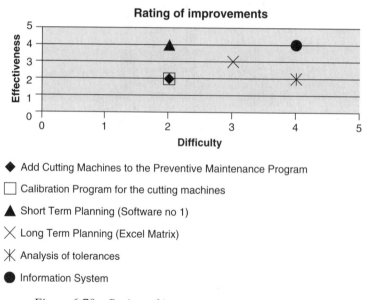

Rating of improvements

◆ Add Cutting Machines to the Preventive Maintenance Program

☐ Calibration Program for the cutting machines

▲ Short Term Planning (Software no 1)

✕ Long Term Planning (Excel Matrix)

✳ Analysis of tolerances

● Information System

Figure 6.70   Rating of improvement (prioritising tool).

than the others because it implies a high difficulty and it has a low effectiveness. Vice versa, the tool shows the highly prioritised improvement as the information system.

One of the problems that emerged in the affinity session with operators was that there is no plan for maintenance and calibration of the cutting machine. The solution could be to include this machine in the existing general plan of maintenance and check periodically the effectiveness of the way it works.

Another important issue is concerning the tolerances. The cutting tolerance is given per standard ($+2/-0$ mm) and it is used for orders when no requirement of a specific tolerance is given. When these standard tolerances are not met, this consequently results in waste of material. But maybe that material could be acceptable for the customer and the waste could, in some extent, be avoided. This issue can be easily overcome by reviewing the tolerances for each customer regarding that particular order.

A very important problem to solve regards the data available in the information system of the company. On one hand the fact that we found some incorrect data (when the yield is $>100\%$) shows that there is something wrong in the system used for registration of data or in the way it is used. On the other hand there is a difference in the accuracy of the information contained in the system regarding the customer orders and the available stock of long tubes produced in the company or bought from other suppliers. Another issue is how the system shows the quantity in stock, regarding the input to the cutting process. Today the system shows only a total quantity per batch, which is, for example 12 pieces with total (added) quantity of 100 m. However, the system does not show the length of each one of the 12 pieces. The solution for this issue is related to a good design of the information system.

The most important issue that emerged during this project is the lack of a structured way of planning and cutting the long tubes in fixed lengths according to customers' specifications. Both at the high level and at the operational level the cutting plan is essentially based on the operators' experience and knowledge. Since there is no systematic way of cutting when orders are received by the operators, they start the cutting based on what they feel is the best solution and has the least waste. So the input to the cutting plan is mostly their mathematical knowledge and experience with the guidance of the production leader of the warehouse. Generally there are two kinds of improvement that should be considered, short term and long term.

- The short-term improvement tries to answer the question of how to cut the tubes that are already in stock. Since there is no information system saying the exact length of the tubes in stock before taking the order, the short-term solution can be to optimise the cutting plan to have the least waste as possible.

- The long-term improvement idea is to forecast the orders before getting them from the customers so as to be able to purchase the right length of the tubes which leads to decreasing the waste later during the cutting phase. The current situation is that a purchase order occurs according to an individual estimation and the experience of the responsible person and this does not always lead to the right length of tubes purchased.

Generally the cutting stock problem can be solved using different methods, heuristic, meta-heuristic or algorithmic methods. Anyway, for an industrial application, the problem can be very large and complex to solve and so it is very common to use specific software that helps to optimise a cutting process. The choice and implementation of software of this kind is strictly connected to the design of the information system. The information that the software needs for the optimisation should be available and easy to extract from the system, for example the number of pieces and the exact length of each tube in stock, to use it as an input for cutting.

After finding different solutions, their costs were evaluated (see Table 6.5). They have been divided into direct costs and indirect costs, where the indirect costs is including, for example, training time of personnel. The direct costs are purchasing of software and the hourly cost of personnel education and maintaining the different solutions on a yearly basis. The day-to-day use of the planning tools are not included, since it is assumed that this will be used instead of their current processes, and taking the same amount of time or less.

Table 6.5   Cost-benefit analysis table.

Improvement	Direct cost	Indirect cost	Total cost	Benefits
Add cutting machines to the preven-tive main-tenance programme	Maintenance cost	Machine downtime	€500	Less intervention to failure
Calibration programme for the cutting machines	Calibration cost	Machine downtime	€500	More accuracy and precision
Short-term planning (software number 1)	€75	€1000/year	€1075	Using the software, the goal of increasing by 1 % will definitely be achieved
Long-term planning (Excel spreadsheet)	€0	Education and maintaining	€750	
Analysis of tolerances	This is a suggestion for improvement activities/projects			
Information system	This is a suggestion for improvement activities/projects			

As said above, one of the possibilities to optimise the cutting planning is to implement the use of software. One piece of software suitable for this purpose was found and downloaded in the trial version (45 days). Some trials with this software were performed to test the applicability of this tool, and its possible effectiveness, compared to the costs of this solution, both the direct and indirect costs. Different aspects are to be taken into consideration:

- cost of the software

- feasibility of the integration of this software in the company's information system;

- extent of changes in the process necessary for a practical implementation;

- ease of use and training for people who are going to use it;

- potential savings in terms of yield (big Y of this project).

In order to test the software, data about customer orders and stock on hand were needed.

For customer orders it is possible to use both historical data from our previous data collection and current orders extracted from the company database.

For stock on hand, in the current state, the data in the system have the following structure: when looking at the system in order to check the feasibility to execute a customer order, the operator sees in the table the following information per lot in the stock: part number (code which identifies the product) and description, place in the stock, quantity in metres, quantity of pieces in the lot, remarks and lot number. The quantity in metres represents the total amount of metres of that product in that specific lot, but it is not possible to have the information of how long a particular tube in that lot is; of course it is possible to go physically in the stock and check the length of the tubes but this information is not actually available in the system. Sometimes in the remarks column it is possible to register if that lot has been already cut into pieces of a certain length. One other important thing to consider before starting with some simulation with the software is that the data about stock on hand are always and continuously updated during the daily work of the company. The only possibility to use this data for a simulation is to assume that we are taking a snap shot of the current state of stock at a certain time and using the data only for this simulation purpose. We cannot pretend that the results could be directly applied after the simulation because meanwhile the real stock has changed many times in parallel with our study.

After all these considerations it is possible to start some simulations of optimum cutting patterns using the following data extracted from the company system:

- order up to June of the customer 1600;

- stock on hand information about the most popular input part numbers.

The results of one simulation carried out considering orders and stock of the product VCE111 showed that it is possible to reduce the waste and increase the yield of cutting when considering at the same time several orders of different length. With the pattern suggested by the optimisation software (see Figure 6.71) the yield obtained is 99.0 %, which is much higher than the average yield in the orders of that product last year, which was 92.1 %.

Another possibility for improve the cutting schedule is to use a tool developed in a previous project at Structo Hydraulics. This previous project had the same aim, to reduce waste, by finding a better way to execute the orders in order to minimise the remaining material by considering, for instance, two orders at the same time. An Excel matrix was built as a tool to use every time there are two orders of two different lengths specified by

Length	Material	Quantity	1D Graphic
6,5	VCE11 1	15	0,599 0,599 0,599 0,599 0,599 0,599 0,599 0,599 0,422 0,422 0,422 0,422
6,5	VCE11 1	1	0,599 0,599 0,599 0,599 0,599 0,599 0,598 0,598 0,422 0,422 0,422 0,422
6,5	VCE11 1	15	0,598 0,598 0,598 0,598 0,598 0,598 0,598 0,598 0,422 0,422 0,422 0,422
6,5	VCE11 1	1	0,598 0,598 0,598 0,598 0,597 0,597 0,597 0,597 0,597 0,597 0,597 0,422
6,5	VCE11 1	1	0,597 0,597 0,597 0,597 0,597 0,597 0,597 0,597 0,597 0,597 0,597 0,422
6,5	VCE11 1	3	0,597 0,597 0,597 0,597 0,597 0,597 0,597 0,597 0,597 0,597 0,597
6	VCE11 1	8	0,597 0,597 0,597 0,597 0,597 0,597 0,597 0,597 0,597 0,597 0,597

Figure 6.71   Optimal pattern from the software optimisation.

the customer. This matrix was further developed during this project, in order to make it more user-friendly, and easily used for the long-term planning, to decide the most optimal input length using forecasts of orders.

The intent of the project team is to implement these two tools for the optimisation of the cutting process in two different perspectives: the long-term and the short-term points of view. For the most popular products, the best sellers found in the Analyse phase, it is important to go deeper when studying the orders in order to figure out some patterns on the data that can allow us to forecast the demand for this product for a long-term optimisation. Combining the use of the tools found with this forecast it could be possible to buy or produce the optimal input length. On the other side, the analysis shows that some products are bought rarely, in some few lengths required by few and smaller customers. For these products it is economically more convenient to have a short-term perspective for the optimisation by trying to schedule the cuts in a way that can minimise the amount of waste as much as possible by utilising the software tool.

6.5.7   Control phase (planning)

The Improve phase could not fully be completed due to the time limit of the course. In the control plan (see Figure 6.72), both the time schedule for the implementations and the control plan schedule are compiled per solution. Table 6.6 shows the control frequencies and the responsibilities. The person responsible for control is also responsible for implementation. The above control plan has not yet been approved by the management, who could slightly change the plan in respect of time frames. The control chart (Figure 6.73) is made on the top five ordering customers. Each point in the chart considers the yield of a different sample size according to the number of order of that specific week. The plan is to implement the planning tool for these five customers, and continue the use of the control chart, showing also the future implementation points in the chart.

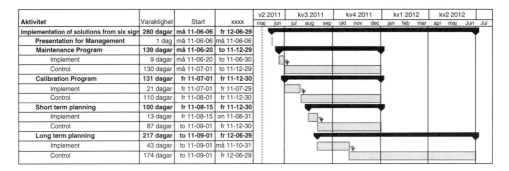

Aktivitet	Varaktighet	Start	xxxx
Implementation of solutions from six sign	280 dagar	må 11-06-06	fr 12-06-29
Presentation for Management	1 dag	må 11-06-06	må 11-06-06
Maintenance Program	139 dagar	må 11-06-20	to 11-12-29
Implement	9 dagar	må 11-06-20	to 11-06-30
Control	130 dagar	må 11-07-01	to 11-12-29
Calibration Program	131 dagar	fr 11-07-01	fr 11-12-30
Implement	21 dagar	fr 11-07-01	fr 11-07-29
Control	110 dagar	fr 11-08-01	fr 11-12-30
Short term planning	100 dagar	fr 11-08-15	fr 11-12-30
Implement	13 dagar	fr 11-08-15	on 11-08-31
Control	87 dagar	to 11-09-01	fr 11-12-30
Long term planning	217 dagar	to 11-09-01	fr 12-06-29
Implement	43 dagar	to 11-09-01	må 11-10-31
Control	174 dagar	to 11-09-01	fr 12-06-29

Figure 6.72   Control plan including implementation of the different solutions.

Table 6.6   Control frequency and responsibilities.

Solution	Control frequency	Responsible
Maintenance programme	1 per month	Production Manager
Calibration programme	1 per month	Production Manager
Short-term planning	1 per week	Black Belt of this project
Long-term planning	1 per week	Black Belt of this project

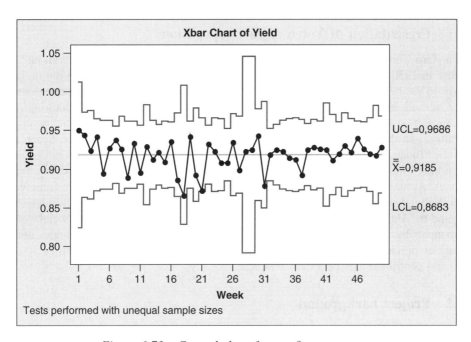

Figure 6.73   Control chart for top five customers.

6.6   Reducing welding defect rate for a critical component of an aircraft engine

Project purpose: to investigate if there is a correlation between repaired castings delivered from the supplier and weld defects occurring in the manufacturing processes of a large structural component of an aircraft engine at Volvo Aero, in order to minimise weld defects and subsequent rework, inspection procedures and overall lead-time, through a revision of the specifications of the castings.

Organisation: VOLVO AERO CORPORATION

Duration of the project: five months (2011)

Black Belt candidate team:
Sören Knuts, Ph.D., Volvo Aero Corporation
Arnela Tunovic and Henric Ericsson – masters students.

Phases carried on and implemented tools: Define (project charter, Gantt chart, 'Y' and 'y' definitions, cause-and-effect diagram, business case, log book on meetings, SIPOC, process map) – Measure (interviews, observation, drawings, work sheets, histograms, Gauge R&R) – Analyse (correlation, regression, process map, cause-and-effect matrix, FMEA, VMEA).

6.6.1   Presentation of Volvo Aero Corporation

Volvo Aero Corporation (VAC) has a long history of producing military aircraft engines, starting in 1930 and producing most of the engines for the Swedish Air Force in Troll-hattan. In the 1970s the strategic decision was made that the company's operations would be broadened, and the technical capabilities that were gained from manufacturing military aircraft engines were put into practice building up expansive commercial aerospace operations. In the 1970s Volvo Aero also joined the Joint Ariane European Space Programme. Today Volvo Aero develops and manufactures components for aircraft engines, gas turbines, military aircraft engines, subsystems for rocket engines, and also has engine services. Around 90 % of all new large commercial aircraft use components from Volvo Aero. Volvo Aero has about 2850 employees, with production facilities in Sweden, Norway and the United States. The motto 'Make it light' aims at reducing the weight of the components by up to 30 %, by for example using unique composite technologies instead of cast as previously. In this way, carbon emissions will be reduced by up to 50 % in 2020 and contribute to a better environment.

6.6.2   Project background

The Six Sigma Black Belt project described in this report is derived from the 'Make it light' initiative. By making the decision of going from manufacturing the large structural components of aircraft engines entirely in cast, to making it consist of smaller parts that

Figure 6.74   Assembling steps of aircraft component.

are welded together (Figure 6.74), a new set of issues arose that need to be considered in order to ensure high quality. The choice of using castings, welding them together with sheet metal and heat-treating the whole part brings difficulties such as defects occurring in the welded zone, as well as around the welded zone and in the castings. The number of defects in welded cast material is generally higher than in welded forging/sheet metal, and determining the cause of the high number of welding defects and how to minimise them is the scope of this project.

The specific product of interest is a structural component produced at VAC. The component is assembled through welding of sub assemblys or sectors. The main advantage of building the product out of separate parts is that different materials can be used for different parts in order to reach the long-term goal of building lightweight structures with a lot of advantages for the end customer. The inner part and the welds connected to it is the main focus area of this project since weld defects when assembling (both in sector assembly and weld assembly) are most common in this region, about 80 % of defects occurring at weld assembly are connected to this area. When the supplier is producing the castings some defects in the material are occurring. The defects are documented and then repaired by a welding operation. The definition of this project is to investigate if there is a correlation between repaired castings delivered from supplier and weld defects occurring in the manufacturing processes at VAC.

6.6.3   Define phase

The CTQ property that has to be characterised in this project is the defect-free weld that has to be delivered from VAC (see Figure 6.75).

Regarding the business case, this has been based and calculated with use of a standardised VAC template. A simplified estimation of business case is shown in Table 6.7.

To avoid diving too deep into details at this early stage of the project, the core team first of all decided to get a clear overview of what and how different types of factors could play an important role when welding in casting material. To identify and get an understanding of possible causes (small x's) and their effect (big Y) of weld defects, the core team developed a cause-and-effect diagram (Figure 6.76). The diagram shows a lot of different perspectives on how weld defects could occur, both from a VAC process perspective (e.g. heat treatment cycles) and also from a supplier perspective (e.g. variations in material structure).

To avoid scenarios where too much time is spent covering certain areas, causing problems to complete other equally or even more important fields, the Gantt chart has

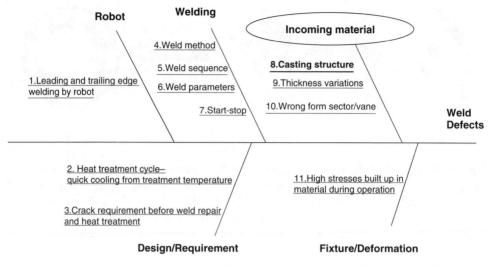

Figure 6.75   Cause-and-effect diagram.

Table 6.7   Business case.

---

100 hardware per year in 10 years: cost per weld repair 10 KSEK.
Cost per weld repair on annual basis = 100 × 10 KSEK= 1000 KSEK
In case we lower the average with one weld repair per hardware.
Total cost on weld repair is lowered on annual basis = 1 × 100 ×
    10 KSEK = 1000 KSEK
In 10 years this means 10 MSEK at least.

---

been used (Figure 6.77). The project outcome, with many meetings at VAC, can be found in Table 6.8.

To get a clear picture of the process of producing the component and get a grip of what factors that comes in to play, a process map (Figure 6.82) was developed. The first level was set as the process of producing the castings at the supplier, the second level was the process of producing the sectors and the third level of the process map was set as the process of reaching the final component. In this early phase of the project it was motivated to first of all clarify the process within VAC and therefore collect more information about the sector and component levels.

When the four different parts included in one sector are delivered from the supplier to VAC they are fixed in the right position and then tack welded together. After the initial tack welding, the sector is placed and welded in the plasma welding equipment. When the sector is welded it goes into the inspection of welds in form of a fluorescence penetrant inspection (FPI) and X-ray.

When all the sectors are approved, they go through a similar process as before, that starts with fixing all the sectors and then tack welding them together. After this we have a weld assembly that is welded in the plasma weld and then sent for inspection in FPI and X-ray.

Company (organization)	Volvo Aero	Unit/Department		7163/DfR
Industrial participant (Black Belt candidate)	Sören Knuts	Telephone/ e-mail		
Project Start Date	20110201	Project completion Date		20110524
Expected impact level		Expected financial impact (savings/revenues)		14% savings of total weld repairing cost.
Project summary	A short description of the project	High number of weld defects are causing quality problems and too long lead-time in the serial production phase of a civil aircraft component. The focus of the project is to evaluate correlation between incoming material defects (previously repaired by the supplier) and defects that appears during fabrication process at Volvo Aero.		
Impacted process	The specific processes involved in the project	Assembling processes of struts involve fixturing, welding, heat treatment and inspection (NDT=None Destructive Testing).		
Benefit to customers	Define internal and external customers (most critical) and their requirements	Production (High Volume) Trollhättan. To reach acceptable capability of identified quality problem.		
Benefit to the business	Describe the expected improvement in business performance	Enhances both in lowering of lead-time and improvement in quality cost. (Potential lowering of 14% of time and cost.)		
Required support	Support in terms of resources (human and financial) required for the project	Travel costs (students)		
Team members (including students BB candidates)	Names of the master students who will take part in the project	Arnela Tunovic Henrik Ericsson		
Other people involved	List componenthnical experts and other people who will be part of the team	NN (design concept), NN (Purchase), NN (weld manufacturing), NN (manufacturing analysis), NN (manufacturing subassembly), NN(manufacturing assembly)		

Specific goals	Define the baselines, your realistic goals for the project and the best case targets for improvement.	Actual value (baseline)	Realistic goal by project end date	Best case goal
		1 000 000 SEK/year	860 000 SEK/year	140 000 SEK/year

DEFINE phase completion date	2011-02-25	MEASURE phase completion date	2011-03-23
ANALYZE phase completion date	2011-04-07	IMPROVE phase completion date	2011-05-05
CONTROL phase completion date	2011-05-20	PROJECT results presentation date	2011-05-25

Figure 6.76   Project charter.

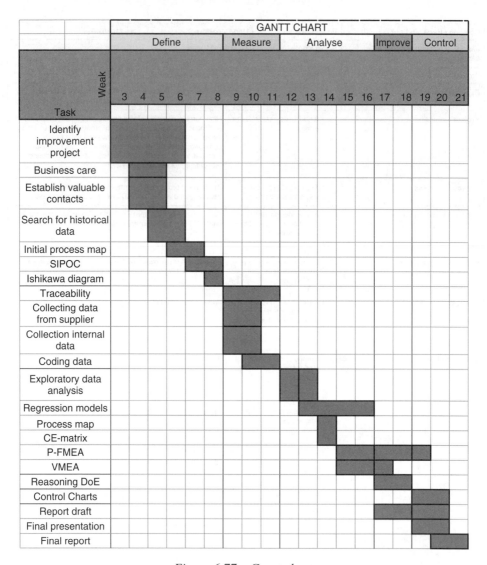

Figure 6.77   Gantt chart.

When the process map was constructed (see Figure 6.78) it was used as the P (Process) step in the SIPOC (see Table 6.9).

6.6.4   Measure phase

To be able to proceed with the Measure phase, first some more details need to be explained about the product. A sector is shown below, and the welds that are performed on it before all sectors are assembled to a complete component. The welds connected to the casting that is of interest for this project are called welds A, B, C and D. Weld A is shown

Table 6.8   Log book on meetings.

Date	Meetings, mainly in Trollhättan
27 January 2011	First meeting -presentation of Volvo Aero and description of problem
9 February 2011	Start of Six Sigma Project; discussion of business case; supplier data, walk the process; project charter
15 February 2011	Work on measurement data; discussion with project on weld indications
16 February 2011	Work on measurement data
17 February 2011	Supervision with Stefano Barone regarding project charter
22 February 2011	Preparing for Friday meeting regarding project definition
25 February 2011	Meeting with project on definition
8 March 2011	Castings versus weld assembly map
9 March 2011	Second walk the process
15 March 2011	Analysis of measurement data on sectors
17 March 2011	Analysis of measurement data on sectors; report to project
28 March 2011	Follow up together with project team
31 March 2011	Material mechanism at welding; project steering committee meeting
11 April 2011	Process map; Cause & Effect (CE) matrix
14 April 2011	CE matrix; FMEA; project steering committee meeting
18 April 2011	Continuous work with the FMEA; VMEA
19 April 2011	Plan report writing; analysis of assembly data
9 May 2011	Report writing; analysis of assembly data; SPC (statistical process control)
12 May 2011	Report writing; analysis of assembly data
13 May 2011	Meeting on casting structure interpretations.
16 May 2011	Report writing; analysis of chemical composition data
19 May 2011	Report writing; meeting with material specialist
26 May 2011	Final presentation to the project team

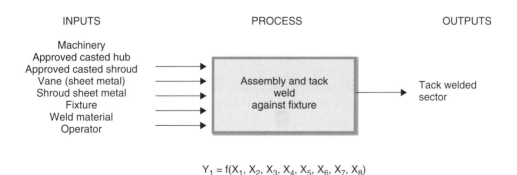

$$Y_1 = f(X_1, X_2, X_3, X_4, X_5, X_6, X_7, X_8)$$

Figure 6.78   First process map lay-out.

Table 6.9   SIPOC.

Supplier	Input	Process	Output	Customer
Company A	Casting (incl defects)	Repairment of casting	Repaired casting (Require-ment: defect free)	VAC
Company A	Repaired casting	Fixturing and tack welding		
Company B	Sheet metal			
VAC	Welding operators	Welding		
			Sector assembly	VAC
	Welding machines	Fluorescence penetrant inspection (complete sector)		
	Fixtures	X-ray inspection (welds and HAZ)		
	Sector assembly	Fixturing and tack welding		
	Welding operators	Welding		
VAC	Welding machines	Flouresence penetrant inspection (complete sector)	Weld Assembly (Require-ment: defect free)	VAC
	Fixtures	X-ray inspection (welds and HAZ)		

in Figure 6.79, it is the weld between the casting and the vane. The heat affected zone (HAZ) is a zone that is running along a weld, 18 mm in width, that becomes heat affected when the weld is made.

With every casting delivered, the supplier has an attached file with data on the casting. This file is called a MAF (Manufacturing Approval Form), and it contains an identification number for the casting, as well as information about the chemical composition of the alloy, the casting date and results from tensile tests among other things. The MAF documents also include drawings of the castings, and where the repairs that have been made on the castings have been noted by hand. Each repair has a number,

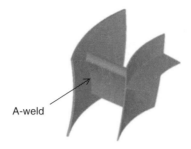

A-weld

Figure 6.79   Weld A is the weld sheet metal to outer casting.

and also includes a list of all the repairs, what the type of defect is that has been repaired, the size of the defect and also the date that the defect had been repaired. An example of supplier data (MAF document) including drawing of defects is shown in Table 6.10 and Figure 6.80.

Table 6.10   Supplier defect data (MAF).

** If Reheat Treatment is required place note under comments **

#	~~ZONE~~ WELDER	WELDER	DATE	DEFECT TYPE	SIZE Per Technique Sheet	VOLUME	COMMENTS
1	C	W 17	7·28·09	Inc	.325√.210√.030		
2	C	W·17	7·28·09	Inc	.115√.115√.025		
3	C	W·17	7·28·09	Inc	.280√.130√.030		

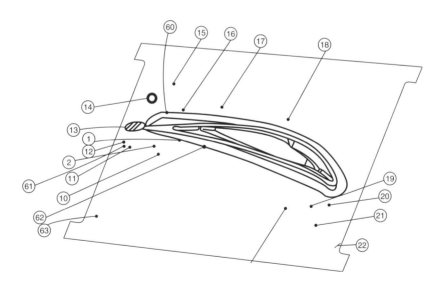

Figure 6.80   Supplier defect data (MAF).

Each weld assembly is named as serial hardware (SHW), and is assigned a number which indicates when it was manufactured. SHW 1 is the first one manufactured, and the production of SHW 30 is under production now. There is not complete defect data about all SHWs that came before SHW 30; data is continuously noted and entered into the system as the components are being processed. At the start of this project, relevant data on the SHWs was available until approximately SHW 15.

The defect data at VAC consists of two parts. One part is drawings of the sectors after they have been welded together, with the defects noted by hand in a similar way as the MAF data. The difference is that there is no list of the defects where the type of defect, size, and so on are noted. Occasionally, there is a comment noted next to the defect detailing what type of defect it is and the size of it, but this was done on far from every defect. From these drawings it is not that difficult to determine approximately where defects are located, since it consists of fairly large drawings with several views provided. Still, the defects were noted by hand, and the accuracy of this may be debated.

The other part of the data concerning defects from VAC consists of drawings of the components after they have been welded together. These drawings are similar to the previous in the respect of defects being noted by hand, and an occasional comment given next to the defect. What was more problematic was that some drawings of the whole component were very small, and the accuracy of the notation of defects might be less than in previous drawings. Also, from some views it was difficult to determine the exact location of the defects. However, the drawings from VAC state in what position and sector the defects are located. The data on defects, both from sector and component, will from now on be called Syglo data.

Starting with the MAF data, if one is able to correlate the defects in that data to the defects from the Syglo data, the exact position of the weld repairs can be noted. A coordinate system with x- and y-axes was established in the drawings, and an x- and y-position of each repair was noted in an Excel spreadsheet for each of the welds A, B and C. The M weld has not had many defects connected to it, and it was therefore left out of the compilation of MAF repairs. Not all defects from the A, B and C welds were noted in the Excel spreadsheet, but only those that were located in the welds and their HAZ. What was not noted was the sizes of the defects, and the reason for this is that they are noted as three-dimensional defects in the MAF data, and there was no such data to compare with at VAC. The next step consisted of entering the Syglo data in an Excel spreadsheet in a similar manner. As mentioned previously, the scales of the drawings from where the defects are noted differ, both for the Syglo data, but also the MAF data, so for now on only the location on the drawings was noted, and the scaling in order to get the data to be of equal scale was left to be done later.

In order to investigate the correlation between repairs at the supplier and defects at VAC, it had to be known which sector and SHW each casting was part of. After the castings come in to VAC, it is not so obvious in what SHW, and what position in a SHW, they are located after processing. The castings had to be traced somehow through their identification numbers. The identification number was set at the supplier, but the same identification number was not used at VAC, so this introduced complications when it came to tracing the castings. A database exists of the SHWs and what in-going components they have. This is a fairly unstructured database though, and it was not possible to trace the location of the castings only based on the information in this database. The team also needed to find 'traceability maps', which were a sort of map of the entire SHWs, with identification numbers on them that helped to encode the position of each casting.

The traceability maps were not that easy to find, since they were not available electronically, and followed the log-book of each SHW as it was being manufactured. At last though, the team got the hold of all traceability maps of the relevant SHWs (up to 21), and could trace each casting to which SHW it was part of, and its position in the SHW.

To establish this traceability was an important step in the Measurement phase, since it had not been done before. This means that the defects repaired in incoming material could be located and compared with defects outcome both at sector assembly as well as component assembly at Volvo Aero.

The capability aim is a weld defect free process. However, it is known that there exist defects in welding of casting material. It was shown how the weld defect number was varying connected to the so-called weld A. It was shown that a large improvement was obtained when going from the Development Hardware DHW4 to SHW1. The level after SHW1 has improved slightly, but is more or less within the process variation.

The gauge R&R has to be critically studied for this project. There is a high dependence on the operators that have performed the recordings. Three main uncertainties in measuring procedure can be identified: accurate number of defects, accurate position of defect and accurate classification of defect. There is also a problem of accurate interpretation of the position of the defect and the weld repair in a casting (see Figure 6.80), as well as interpreting the position of the defect after sector assembly.

6.6.5   Analyse phase

The repairs have been noted using the supplier's drawings, and the defects at VAC have been noted using VAC's drawings, but these drawings were not comparable and could lead to erroneous interpretations. For this reason it was decided that only the number of defects for each weld and its HAZ has to be correlated. The first step taken towards making a correlation is to make a chart of the number of repairs needed, both from the supplier and VAC. One chart for each SHW that has a sufficient amount of data accessible is made. The x-axis shows at what position a specific sector is located on the component; this is done in order to connect the specific sector to its specific castings MAF. The y-axis indicates the number of repairs needed on the casting. As an example, if the C welds in SHW 5 is about to be plotted, the A weld for each Sector in SHW 5 is plotted on the x-axis. The number of repairs needed for each A weld and its HAZ, both from supplier and at VAC, is then plotted on the y-axis. This is done in order to get an overview of the repairs needed, if the number is overlapping, and if we could expect a correlation later on. Figure 6.81 shows the correlation between number of weld repairs needed on casting and the number of weld repairs needed in the A weld, middle, for SHW5.

No significant statistical correlation was found.

Even if there seems to be no correlation when it comes to specific welds before and after processing, there might still be a factor that influences the possibility of both these defects to occur. The next step was therefore to investigate if there was such a hidden factor, and in that case what that factor could be. Firstly, it was verified if there was a correlation between the total number of repaired defects on a casting from the supplier, and the total number of defects on the same casting after being processed at VAC. A positive correlation was found. Since there was no correlation found to the location of the repairs of the castings at the supplier, there are other factors that are causing defects at VAC. In

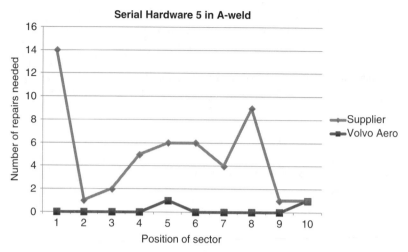

Figure 6.81   Correlation between number of weld repairs needed on casting in HAZ and number of weld repairs needed at assembly in HAZ.

this case a high number of weld repairs at the supplier could hint that there will also be a higher rate of weld defects in VAC production, even though it is an indirect connection.

First of all a complete process map was prepared (Figure 6.82), then a cause-and-effect matrix was built (Figure 6.83).

Then a tool named potential failure mode and effect analysis (p-FMEA) was used (see Figure 6.84).

When all the processes are inserted in the tool, the following steps are about analysing these processes in order to understand what can cause failure in the process and what effects a potential failure can lead to. The effects and causes are then graded by severity and occurrence in order to take a decision on what potential causes to improve. The second half of the p-FMEA is about the actions taken to avoid failures, but this will not be handled at the next stage of the project.

A variation mode and effect analysis was conducted to determine in which direction the improvement efforts should be directed (Figure 6.85). A defect-free welded material was chosen as the key product characteristic (KPC). The initial sub-KPCs were retrieved from the Ishikawa diagram from the Define phase and some others were added after some discussions with the welding process expert at VAC. Furthermore, the main noise factors to the sub-KPCs were identified, and the sensitivities and variation sizes were determined.

The VRPN was then calculated for each sub-KPC and each noise factor. The sub-KPCs contributing with most sensitivity to having a defect-free welded material are listed in the pie chart of Figure 6.86 The sub-KPC contributing with highest sensitivity to having a defect-free welded material is the incoming material quality, followed by welding parameters for robot and manual, having equal VRPNs. Casting structure is the noise factor scoring the highest VRPN; according to the material expert at VAC, there can be hundreds of parameters affecting material structure, everything from the compositions of materials included in the casting, to heat treatments that the casting have undergone and how much built-in strain the casting detail has. The built-in strain can have several reasons, among

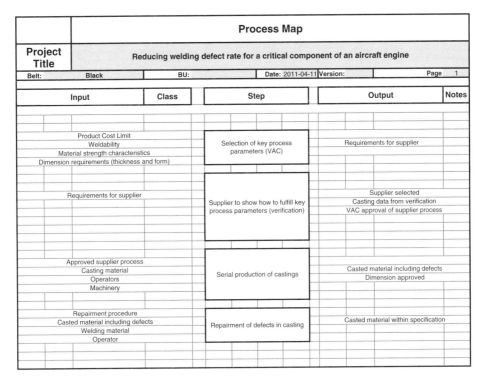

Figure 6.82   *Process map (first part of process).*

	Key Process Outputs	Defect free part	On time delivery	Cost		
Process Step	**Process Input**					**Total**
Selection of key process parameters (VAC)	Product Cost Limit	9	3	9		210
	Weldability	3	0	9		120
	Material strength characteristics	3	0	3		60
	Dimension requirements (thickness and form)	3	0	1		40
						0
						0
Supplier to show how to fulfill key process parameters (verification)	Requirements for supplier	9	9	9		270
						0
						0
Serial production of castings	Approved supplier process	9	9	9		270
	Casting material	9	3	9		210
	Operators	3	1	1		50
	Machinery	9	9	9		270
						0
						0
Repairment of defects in casting	Casted material including defects	9	3	3		150
	Welding material	9	0	1		100
	Operator	3	1	1		50
	Repairment procedure	9	1	3		130

Figure 6.83   *Cause-and-effect matrix (first part of process).*

Process Step	Input	Potential Failure Mode	Potential Effect(s) of Failure	Severity	Potential Cause(s)/Mechanisms of Failure	Occurrence	Current Process Controls Detection	Detection	RPN
Selection of key process parameters (VAC)	Product Cost Limit	Underestimation of cost	Cost too high	7	Casting process contains too many uncertain parameters	7	Previous casting experience?		
Supplier to show how to fulfill key process parameters (verification)	Requirements for supplier	Hard to fulfill material requirements	Cost too high due to process corrections	7	Casting process contains too many uncertain parameters	7	Verification program?		
				7	Criteria for defect material uncertain	7			
Serial production of castings	Casting material	Defect material	Cost too high due to process corrections	7	Heterogenous material: dross, variation in material structure.	5			
Serial production of castings	Machinery	Defect material	Cost too high due to process corrections	7	Rest material from casting process, e.g. the mold material	5			
Repairment of defects in casting	Repairment procedure	Defect material	Cost too high due to process corrections	7	Impurities in reparation procedure	5			

Figure 6.84   p-FMEA (first part of process).

VMEA

Performed by:	Arnela Tunovic, Henrik Ericsson & Sören Knuts
Date:	18/04/2011
System:	
Function:	

KPC	Sub-KPC	KPC Sens to Sub-KPC	NF	Sub-KPC Sens to NF	NF Variation Size	VRPN (NF)	VRPN (Sub-KPC)
Defect free welded material	Incomming material quality	10	Number of repaired defects	3	10	90000	1352500
			Thickness variation	5	10	250000	
			Casting structure	10	8	640000	
			Geometrical variation	5	10	250000	
			Heat treatment	7	5	122500	
	Welding parameters (Auto)	10	Welding material (A)	6	3	32400	999800
			Weld joint complexity (A)	10	7	490000	
			Impurity in welding zone (A)	4	5	40000	
			Power (A)	9	3	72900	
			Gap (A)	9	3	72900	
			Alignment (A)	9	3	72900	
			Angle (A)	9	3	72900	
			Flow (A)	9	3	72900	
			Velocity (A)	9	3	72900	
	Welding parameters (Manual)	10	Welding material	6	3	32400	999800
			Weld joint complexity	10	7	490000	
			Impurity in welding zone	4	5	40000	
			Power	9	3	72900	
			Gap	9	3	72900	
			Alignment	9	3	72900	
			Angle	9	3	72900	
			Flow	9	3	72900	
			Velocity	9	3	72900	
	Welding equipment	10	Robot performance	8	5	160000	163600
			Equipment does not fit material	2	3	3600	
	Welding operator	10	Judgement uncertainties	8	6	230400	640000
			Performance	8	8	409600	
	Heat treatment	6	Temperature	10	3	32400	207684
			Time	9	7	142884	
			Speed of heating/cooling	10	3	32400	
	Aging	4	Temperature	10	3	14400	40464
			Time	9	3	11664	
			Speed of heating/cooling	10	3	14400	
	NDT (Non destructive testing)	3	Operator judgement	81	10	57600	72324
			Accessability of fluid	8	5	14400	
			Equipment	2	3	324	

Figure 6.85   VMEA.

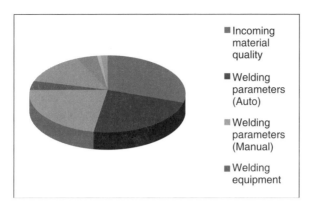

Sub-KPC:	Incoming material quality
Noise factors:	▪ **Casting structure,** VRPN: 640 000 ▪ **Thickness variation,** VRPN: 250 000 ▪ **Geometrical variation,** VRPN: 250 000

Figure 6.86   Pie chart of sub-KPCs (a) and table of noise factors (b).

them are welding in the material without having a subsequent heat treatment, or when straightening the material without heating it, to fulfil the design specifications.

Weld joint complexity (for robot as well as manual) is the noise factor acquiring the second highest VRPN, and is the noise factor explained by complex weld joints in, for example, there exists an ellipse weld of the shroud. The leading and trailing edge of Weld A, see Figure 6.79, is also a complex geometry, and has been identified as a particularly problematic area for welding, since it involves welding over an edge, with several starts and stops as a result. Performance is another noise factor scoring a high VRPN, and was identified as a factor that has high variety among welders. The incoming material quality is going to be the main target for further investigation in this report.

The measurability of the casting structure and the ability to detect if there is something in the structure of the material that might be causing the defects are not so obvious. In the Measure phase, a compilation was made of the data that is available on the mount castings from SHW 1 to SHW 25. Figure 6.87 is a time series plot of when the mount

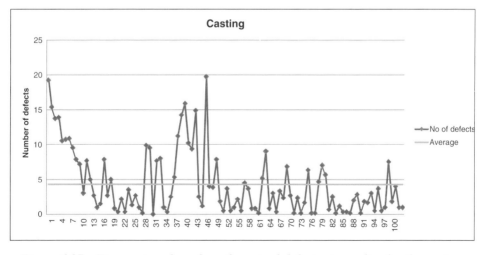

Figure 6.87   Time series of number of repaired defects at supplier for the casting.

hubs were casted; the number of repairs made on the castings are plotted in time series to see how the process has been changed over time. The number of defects is chosen as a measure, and this can be an indication of how well the material quality is with respect to weldability. The process has a lot of defects in the beginning, and a lot of variation in the amount of defects. The defects decrease somewhat over time, but there are still great variations. The variations are not only large when it comes to when the castings were casted in time; there are also great variations within a specific cast. This is an indication that the process is not completely stable. The supplier data on casting material also shows some data of the chemical composition. By 'some data' we mean that not all chemical components are specified by the weight percent; some just state that they are under a certain percent within material specification. In the Measure phase, a compilation of material data was made for each of the mount castings from SHW 1 to SHW 21 that has available data. A regression analysis is performed in Minitab, where for each mount casting the amount of one chemical element at a time is correlated to the amount of defects occurring at the supplier on that mount casting. To fully understand the data the supplier has to be involved in the analysis.

6.6.6   Improve phase (ideas and intentions)

The most preferred way to validate how different factors influence the result of producing defect-free castings would be to do a design of experiment (DOE). This would make it possible to evaluate which factors that are the most influential, and how to correct the process so a real improvement can be seen. It should be noted regarding the concept selection and the total product cost related to a fabricated solution, that is small castings welded into a larger structure of castings and sheet metal parts, that there exist no defect-free structures, due to the inhomogeneous material properties that castings consist of. Therefore a capability number should be given to each weld that is based on a real outcome. This will give a fairer balance between materials in the future concept selection, where also other material combinations like forgings, as well as different suppliers of casting, where each supplier has to show their capability of the process.

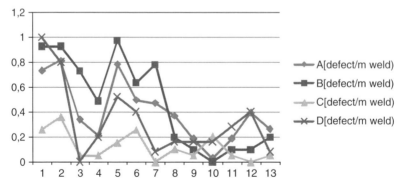

Figure 6.88   Number of defects per meter in a time series of different SHW, and a comparisons between different welds.

6.6.7   Control phase (ideas and intentions)

In this project there was no availability of follow up implementations, but in the future, the supplier shall be directly involved, and take control of the handling of his own data. In particular, what is proposed to the project and the supplier is to perform U-chart measurements. Another process measure (key process indicator, KPI) would be to transform measured values into defects per meter weld (see Figure 6.88).

6.7   Attacking a problem of low capability in final machining for an aircraft engine component at VAC – Volvo Aero Corporation

Project purpose: to find the main sources of variation in the manufacturing process of a Turbine Exhaust Case (TEC), in order to increase capability in final profile tolerance requirements and to assure optimal conditions before starting the final machining phase.

Organisation: VOLVO AERO CORPORATION

Duration of the project: five months (2011)

Black Belt candidate team:
Johan Lööf, Ph.D, Volvo Aero Corporation
Christoffer Löfström – masters student; Giovanni Lo Iacono – PhD student.

Phases carried out and implemented tools: Define (project charter, Gantt chart, 'Y' and 'y' definitions, Ishikawa diagram, SIPOC) – Measure (brainstorming, observations, pictures, Ishikawa diagram, geometric optical measurements) – Analyse (correlation, regression, normal probability plot, capability analyses, control charts).

6.7.1   Presentation of Volvo Aero Corporation

Volvo Aero Corporation is a company located in Trollhättan, Sweden, that develops and manufactures components for commercial and military aircraft engines and gas turbines. In this area, we have specialised in complex structures and rotating parts. We have market-leading capabilities in the development and production of rocket engine turbines and exhaust nozzles.

The main challenges for the flight industry today are to reduce the emissions of CO_2. The lighter an aircraft engine is, the less fuel it consumes for any given flight. That is why we focus on developing lightweight solutions for aircraft engine structures and rotors. Our optimised fabrication of fan/compressor structures and turbine structures, realised through our advanced computer weld modelling, makes these structures lighter than conventional single-piece castings. Moreover, we have also developed our fabrication processes to high levels of automation in order to reduce costs, improve robustness and optimise fabrication concepts, especially when manual assembly is not feasible.

6.7.2   Project background

In our current production it is hard to reach capability goals in the final profile tolerance requirements on a structural component. This component consists of a number of parts that are welded together in a number of steps and then machined to its final form. The machining process consists to a big part of removing (e.g. shaving off) excess material to make the whole component meet tolerance requirements and to provide surface treatments needed before delivery. Today there are problems with variation of geometric deviations in different steps of the manufacturing process which leads to long lead times in the final machining phase to be able to meet the customer requirements of maximum nominal deviations in the finished products. This may in the worst case even lead to scrapping of a whole component.

6.7.3   Define phase

The manufacturing process consists of a large series of sequential operational steps that can be grouped together as operations, with measuring steps in between. A number of incoming parts are measured and then placed into a fixture and tack-welded together. Another measurement is then carried out to find out if the parts are still in their planned positions after the welding. If they are in their correct positions the parts are permanently welded together, if not their positions are corrected and a new measurement performed until satisfactory results are achieved before moving on to the next step in the process. The operation steps contain of adding material with different welding techniques, measuring, possibly aligning, and cleaning and preparing surfaces for the next material to be added. At the end of the process, the whole assembly is X-rayed to see that welds are meeting requirement standards and if necessary weld-repaired, before the assembly is heat-treated and sent off to final machining. This is where the process stops and the final machining group take over as an internal customer.

The big Y's are defined in the measurement operation after the heat treatment. In this project the focus has been set on four critical areas on the product that have to be in statistical control before entering the final machining phase. These four areas are measured in each of the seven measurement steps.

As input into the process there are the incoming parts supplied by an internal supplier at VAC, the welding functions supplied by Production, and the measuring and controlling by the Control and Measure departments. The numerical requirements on the inputs are geometrical tolerance requirements on the incoming parts. The measurement requirements are fulfilled in view of accuracy and reliability. When it comes to the input from the operators' performance we have marked them as grey in the SIPOC (see Figure 6.89) since they are outside the limitations in the project, but we still included them in the graphics for comprehensiveness. The outputs of the process are the weld assembly with the numerical requirements of meeting geometrical profile tolerances fulfiled.

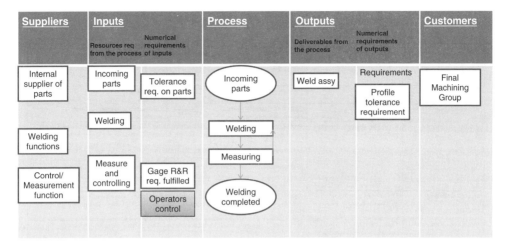

Figure 6.89   SIPOC diagram.

Other outcomes from the Define phase are a project charter (Figure 6.90), a Gantt chart and a first-level Ishikawa diagram based on brainstorming that will be more detailed and further sorted out after forthcoming meetings with the final machining manager and the quality managers of the process.

6.7.4   Measure phase

The measurement steps are carried out by GOM and CMM (coordinate measuring machine) where data is recorded and kept in the system. While the CMM is only carried out at a few single stages, the GOM measurements cover the whole process measuring in all stages that are considered potentially critical and are therefore our natural choice of data to focus on. GOM is a measurement system based on a camera that records hundreds of still images of an object that are then put together in the computer system into a single 3D image of the object. The object's geometrical deviations from a reference model (the nominal values) are colour coded and the image provides information on the magnitude and direction of the deviation.

The detailed Ishikawa diagram (Figure 6.91) constructed during this phase highlighted that the main problem of variation seems most likely to be due to the welding operations that, when performed, transfer a significant amount of energy in form of heat into the surrounding materials. These materials are usually both of different compositions and thickness, which leads to different degrees of expansion/contraction and different cooling times that results in stress building up in the material when cooling down back to room temperature. This stress causes the still not cooled material to move somewhat which in its turn leads to geometrical deviations. The variations are attempted to be controlled and held as low as possible by trying different welding techniques that are more focused and concentrate the heat transfer into a much smaller area leading to lower temperatures in surrounding area and less material stress than compared with conventional welding techniques. This has reduced the effects but some movements still remain.

Company (organization)	Volvo Aero	Unit/Department		7163/DfR
Industrial participant (Black Belt candidate)	Johan Lööf	Telephone/e-mail		
Sponsor & process owner (Manager Production)	(masked data)	Site or location		VAC-THN
Project Start Date	20110201	Project completion Date		20110524
Expected impact level		Expected financial impact (savings/revenues)		Decreased lead time for final machining.
Project summary	A short description of the project	Problems with low capabilities in the final machining process causing problems in the ramp up phase of structural component manufacturing. In order to achieve a better outcome of the process, we need to investigate how big the geometric deviations are and where they occur before going into the final machining process. The final goal of this project is to find main root causes to the geometric deviations such that we can make improvements in order to assure optimal conditions before starting final machining.		
Impacted process	The specific processes involved in the project	Assembly processes of complete product starting with parts from internal supplier. This includes fixturing, welding, heat treatment and inspection (NDT=None Destructive Testing).		
Benefit to customers	Define internal and external customers (most critical) and their requirements	Production Trollhättan. To reach acceptable capability of identified quality problem.		
Benefit to the business	Describe the expected improvement in business performance	Enhances both in lowering of lead-time and improvement in quality cost.		
Project delimitations	What will be excluded from the project	The project will concentrate on a specific area on the complete product in order to make it possible to get a result within this course.		
Required support	Support in terms of resources (human and financial) required for the project	Travel costs (students)		
Other people involved	List technical experts and other people who will be part of the team	Machining experts, Component owners, Producibility leader, Quality leader, Welding experts		
Specific goals	Define the baselines, your realistic goals for the project and the best case targets for improvement.	Actual value (baseline)	Realistic goal by project end date	Best case goal
		(masked data)	(masked data)	(masked data)
DEFINE phase completion date	2011-02-24	MEASURE phase completion date	2011-03-29	
ANALYZE phase completion date	2011-04-21	IMPROVE phase completion date	2011-05-05	
CONTROL phase completion date	2011-05-20	PROJECT results presentation date	May 25th, 2011	

Figure 6.90   Project charter.

Then there is another difficulty that also stems from heating the material. The weld assembly arrives for the final machining after the heat treatment. Here the assembly may be well within tolerance limits before the heat treatment but thanks to both the built-in stress from earlier welding operations and the new heat and cool cycle the assembly may get outside tolerance limits.

Due to our limitations we have basically two options when approaching the process. The first one is to investigate how much the incoming parts are affecting the output of the process by analysing the geometric variations of these parts and measuring their contribution to the total process variation. The second approach is to focus on the measuring procedure and on how the parts are placed into the fixture before they are welded together. The adjustments are done manually, and specific guidelines help operators to place the

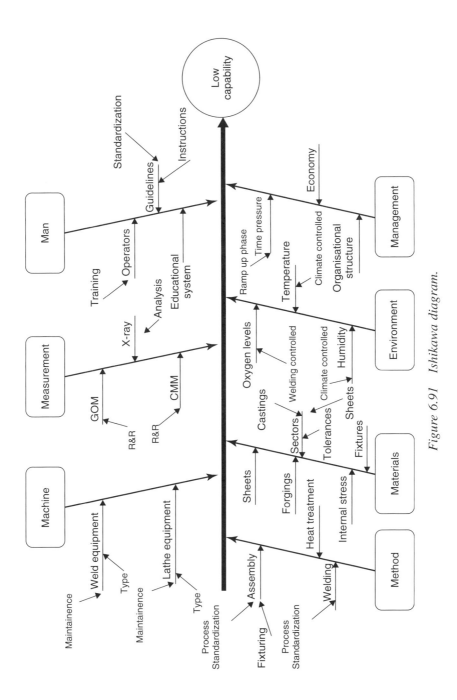

Figure 6.91   Ishikawa diagram.

parts in the correct position. Thanks to these adjustments a whole component is well fitted within tolerance limits but can be slightly shifted from the nominal position that the measurement system has as its reference. These shifts will be recorded and, consequently, taken into account into coming operations steps. That may cause operational procedures that are unnecessary or are wrong.

From the process meetings there are some details of the component that have been identified as most critical to fulfil the output tolerance requirements. These details (see Figure 6.92), so called 'bosses', are easy to identify and unlikely to deform geometrically (i.e. compared to the surrounding material). The four bosses have different functions. In the centre of each boss there is a hole; if the boss is not within its tolerance limits there will not be enough material surrounding the hole to fulfil the material strength requirements.

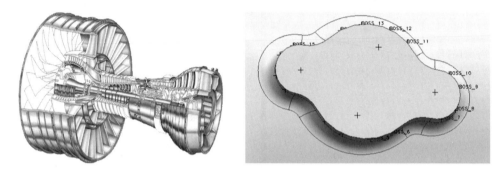

Figure 6.92   The aircraft engine (a), a boss on an aircraft component (b).

Each boss has been divided into 16 subareas where are located the points where GOM measurements are made (see Figures 6.92 and 6.94).

This type of GOM data is stored in the system as colour-coded pictures that give qualitative information on the magnitude and direction of deviation from nominal values. Figure 6.93 shows the measured points on a boss and a colour-coded GOM data

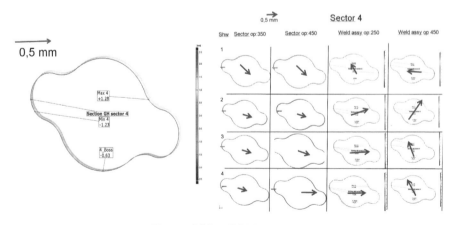

Figure 6.93   GOM pictures.

picture. GOM pictures do not contain quantitative information of deviation or direction of deviation in any fixed points (except for maximum and minimum deviations, and the deviation of the point 5 on the bosses). So, data have to be ordered and measured by specialised personnel in order to analyse specific points on a boss. GOM pictures show a nominal positioned boss silhouette in black, the actual boss position silhouette is placed on top of it. The actual position silhouette holds different colours depending on the size and direction (i.e. positive or negative direction) of the deviation. On the right side of each picture there is a coloured scale, the pictures concerning the same boss are then put together in series (see Figure 6.93).

6.7.5 Analyse phase

Figure 6.93, on the right, shows how one boss moved during four different process operations on four different components. A visual analysis highlights a clear trend both in the movement direction and in the magnitude of the bosses in the sector level.

The process starts with the incoming parts. Some of these incoming parts are identical and have a boss placed on the outer surface. The first analysis consisted in measuring the capability of one point on the boss in a certain direction, illustrated by the arrow in Figure 6.94. There is a tolerance requirement on the position of the boss of ± 1.5 mm (general) in that direction.

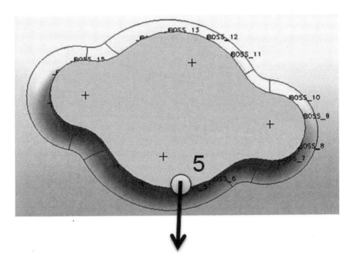

Figure 6.94   Critical direction of a point on the boss.

In order to deepen the knowledge on the distribution of data, capability and normality analyses were performed (Figure 6.95): the process is normally distributed and is stable but with a drift from the nominal value and a low capability (see for example Figure 6.95).

The next step was to investigate on how the boss performs not only in the Z direction but also in the X direction by a correlation analysis. Analyses show that the bosses all

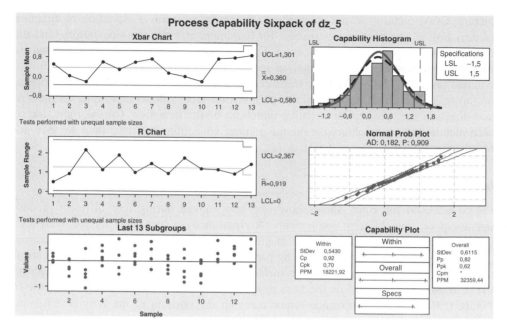

Figure 6.95   Capability analysis.

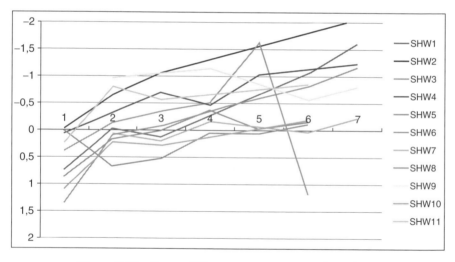

Figure 6.96   Seven different measurements of one boss.

rotate in the same way and the process is stable, since stochastically independent events (the related manufacturing processes) head to the same kind of performance. There is not a correlation between the maximum/minimum displacements across the series.

In Figure 6.96, the position of one boss can be followed through the seven different measurement operations.

After having measured the capability of the incoming parts and the outcome in the big Y, the next step was to investigate which process contributes the most to the final variation.

There are seven different phases that affect the variation in the big Y: firstly, the position of the bosses, then the six phases allocated between the measurement operations. In order to calculate the contribution of each phase to the variation, the differences of data resulting from consecutive measurements (one in correspondence of each phase) were calculated. In Figure 6.97 the capabilities associated with each of the six phases are shown. In order to calculate Cpk values, a tolerance of ±1 mm has been defined. For the boss considered, the main contributors to the variation are the processes located between 'Op1' and 'Op2' measurements. Certainly, also the incoming position of the boss affects the outcome significantly, since the capability value is 0.7 and the tolerance limit is ±1.5 mm.

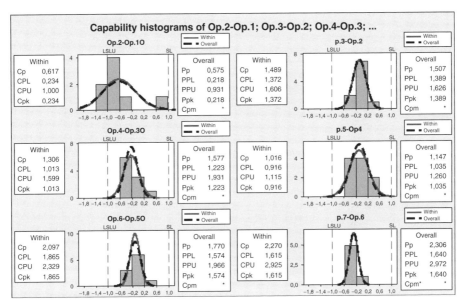

Figure 6.97   Capability associated to the seven process phases.

6.7.6 Improve phase (ideas and intentions)

Analyses concerned only six components that exceeded the measurement stage before starting the final machining phase, so it is necessary to collect data concerning the other components in order to analyse, in a more complete way, the real situation about the big Y. The main suggestion for improvement is to focus the attention on the incoming parts because relevant improvements can be achieved in them. Firstly, it should be appropriate to perform a well defined p-FMEA on the manufacturing process of the incoming parts in order to find the main sources of variation, and increase the process capability. A DOE (design of experiments) will allow an understanding of how robust the process is, as soon as the noise factors in the incoming parts (sub-assembly) welding phase are under control.

Index

7QC tools, seven quality control tools
14, 23–4

acceptance control 183
  acceptance number 184
  rejection number 184
affinity diagram 26–8, 291–3, 339
AHP (analytic hierarchy process) 122–9
  consistency index 126
  consistency ratio 126
  pairwise comparison matrix 125,
          127–8
  random index 126
  semantic scale 124
ANOM (analysis of means) 291
ANOVA (analysis of variance) 218–21,
    276
  ANOVA table 220–1
  F ratio 220
  mean squares 220
  p value 220
  sum of squares 219
AQL acceptable quality level 189
  see also operating characteristic curve
Atlet AB 297
autocorrelation 169
average 50
  average run length see control chart
  weighted average 51

bar chart 48
Bayes theorem 68
benchmarking 13–14, 18, 137

Bernoulli model 74, 75
  Bernoulli proportion model 179
beta model 102–3
binomial coefficient 70
binomial model 75, 77, 179
bivariate data 158, 160
  bivariate Gaussian distribution 165–7
  bivariate histogram 159
Bill Smith 7–9
black belt(s)
  see Six Sigma human resource(s)
  see Six Sigma project(s)
BLUE (best linear unbiased estimator)
    150
box-whiskers plot 54
  double box-whiskers plot 55
brainstorming 23–8, 319–20
  5M criterion 24, process steps
          method 26
  cause listing method 26

C_p, C_{pk}, C_{pm}, C_{pmk}, see process
    capability
CA conjoint analysis 19
capability see process capability
catapult 235
  see also DOE
cause and effect diagram 12, 17, 23–6,
          193, 280, 294, 304, 324–5,
          340, 352, 369
cause and effect matrix 341
cause-effect relation see correlation
central line see control chart

Statistical and Managerial Techniques for Six Sigma Methodology: Theory and Application, First Edition.
Stefano Barone and Eva Lo Franco.
© 2012 John Wiley & Sons, Ltd. Published 2012 by John Wiley & Sons, Ltd.

champions *see* Six Sigma human
    resource(s)
character 43
check sheet 24
chi-square distribution 152, 219–20
classes of values *see* variable
coefficient of variation 57
coefficient of determination 151
combinations 70
comfort assessment 250
common cause(s) *see* variation
condition monitoring 253
confidence interval 90
    confidence level 90
    pivotal quantity or ancillary function
        90
conjoint analysis 246
consistent estimator 89
constraint(s) 95
    sequence constraints 98
continuous improvement 10–12
control chart(s) 14, 24, 171, 174, 310,
        349
    average run length of 176–7
    c chart 180
    for the average 175
    for the standard deviation 177
    individual and moving range charts
        182
    p chart 179
    phase I, phase II 178
    setup 178
    u chart 181
    with or without prescriptions 176
control limit(s) 174
    variable control limits 180
    see also control chart
control factor(s) 284
    see also DOE
    see also variation
control plan 349
correlation 61, 160
    cause-effect relation 162
    correlation analysis 17, 19
    correlation coefficient 160–2
    correlation matrix 168

cost(s)
    cost-benefit analysis 346
    cost-time relationship 104
    productive, unproductive 11
    see also CPM
covariance matrix 168
    determinant of 168
    eigenvalues of 168
CPM (critical path method) 98, 104–9
    acceleration cost 104
    activity slack(s) 106
        see also PERT
    crash time 104
    direct cost(s) 104, 108
    indirect cost(s) 104
    node slack(s) 106
    normal time 104
    utility cost(s) 105
    worst time 104
    see also project network diagram
CTQ (critical to quality) characteristic(s)
        13, 16, 18–19, 285
customer satisfaction 3, 37
    customer expectations 3
    customer's risk *see* operating
            characteristic curve
    see also Kano model
    see also QFD
cycle time 3

data collection 44
    data array 44, 46
    data collection sheet 44–5
degradation *see* OBD
defect rate 187
Deming cycle 1–2
design parameter(s) *see* robust design
design setting *see* robust design
deterministic component of a statistical
        model 146, 169
diagnostic index(es) 254
dichotomous variable 156
dispersion index 50
dispositions 69
distribution function
    distribution fitting 84

empirical distribution function 48
estimation of 84
DFFS design for Six Sigma 15–17
DMADV (define, measure, analyse, design and verify) 15
DMAIC (define, measure, analyse, improve and control) 12–15, 267–8
DMARIC (define, measure, analyse, redesign, implement and control) 18–20
DOE (design of experiments) 14, 17, 19, 212–43, 364
   blocking 234
   computer experiments 217
   control factor(s) 213, 244
      control domain 224, 226
   experimental effort 250
   experimental plan 216, 219, 225–6, 229, 237–8, 248, 251, 254
   experimental risk factor(s) 242
   experimental treatment 213, 245
   experiments with the catapult 235–43
   factorial designs 222
   factorial effect(s) 218, 224, 228, 240
      confounding or aliasing 232, 238
      hierarchy of 231
      redundancy of 231
      sparsity of 231
   fractional factorial designs 232
      half-fractions 233
      resolution 233–4
   inner and outer array 251, 258
      see also robust design
   interaction effect(s) 224
   main effect(s) 224, 252
   OFAT One Factor At Time experiments 214
   orthogonality 228
   physical experiments 255
   randomisation 215
   replications and repetitions 217
   response 212
   sequential approach 231
   split-plot technique 239

variance homogeneity 242
dot plot 46, 47
DPMO (defect per million opportunities) 7, 12
Durbin-Watson test 169

EDA (exploratory data analysis) 41
empirical frequency distribution 46, 49
   dispersion of 50
   location of 50
   shape of 50
   symmetry of 58
   unimodal, bimodal, multimodal 53
ergonomics 249
error
   error term 145–7, 218
   error variance 152, 155
   random error 145
event 63
   complementary event 64–5
   concomitance of 64, 66
   event space 64
      event space partition 68, 70
   impossible event 64
   incompatible events 64, 66
   intersection of 64
   union of 64–5
   Venn diagram 64–5
   see also FMEA
   see also FTA
executive leadership see Six Sigma human resource(s)
experiment 64
   see also DOE
experimental error 193
   see also DOE
exploratory analysis 288
exponential model 80, 81, 85–6

F ratio see ANOVA
false alarm 176
failure see fault
fault 109
   see also FTA
   see also FMEA
   see also VMEA

fact 63
final tollgate review 288
fishbone diagram *see* cause and effect
            diagram
Fisher index see
            skewness
five whys 271
flow chart 15–19, 32–4
flow-down tree 302
flow-tree-block diagram 210
FMEA (failure mode and effect
            analysis) 17, 110–14, 362
    FMECA 114
    event 110
    process FMEA 110
    project FMEA 110
    see also VMEA
frequency 43
    absolute frequency 46
    empirical frequency 46–7
    relative frequency 46
FTA (fault tree analysis) 110, 114–22
    cut-set 115
    dual tree 120
    event symbol(s) 116–17
    Mocus algorithm 119
    path-set 120
    quantitative FTA 120
    top event 114–15

gamma function 82
Gantt chart 30, 31, 98, 271, 301, 317,
            334, 354
Gaussian
    distribution 6, 78, 79, 103–4
    probability plot 86–8, 223
geometric model of probability
            distribution 177
GR&R gauge repeatability and
            reproducibility 16, 308
green belts *see* Six Sigma human
            resource(s)

histogram 24, 48-9, 290, 342
    normalised histogram 48
HOQ *see* QFD

Hotelling-Solomon index *see* skewness
hypergeometric formula 186
hypothesis test 91–3, 218
    alternative hypothesis 92
    null hypothesis 92
    power 92
    significance level 92
    test statistics 92
    type I risk 92
    type II risk 92

ICOV (identify, characterize, optimise
            and validate) *see* IDOV
IDEF (integration definition for
            function modelling) 32, 95–7
IDOV (identify, design, optimise and
            validate) 15
IIDN (independent and identically
            distributed as normal) *see*
            ANOVA
indicator(s) 181
inference 62, 82
inferential procedures 83
input(s) 95
    see also process
interaction effect *see* DOE
interval estimation 90
interval plot 289
IQR (interquartile range) 55
Ishikawa diagram *see* cause and effect
            diagram

Kano model 16, 37–9
    product attribute(s) 37–9
kansei engineering 245
KJ analysis 323
    see affinity diagram
KPC (key process characteristics)
            291–3, 363
KPIVs *see* variable
KPOVs *see* variable
KSC (key system characteristic(s)) 190
    KSC breakdown 203
    see also VMEA
kurtosis 58

lag 169
latent variable 157
least squares 148, 155
Likert scale 248, 343–4
linear regression 145–55
   multiple linear regression 146, 155
   simple linear regression 146
   see also regression
location index 50
load-strength scheme 201
logistic regression 156–7
   ordinal logistic regression 157, 246
logit 156
   cumulative logit 157
loss function *see* robust design

MAIC (measure, analyse, improve and control) 8, 12
main effect *see* DOE
management by facts 10
MBBs (master black belts) *see* Six Sigma human resource(s)
measurement error 145
measurement scale 43
   interval scale 44
   nominal scale 44
   ordinal scale 44
   ratio scale 44
measurement system analysis 259–65, 307, 313
   accuracy and precision 260
   actual value 259
   appraiser 264
   direct vs. indirect measurement 261
   gauge repeatability and reproducibility 260–5
   measurement system discrimination 259
   measurement uncertainty 259
   true value 259
median 52
method of moments *see* VMEA
mode 52
   modal class 53
moments generating function 73
Michael Harry 8–9

modality 43
monitoring 171
MUDA 10
multiple linear regression *see* linear regression
multivariate statistics 157–70
   multivariate dataset 158

natural variability interval 172
network diagram *see* project network diagram
noise factor(s) 292
   noise factor assessment 206
   see robust design
   see variation
   see also VMEA
normal distribution *see* Gaussian distribution
normal equations 149, 155
   matrix form of 155

OBD (on-board diagnostics) 252–8
odds 156
   odds ratio 156
   proportional odds 157
OFAT (one factor at time) experiments *see* DOE
online process control 174
operating characteristic curve 185
   customer's risk 187–8
   ideal operating characteristic curve 185
   supplier's risk 187–8
   type A and type B 186–8
ordered observations 52, 84
organization success 4
   innovativeness 4
outlier 52
   outlier detection 54
output *see* process
   output characteristic(s) *see* QFD

p value *see* ANOVA
P diagram 31, 191, 253, 287, 318
Paper helicopter 192, 263
parameter estimation 17

Pareto chart 12, 19, 24, 61, 63, 278–9, 290, 326, 342
Pearson index *see* kurtosis
permutations 69
PERT project evaluation and review technique 30, 98–104
  critical path 101–2
  earliest time 101–2
  latest time 101–2
  slack 102
    see also CPM
  see also project network diagram
pivotal quantity *see* confidence interval
point estimation 88
Poisson model 76, 78, 180
Poka-yoke 330
polychotomous variable 157
population 43, 82
preference uncertainty theory 129
probability 63
  calculation rules 66
  composite probability rule 67
  conditional probability 66, 67
  definitions of 63
  density function 71
  distribution function 71
  mass function 71
  plot 84
  total probability rule 68
probability density function 5
process 1–2
  decision making processes 122
    see also AHP
  managing by 1, 11
  process capability 7, 13–14, 17, 172–3, 372
  process capability index(es) 172, 308–9, 373
  process control
    offline 181
    online *see* control charts
  process description 95–7
  process design 31, 274–5
  process diagram *see* P diagram
  process drift 175
  process in control 6–7, 176

process mapping 30–1
process map 273, 286, 303, 320, 336, 361
process monitoring 171
process performance 2–3
  process performance triangle 3
stationary process 171, 176
variation in 5–6
see also Six Sigma
see also variation
product
  product attribute 246
  product profile 246
  see also Kano model
  see also variation
project 98
  manage a 98–109
  project network diagram 99–100, 105–7
  project charter 270, 285, 299–300, 316, 332–4, 353, 368

QFD (quality function deployment) 13, 16, 135–42, 191
  QFD matrix(es) 135, 139–42
  HOQ (house of quality) 136
quantile 53, 91
  percentile 53
  quartile 53
QQ (quantile-quantile) plot 86, 88

R&R (repeatability and reproducibility) *see* gauge repeatability and reproducibility
r.v. random variable 70
  continuous 71
  discrete 71
  expectation of 72
  models of 74
  moments of 73
  standard deviation of 73
  transformation of 146
  variance of 72
  see also probability density function
  see also probability distribution function
  see also probability mass function

random function 171
random sample 83
randomisation *see* DOE
range of variation 55
rational subgroup 175, 179
   see also control chart
reliability 109–10
regression analysis 17, 19,
   prediction by 153
   regression line 148, 151
residual 152
   analysis of residuals 153, 221, 241
   estimated residuals 154
   residual variation 152
resource(s) 95
respondent preferences 246
response *see* DOE
response latency model 129–35
response surface 195
   contour plot 252
robust design 17, 31, 189–258
   cross array 258
   design parameter(s) 190
   design setting 191
   loss function 197–9
   noise factor(s) 191
   nominal-the-best,
         the-smaller-the-better,
         the-larger-the-better
         problems 191
   robust calibration 253
   robust thinking 329
   robustness indicators 197, 255
   S/N signal/noise ratio 199–200
   signal factor(s) 191
   signal-response systems 192
   system, parameter, tolerance design
         190
   systemic robust design 209

S/N signal/noise ratio *see* robust design
SAAB Microwave Systems 269
Sahlgrenska and Östra Hospitals 284
sample 43, 82–3
   sample average 88
   sample statistic(s) 84

sample variance 89
   corrected sample variance 89
sampling plan 184
   simple, double, multiple sampling
         plan 184
scatter plot 24, 61, 62, 148, 160,161
sensitivity 196
   sensitivity assessment 206–7
sequential sampling 184
service
   definition 33
   service quality determinant(s) 35
   see ServQual model
ServQual model 33–7, 243
shape index 50
   see also kurtosis
   see also skewness
signal factor(s) 317–18
   see robust design
SIPOC (suppliers, inputs, process,
        outputs and customers) 16,
        31–2, 96, 273, 287, 302, 319,
        336, 356, 367
Six Sigma
   definition(s) 9–11
   evolution 9
   human resource(s) 20
   Lean Six Sigma 9
   meaning 6–7
   origin(s) 7–9
   philosophy 5, 11, 20
   process 6–7
   programme(s) 10–11
   project(s) 11–12, 15, 20, 267–373
Skaraborg Hospital 314
skewness 57
slack *see* PERT
sorted sample 84
spatial series 61
SPC (statistical process control) 17
special cause(s) *see* variation
specification limits 171, 198
SS (sum of squares) *see* ANOVA
standard deviation 6, 56
standardization of a random variable
        79, 161

standard Gaussian random variable 79
state of knowledge 66
statistical control 184
statistical model 146, 217
statistical software 44
statistical unit 43
stochastic 41
    stochastic independence 67
stratification, analysis by 24, 59
Structo Hydraulics AB 330
supplier's risk *see* operating
                characteristic curve
survey 43

Taylor series expansion 198, 203
TESF (teaching experiments and
            student feedback) 243
time series 60
    time series plot 276, 281, 363
tolerance 345, 370
    interval 6
    limit 6
    see also robust design
traceability 358–9
trend 60
    linear trend 148
TQL (tolerable quality level) 189
    see also operating characteristic curve
TQM (total quality management) 20
    see also 7QC tools
TSS (Transactional Six Sigma) 18

unbiased estimator 88
uncertainty 62

variable
    categorical variable 43
    continuous variable 43
        classes of values 48
    discrete variable 43
    input variable(s) 13
    KPIVs (key process/product input
            variable(s)) 17

KPOVs (key process/product output
            variable(s)) 17
    numerical variable 43
    response variable(s) 13
    see also random variable
variance 55
    variance homogeneity *see* DOE
variation 2, 5–6
    common cause(s) of 6, 174
    control factor(s) 5
    external, internal sources of 193
    in-use variation 193
    noise factor(s) 5
    sources of variation 191, 194
    special cause(s) of 5, 14, 174
    unit-to-unit variation 193
    unwanted variation 190
    variation channeling 253
    variation in process 5–6
    variation in product 5–6
    variation risk management 209
    variation risk priority number *see*
            VMEA
    see also robust design
Venn diagram *see* event
virtual environment 249
VMEA (variation mode and effect
            analysis) 200–9, 327–8, 362
    method of moments 202
    variation risk priority number 201–2
    VMEA freeware 209
Volvo Aero Corporation 350, 365

WBS (work breakdown structure)
            29–30
Weibull model 80–1
white noise 169

yield 3, 335